Deferring Development

Evolutionary Cell Biology

A Series of Reference Books and Textbooks

PUBLISHED TITLES

Cells in Evolutionary Biology: Translating Genotypes into Phenotypes – Past, Present, Future

Edited by Brian K. Hall and Sally A. Moody

Deferring Development: Setting Aside Cells for Future Use in Development and Evolution

Edited by Cory D. Bishop and Brian K. Hall

For more information about this series, please visit: www.crcpress.com/Evolutionary-Cell-Biology/book-series/CRCEVOCELBIO

Deferring Development

Setting Aside Cells for Future Use in Development and Evolution

Edited by

Cory D. Bishop

Associate Professor in the Department of Biology at St. Francis-Xavier University, Antigonish, NS, Canada

Brian K. Hall

University Research Professor Emeritus at Dalhousie University, Halifax, NS, Canada

CRC Press
Taylor & Francis Group
Boca Raton London New York

CRC Press is an imprint of the
Taylor & Francis Group, an **informa** business

CRC Press
Taylor & Francis Group
6000 Broken Sound Parkway NW, Suite 300
Boca Raton, FL 33487-2742

Printed on acid-free paper

International Standard Book Number-13: 978-1-138-33428-1 (Hardback)

Visit the Taylor & Francis Web site at
http://www.taylorandfrancis.com

and the CRC Press Web site at
http://www.crcpress.com

Contents

Series Preface

In recent decades, the central and integrating role of evolution in all of biology was reinforced, as the principles of evolutionary biology were integrated into other biological disciplines, such as developmental biology, ecology, and genetics. Major new fields emerged, chief among which are Evolutionary Developmental Biology (or Evo-Devo) and Ecological Developmental Biology (or Eco-Devo).

Evo-Devo, inspired by the integration of knowledge of change over single life spans (ontogenetic history) and change over evolutionary time (phylogenetic history), produced a unification of developmental and evolutionary biology that is generating many unanticipated synergies. Molecular biologists routinely employ computational and conceptual tools generated by developmental biologists (who study and compare the development of individuals) and by systematists (who study the evolution of life). Evolutionary biologists routinely use detailed analysis of molecules in experimental systems and in the systematic comparison of organisms. These integrations have shifted paradigms and answered many questions once thought intractable. Although slower to embrace evolution, physiology is increasingly being pursued in an evolutionary context. So too is cell biology.

Cell biology is a rich field in biology with a long history. Technology and instrumentation have provided cell biologists the opportunity to make ever more detailed observations of the structure of cells and the processes that occur within and between cells of similar and dissimilar types. In recent years, cell biologists have increasingly asked questions whose answers require insights from evolutionary history. As just one example: How many cell types are there and how did these different cell types evolve? Integrating evolutionary and cellular biology has the potential to generate new theories of cellular function and to create a new field, which we term "*Evolutionary Cell Biology*."

A major impetus in the development of modern Evo-Devo was a comparison of the evolutionary behavior of cells, evidenced in Stephen J. Gould's 1979 proposal of changes in the timing of the activity of cells in development (heterochrony) as a major force in evolutionary change and in Brian Hall's 1984 elaboration of the relatively small number of mechanisms used by cells in development and in evolution. Given this conceptual basis and the advances in genetic analysis and visualization of cells and their organelles, cell biology is poised to be transformed by embracing the approaches of Evo-Devo as a means of organizing and explaining diverse empirical observations and testing fundamental hypotheses about the cellular basis of life. Importantly, cells provide the link between the genotype and the phenotype, both during development and in evolution. No books that capture this cell focus exists. Hence the proposal for a series of books under the general theme of

"Evolutionary Cell Biology: Translating Genotypes into Phenotypes," to document, demonstrate, and establish a long-sought level in evolutionary biology, viz., the central role played by cellular mechanisms in translating genotypes into phenotypes in all forms of life.

Brian K. Hall
Sally A. Moody

Preface

Deferring Development: Setting Aside Cells for Future Use in Development and Evolution is a contribution to the series of books on *Evolutionary Cell Biology: Translating Genotypes into Phenotypes.*

A cyclical return to a unicellular state called a zygote is a near-universal feature of multicellular life. Zygotes, by definition, are totipotent (can make all cells of the body), and therefore, embryonic development is defined by a stepwise and hierarchical decrease in potential for cellular function. Yet, multicellular organisms usually live longer than their cells do, cells and tissues can be lost through injury, and many organisms have complex life cycles. Therefore, developmental potential must be maintained long after embryogenesis is complete.

The patterns and mechanisms by which developmental potential is maintained throughout the ontogeny and evolution of multicellular – what we refer to as *deferred development* – are a major theme of this volume. The most prevalent, but perhaps not ancestral mode by which this is accomplished in animals, is via physical segregation (setting aside) of lineages of pluripotent stem cells. Such cells can persist for decades, can self-renew, and depending on the circumstance, can form germ cells from somatic ones. Cells with such properties in different organisms are neoblasts in planarians, meristematic cells in plants, interstitial stem cells in cnidarians, and archeocytes in sponges.

In the context of complex animal life cycles, as in echinoderms and ribbon worms, the term "set-aside cells" was coined for cells set aside in echinoderm larvae to produce many of the features of the adult. However, the term set-aside is used in other contexts and, as outlined in Chapter 1, "set-aside cells" were not defined by a coherent set of properties. To accommodate all instances in which developmental potential is maintained beyond embryogenesis, we propose the generic term "*deferred-use cells.*" Whereas stem cells are a major category of deferred-use cells, the range of cell types, organisms, developmental, and evolutionary situations discussed in these chapters demonstrates the importance of deferred-use cells for all forms of multicellular life. Indeed, we can ask whether deferred-use cells are essential for the existence of multicellular organisms.

All the authors of the chapters herein are active researchers and scholars in one or more aspects of deferred-use cells. They represent the best thinking on these cells from the disciplines of cell, developmental and evolutionary biology, developmental and molecular genetics, stem cell biology, life history research, and studies that integrate two or more of these fields. By using this diverse knowledge, *Deferring Development* evaluates deferred-use cells to ask:

1. whether they share fundamental similarities;
2. whether they are uni-, multi-, pluri- or totipotential;
3. whether all occupy a specialized niche in which they function;

4. whether they participate in day-to-day development, maintenance, and repair, or
5. whether they are truly set aside for future use, either later in a single generation or for future generations (as germ cells so obviously are).

Cells considered under the broad category of deferred-use cells, the concept of the niches in which such cells reside and function, and the concept of deferred-use molecules and their relation (if any) to deferred-use cells are discussed in evolutionary, life history, developmental, and cell cycle contexts in Chapters 1, 2, and 3. Chapter 1 describes the diverse ways in which development may be deferred and then, to spur comparisons of evolutionary patterns, proposes an *ontology of deferred development*, organized into six categories. Three evolutionarily independently derived patterns of maximal indirect (i.e., deferred) development are used to exemplify this ontology. Chapter 2 descends a level in the biological hierarchy and explores the concept of *deferred-use molecules* that are employed during animal oogenesis (and apparently much less so in plants) to form a prepattern in early embryos. Development is thus deferred across generations. Chapter 3 elucidates the intricate regulatory controls that interact to balance the cell cycle with pluripotency, cell fate decisions, and differentiation. Contrary to common wisdom, cell cycle regulators may actively affect cell differentiation; variation in the time that a cell spends in different phases of the cell cycle correlates with self-renewal, fate, and differentiation decisions. Deferred-use cells often display specialized modes of cell division—the asymmetrical division of stem cells is a prime example—that evolved coincident with the cell cycle and the evolution of mitotic division (Chapter 3).

Other important questions considered in subsequent chapters are: Did deferred-use cells originate before their niche, in parallel with the niche, or in response to the niche? Do these cells or does the niche control interaction with the niche and by what mechanism(s)? Some deferred-use cells do not occupy a specialized niche; *neoblasts* (reserve cells) in planarians discussed in Chapter 8 are a prime example. Further, deferred-use cells can be specified maternally or zygotically, and they can remain in a proliferative compartment for many years or generations. They can de- and then redifferentiate. Some deferred-use cells—such as those in insect imaginal discs—can transdifferentiate into a different class of deferred-use cells. Indeed, life cycle transitions and regeneration are mediated by transdifferentiation to regulate cell fate in non-Bilateria such as cnidarians and sponges, indicating that transdifferentiation may be an ancient feature of animals that is less common within Bilateria (Chapters 1 and 4).

The five chapters in Section II examine the origin of deferred-use cells and their niches within phylogenetic contexts. The rich and diverse evolution of both deferred-use cells and niches is revealed by the evolution of cellular reprogramming in animals (Chapter 4) and the origin of such cells and their niches in phylogenetically disparate lineages of macroalgae (Chapter 5), as meristems in vascular land plants (Chapter 6), as multipurpose stem cells (neoblasts) for body maintenance, growth, repair, and regeneration in planarians (Chapter 7), and as more specialized satellite cells to repair and replace muscle (Chapter 8).

Like other lineages of multicellular organisms, vascular land plants evolved stem cells and niches as meristems to produce shoots and leaves using molecular mechanisms that are conserved in plants. In vascular plants and macroalgae pluripotency is not assigned and maintained in a lineage-based manner as it often is in animals, yet all need to maintain pluripotency beyond embryogenesis and commonly do so using stem cells and niches. An examination of the evolution and development of these and other deferred-use cells and their niches from all independently evolved multicellular organisms will surely help distinguish their shared, derived properties from homoplasies.

The six chapters in Section III examine important case studies of deferred-use cells in development and evolution: Their sustained pluripotency throughout life as adult pluripotent stem cells as a source for evolutionary innovation, and their response to hormones and environmental factors in the evolution of allometry are the subjects of Chapters 9 and 10. Their roles in the evolution of adaptive immunity are discussed in Chapter 11—the adaptive immune system utilizes set-aside cells to enable host organisms to respond faster and more specifically to a second or subsequent encounter with a pathogen, all under the control of a key transcriptional regulator FoxO. Cancer can be a cost of not having set-aside cells, which is the topic of Chapter 12.

Microbes, surging in relevance to several areas of human health, are important players in the formation of stem cells niches: indigenous and pathogenic microbes are located in the niches from which they modulate stem cell activity in the regulation of proliferation and differentiation (Chapter 13). The demonstration that animals are not individuals in the traditional sense of that word has been extended to symbiosis between two or more organisms in what is known as a *holobiont*, in which interactions between the different genetic lineages are critical for the development and maintenance of the entire metaorganism. Stem cells play important roles in the ecological interactions between multispecies assemblages in holobionts (Chapter 14). Such enormous diversity prompted the development of the hierarchical view of deferred development and deferred-use cells outlined in Chapter 1.

CDB gratefully acknowledges support from an NSERC of Canada Discovery Grant.

Cory D. Bishop
Brian K. Hall

Editors

Cory D. Bishop, Associate Professor in the Department of Biology at St. Francis-Xavier University, Antigonish, NS, Canada, received his PhD in Evolutionary Developmental Biology from Simon Fraser University, where he studied cellular and molecular regulation by nitric oxide signaling of metamorphosis in Deuterostomes. Throughout that work, and into his postdoctoral years, he maintained an interest in comparative embryology, larval biology, and life cycle evolution, especially as it related to the capacity of larvae to defer metamorphosis in the face of suboptimal conditions. He and colleagues developed methods to inject oligonucleotides into sea urchin juvenile rudiments, thus providing a method to experimentally investigate later stages that form inside the larval body. As a member of the Centre for Biofouling Research at St. Francis-Xavier University, he has applied his interest in metamorphosis to developing antifouling strategies for invasive species and, more recently, has focused his attention on a symbiotic relationship between unicellular green algae and embryos of several amphibians.

Brian K. Hall, University Research Professor Emeritus at Dalhousie University in Halifax, NS, Canada, was trained in Australia as an experimental embryologist. His research concentrated on the differentiation of skeletal tissues, especially, how epithelial-mesenchymal signaling initiates osteogenesis and chondrogenesis through the formation of cellular condensations. These studies led him to earlier stages of development and the origin and function of skeletogenic neural crest cells. Comparative studies using embryos from all five classes of vertebrates provided a strong evolutionary component to his research. These studies, along with analyses of the developmental basis of homology, played significant roles in the establishing of evolutionary developmental biology. A Fellow of the Royal Society of Canada, Foreign Fellow of the American Academy of Arts and Science, and recipient of a Killam Prize, he was one of eight individuals awarded the first Kovalevsky Medals in 2001 to recognize the most distinguished scientists of the twentieth century in comparative zoology and evolutionary embryology.

Contributors

Ehab Abouheif
Department of Biology
McGill University
Montreal, Quebec, Canada

Jaume Baguñà
Department de Genètica
Universitat de Barcelona
Barcelona, Spain

Cory D. Bishop
Department of Biology
St. Francis Xavier University
Antigonish, Nova Scotia, Canada

Thomas C.G. Bosch
Zoological Institute
University of Kiel
Kiel, Germany

Kurt Buchmann
Laboratory of Aquatic Pathobiology
Department of Veterinary Disease Biology
University of Copenhagen
Copenhagen, Denmark

Nicolas Buchon
Cornell Institute of Host-Microbe Interactions and Disease
Department of Entomology
Cornell University
Ithaca, New York

Erin L Davies
Howard Hughes Medical Institute
Stowers Institute for Medical Research
Kansas, Missouri

Raffaele Dello Ioio
Biology and Biotechnology Department
University of Rome "Sapienza"
Rome, Italy

Moira E. Galway
Department of Biology
Saint Francis Xavier University
Antigonish, Nova Scotia, Canada

David J. Garbary
Department of Biology
Saint Francis Xavier University
Antigonish, Nova Scotia, Canada

J. Manuel Hernández-Hernández
Sprott Center for Stem Cell Research
Ottawa Hospital Research Institute
Ottawa, Ontario, Canada
and
Department of Cellular and Molecular Medicine, Faculty of Medicine
University of Ottawa
Ottawa, Ontario, Canada

Andreas Heyland
Department of Integrative Biology
University of Guelph
Guelph, Ontario, Canada

Jason Hodin
Friday Harbor Labs
University of Washington
Friday Harbor, Washington

David K. Jacobs
Department of Ecology and Evolutionary Biology
University of California Los Angeles
Los Angeles, California

Steven L. Klein
Developmental Systems Cluster
Division of Integrative Organismal Systems
Biology Directorate, National Science Foundation
Alexandria, Virginia

Sophie Koch
Department of Biology
McGill University
Montreal, Quebec, Canada

Alexander Y.T. Lin
Centre for Research and Advanced Studies of the National Polytechnic Institute, Genetics and Molecular Biology Department
Mexico City, Mexico
and
Department of Cellular and Molecular Medicine, Faculty of Medicine
University of Ottawa
Ottawa, Ontario, Canada

Sally A. Moody
Department of Anatomy and Cell Biology
George Washington University
Washington, District of Columbia

Laila Moubayidin
Crop Genetics Department
The John Innes Centre
Norwich Research Park
Norwich, United Kingdom

Peter Nagy
Cornell Institute of Host-Microbe Interactions and Disease
Department of Entomology
Cornell University
Ithaca, New York

Nagayasu Nakanishi
Department of Biological Sciences
University of Arkansas
Fayetteville, Arkansas

Siim Pauklin
Botnar Research Centre
Nuffield Department of Orthopaedics, Rheumatology and Musculoskeletal Sciences
University of Oxford
Oxford, United Kingdom

Morten Ritso
Sprott Center for Stem Cell Research
Ottawa Hospital Research Institute
Ottawa, Ontario, Canada
and
Department of Cellular and Molecular Medicine, Faculty of Medicine
University of Ottawa
Ottawa, Ontario, Canada

Michael A. Rudnicki
Sprott Center for Stem Cell Research
Ottawa Hospital Research Institute
Ottawa, Ontario, Canada
and
Department of Cellular and Molecular Medicine, Faculty of Medicine
University of Ottawa
Ottawa, Ontario, Canada

Elena Salvi
Biology and Biotechnology Department
University of Rome “Sapienza”
Rome, Italy

Alejandro Sánchez Alvarado
Howard Hughes Medical Institute
Stowers Institute for Medical Research
Kansas, Missouri

Darryl Shibata
Department of Pathology
University of Southern California
Los Angeles, California

Section I

Deferred-Use Cells and Niches

1 Deferred-Use Cells in Development and Evolution

A Life History Perspective

Cory D. Bishop
St. Francis Xavier University

Andreas Heyland
University of Guelph

Jason Hodin
University of Washington

CONTENTS

1.1 ONTOGENY AS A TIME-STRUCTURED PROCESS OF CELLULAR INTERACTION THAT CULMINATES WITH REPRODUCTION

> The cell is evolution's most magnificent achievement and embryonic development is merely a baroque elaboration.
>
> **Lewis Wolpert (1999, p. 1)**

Multicellular life has resulted in the evolution of ontogenies that generally begin from a single cell every generation (Buss 1987; Grosberg and Strathmann 1998; Grosberg and Strathmann 2007). Plausible reasons for this commonality are that a cyclical return to a unicellular state evolved independently in multicellular taxa to purge cytoplasmic or genetic parasites (Grosberg and Strathmann 1998) and mutations (Muller 1964). The evolution of this broadly shared pattern of reproductive mode also invokes the evolution of the following:

i. Mechanisms for specifying divergent fates. If only gametes carry genes across generations, then a key specification event in a nascent obligate multicellular organism was "germ" versus "not-germ" (Swartz and Wessel 2015), where germ refers to cells having the developmental potential to form gametes. Such not-germ cells would function in resource acquisition, motility, as a defensive structure, or some other nonreproductive function sensu stricto.
ii. Delayed cellular function due to a prereproductive ontogenetic process (see Figure 1.1). One key example would be that the appearance of differentiated germ cells is deferred in time relative to the differentiation of nonreproductive (not-germ) cell types; this is true whether germ cells are formed by intercellular signaling, by preloading of maternal determinants, or, more rarely in animals, by somatic embryogenesis (e.g., in sponges) (Korotkova 1970). Importantly, the notion of delayed cellular function is

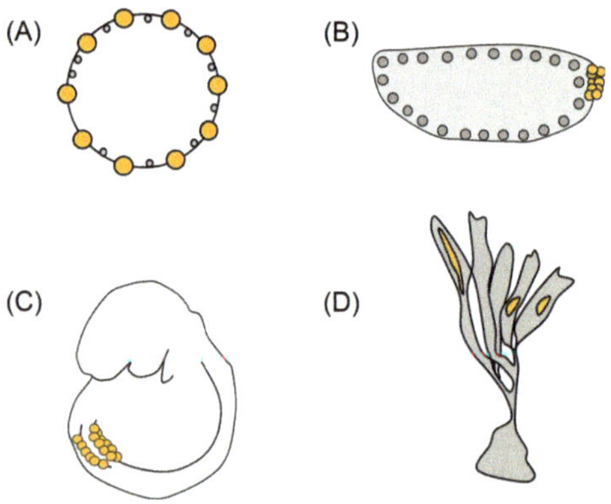

FIGURE 1.1 Schematic of timing of germ cell (gold shading) specification in some examples of disparate multicellular taxa. (A) The green alga *Volvox carteri* consists of 2000–4000 small somatic cells and 16 gonidia (from Kirk 2001). (B) The fruit fly *Drosophila melanogaster* produces a syncytium of nuclei during development; the nuclei that arrive at the posterior end are the first to be enveloped by membrane and thus form "pole cells": the primordial germ cells of the fly. (C) Mice and other mammals produce primordial germ cells in the embryo, which migrate to the genetical ridge during development and stay mitotically dormant until sexual maturation. The diagram depicts stage E10.5 of mouse development, when germ cells are in the process of migrating to the genital ridge from the gut. (D) Mature bull kelp (*Nereocystis luetkeana*) sporophytes produce dark brown patches called sori on their blades. The sporangia within each sorus contain the spores.

true for any functional distinction among cell types, "germ versus not-germ" or otherwise.

Therefore, cell division in multicellular organisms must be linked through gene regulatory mechanisms with the establishment of cell identities, morphogenesis, growth, and regeneration. Organisms that have somatic prereproductive growth—such as multicellular plants, kelp, fungi, and some animals such as sponges and cnidarians—diverge from most animals, in having no specific cells whose function is specified as germ early and then differentiate later (Extavour and Akam 2003). Nevertheless, nonanimal multicellular taxa do share the general characteristics of deferred reproduction with most animals: formation of spores, pollen, or ovules after an often-lengthy period of purely somatic growth. Furthermore, germ cells in these nonanimal groups are typically embedded in and supported by somatic reproductive tissues: sori, fruiting bodies, flowers, gametangia, etc.

Taken together then, a uniting feature of multicellular ontogenies may be usefully described as "*a plan within a plan*," in which the ultimate plan, reproduction, is embedded within the chronologically prior plan: ontogeny and maintenance of somatic and germ cells to support reproduction. The subsequent evolution of complex life cycles, in which a larva or some other intervening, distinct multicellular life stage—such as the encrusting stage in some upright algae (e.g., Santelices and Alvarado 2006)—fits naturally into the above expression of ontogeny, as a "plan within a plan within a plan." Although this phrasing is unwieldy,[1] it places the evolution of complex life cycles naturally in the same conceptual framework as the evolution of multicellularity: *ontogeny as a time-structured process of cellular interaction that culminates with reproduction.*[2]

In this introductory chapter, we place deferred developmental programs, and the cells at their foundation, in the context of complex life cycles and, to a lesser extent, regeneration. While we focus mainly on animals and their life cycles, we endeavor to note how the concepts we emphasize could apply across kingdoms, and hence, across independent origins of multicellularity and ontogeny. We intentionally avoid discussion of processes and mechanisms, deferring to other chapters in this volume for those topics.

1.2 DEVELOPMENT: A BALANCING ACT BETWEEN PROLIFERATION, PLURIPOTENCY, AND FUNCTION

Whether simple or complex, all ontogenies consist of the following sequentially structured processes that function over several different levels of organization (i.e., cellular, molecular, and biochemical):

i. the hierarchical and stepwise fashion with which multicellular bodies develop from one or a small number of cells;

[1] All the more so for organisms, such as parasitic animals with derived "hypermetamorphic" life histories (see Truman and Riddiford 2002), and different types of algae with all manner of hypercomplex life cycles (see, e.g., Lee 2008).

[2] Of course, ontogeny may continue after reproduction commences until the organism dies.

ii. the need to integrate development of a later life cycle stage with a former one; and
iii. the need to replace or repair parts.

Inherent in this conception of time-structured ontogeny is the balance that must be struck between cell division on the one hand, and cell differentiation and function on the other hand. In this context, one bona fide cell biological constraint inherited by the Metazoa is that, in the lineage of protists that gave rise to animals, the microtubule-based cytoskeleton was functionally constrained such that cell division and cell function were mutually exclusive states (Buss 1987). The consequences of this cell biological constraint for macroevolutionary patterns in metazoan ontogeny and regeneration appear to be substantial: differentiation and proliferation cannot occur simultaneously and therefore require molecular mechanisms to regulate transitions in and out of each of those states (Chapter 3). One widespread cellular mechanism for circumventing this constraint is for the developing organism to harbor uni- or pluripotent stem cells, which can be deployed as needed for growth, repair, and replacement of cells. By contrast, it is much rarer for developing animals to have the ability for one differentiated cell to transdifferentiate directly into another with a very different function (see below). Remarkably, however, new evidence from sponge cellular transcriptomes suggests that early animals likely had the ability to readily transition between differentiation states (Sogabe et al. 2019).

One central goal of regenerative medicine and stem cell research is to induce the formation of any cell type from any other cell type. Whereas, "In principle, anything can be changed into anything else by altering the combination of transcription factors" (Slack and Tosh, 2001, p. 1), what makes universal cellular reprogramming such a grand challenge is that, in bilaterian animals at least, development itself seems rarely to accomplish it. Because stem cell biologists *are* inexorably unlocking the mysteries of "stemness" and the minimal set of gene regulatory modifications required for cellular reprogramming (Takahashi and Yamanaka 2006; see Soldner and Jaenisch 2018 for review), we can conclude that a mode of ontogeny in which all cell types can give rise to all other cell types (totipotency) is not fundamentally constrained in animals. Indeed, maintenance of pluripotentiality throughout life by transdifferentiation may *define* early animals (Nakanishi and Jacobs, Chapter 4; Sogabe et al. 2019). Vascular plants and macroalgae are different than most animals in this respect in that differentiated cells can retain totipotency throughout life (see Chapters 5 and 6).

Retaining totipotency through cellular reprogramming would seem to be a rather useful ontogenetic mode. Nevertheless, outside of sponges, cnidarians (see Chapter 14), and ctenophores, this form of ontogeny—including during regeneration (Vervoort 2011)—is restricted to relatively few described examples in animals.[3] The most widely shared mode of regeneration is the use of stem cells

[3] See Slack and Tosh (2001) for reasons why it is difficult to document such cases, suggesting we might underestimate the frequency of transdifferentiation or transdetermination.

(Lai and Aboobaker 2018), which in many ways are comparable to germ cells (e.g., germline multipotency program, Juliano et al. 2010). The cell biological constraint in development identified by Buss (1987) and the universal requirement of pluripotent cells for regeneration purposes pointed out by Alvorado (2000) both converge on the notion that, in animal and perhaps all multicellular life, a source of pluripotent cells is an essential commodity. The hypothesis that postembryonic transdifferentiation potential may have been higher in basal animals and subsequently "traded in" for deferred-use cells (e.g., stem cells or neoblasts in planarians) among descendants suggests an undefined trade-off in modes by which pluripotency is maintained in ontogeny. See Chapter 12 for a treatment of the relationship between "set-aside stem cells" and cancer.

Thus, *if one considers ontogeny a time-structured process where different cell types differentiate at different times, then the notion of deferred development and deferred-use cells is both an emergent phenomenon of ontogeny itself and an evolutionary strategy for maintaining developmental potential after embryogenesis.* Furthermore, regeneration can be considered a recapitulation of a time-structured ontogeny (not necessarily of the specific developmental processes themselves), once again involving cells whose potentials are greater than their fates; in this case, those potentials having been deferred until a portion of the organism is lost and requires healing and regrowth.

In the subsequent sections, we explore the concept of deferred development and deferred-use cells in the context of life cycles. Due to the diversity of perspectives contained in other chapters in this volume, we have not attempted to be broadly inclusive of the remarkably diverse life cycles in nonanimal multicellular organisms, nor even of those in animals. Rather, our goal is to use selected examples from animal development to both (i) identify similarities and differences in how deferred developmental programs and the cells at their foundation build bodies in a time-structured manner and (ii) explore how ontogenies produce a later, phenotypically distinct body from an earlier one.

1.3 DEFERRED DEVELOPMENT IN THE CONTEXT OF COMPLEX LIFE CYCLE EVOLUTION

The price of metaphor is eternal vigilance

-attributed to Norbert Weiner and Arturo Rosenbluth (Lewontin 2001, p. 1)

Multicellular bodies at reproductive size vary by ~5 orders of magnitude in linear dimensions, by over 10 orders of magnitude in volume, and by over 14 orders of magnitude in mass. In all cases other than for very small bodied adults (e.g., meiofauna and nematodes in animals), ontogeny must therefore contend with the challenge of patterning growing bodies. Because of selection on final size at reproductive maturity, ontogeny must also scale with evolutionary increases in body size. Werner (1988) pointed out that scaling relationships define absolute size limits over which a particular ecological niche can be exploited by a particular body design. Therefore, if performance requirements during ontogeny are not the same as those of the size of an organism at reproductive maturity, then life history strategies must evolve to accommodate the mismatch. Cohen (1985) identified three evolutionary responses that animals display, in various combinations, to contend with how to make bodies that can be orders of magnitude larger than their embryos (Figure 1.2):

i. coloniality (e.g., bryozoans and hard corals) where the same body unit is multiplied to create more mass;
ii. increased maternal provisioning (e.g., via evolutionary increases in egg size or other types of postzygotic provisioning) increases the size of the initial free-living ontogenetic stage compared to the ancestral state; or
iii. evolution of complex life cycles.

In animals, this latter tactic is the most widespread (Thorson 1950). Whereas the more commonly known examples of complex life cycles in animals are terrestrial—tadpole to frog and caterpillar to butterfly—the most dazzling array of complex life histories among animals are surely found in the ocean. Coastal marine waters especially are replete with diverse larval forms that disperse for a time in the plankton

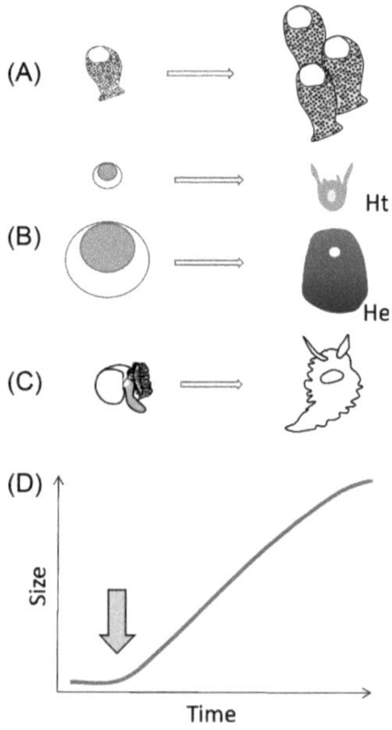

FIGURE 1.2 Three evolutionary strategies to make larger bodies in metazoans. (A) Evolution of coloniality, exemplified by solitary and colonial bryozoans; (B) evolutionary increase of maternal investment, exemplified here by the evolution of nonfeeding development in echinoids (Ht, *Heliocidaris tuberculata*; He, *Heliocidaris erythrogramma*; see Raff 1996); (C) evolution of complex life histories, exemplified by the veliger to juvenile sea slug transition; (D) schematic representation of size increase as a function of time for larval and colonial life histories. The arrow indicates the transition either to a juvenile or a colonial state.

before undergoing metamorphoses into what are often very different looking adults in the benthos. In general, the maximum size of these larval forms (and hence the size of the young juveniles that emerge after metamorphosis) is much smaller than that of the corresponding reproductive adults. How did ontogeny evolve in response to the evolution of large-bodied descendants from small-bodied ancestors?

A prominent attempt to place the evolution of large-bodied animals in the context of developmental innovations is the *set-aside cell* hypothesis of Eric Davidson, Kevin Peterson, and Andrew Cameron (Peterson et al. 1997). Prior to the formal elaboration of this hypothesis, Davidson et al. (1995) proposed the term "maximal indirect development" to reflect not only the significant mismatch in phenotypes between larvae of some marine invertebrates and their adult forms but also their nearly temporally mutually exclusive ontogenies. Larval forms in taxa with proposed maximal indirect development are small, generally less than 1 mm in length, but their adults could be orders of magnitude larger.

The set-aside cell hypothesis suggested a singular and bold resolution to this apparent paradox, involving a key evolutionary novelty in a hypothesized proto-animal with a small-bodied adult. This key novelty was *the proposed origin of groups of cells or tissues that are formed in the embryo but are "set aside" as a rudiment of tissues that remain undifferentiated until the juvenile stage.* In other words, the set-aside cell hypothesis posited that the evolution of large complex animals from small ancestors was facilitated by the developmental innovation of what the authors termed "set-aside cells," which (along with attendant genomic regulatory mechanisms) patterned these rudimentary tissues on a regional, as opposed to a cell-by-cell basis.

Of particular concern to Davidson and colleagues were marine animal phyla characterized by complex life histories involving a larva undergoing a radical metamorphosis into a very different looking adult, such as the sea urchin and ribbon worm depicted in Figure 1.3. Davidson and colleagues considered the ancestral small-bodied adults to be homologous to modern larvae and concluded that the macroscopic adult forms that define the modern phyla arose later and were thus tacked on to the end of ontogeny. Hence the set-aside cell idea was essentially a macroevolutionary hypothesis that spanned innovations in cell behavior and function to novel complex genomic regulatory processes to the origin of large animals themselves.

The set-aside cell hypothesis and the related body of work from which it was derived was a bold attempt at a synthesis of comparative developmental biology and macroevolution, at once accounting for similarities in larval forms among many phyla, the shared use of patterning mechanisms to build adult bodies and the Cambrian explosion itself. Nevertheless, varied critiques soon followed, including, but not limited to, the failure of the set-aside cell hypothesis to meet the criterion of Darwinian plausibility (Wolpert 1999); to infer the likely ancestral metazoan life cycle and critically evaluate typological characterizations of larvae (Jenner 2000); to accurately depict the capacity of larval cells for extended cell division under different conditions, such as phenotypic plasticity (Strathmann 2000); and to contend with the conclusion that adult body plans and possibly the patterning mechanisms that build them (e.g., Hox genes) must *per force* be homoplasies (Raff 2008). Each of these critiques represent significant challenges to the set-aside cell hypothesis as an explanation for broad patterns in animal evolution and development.

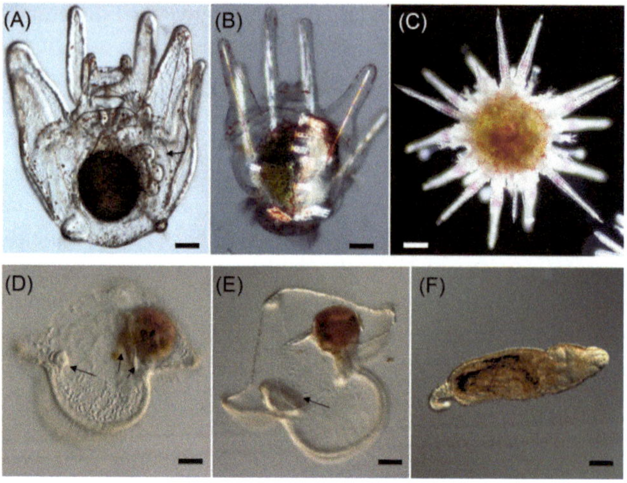

FIGURE 1.3 Sea urchins (phylum: Echinodermata) and ribbon worms (phylum: Nemertea) produce juvenile bodies from phenotypically distinct larval ones. Developmental stages of the purple sea urchin *Strongylocentrotus purpuratus* (A–C) and the ribbon worm *Maculaura alaskensis* (D–F). Both species undergo a dramatic metamorphic transition from the larva (pluteus—A, B; pillidium—D, E) to the juvenile (C, F). The juvenile rudiment (arrow in a) in the sea urchin larva begins to develop in parallel to the larval body and emerges at settlement after metamorphosis is completed (C). In nemerteans, invaginated discs (cephalic, cerebral organ and trunk; arrows in D and E) emerge in the pilidium larva and eventually form the juvenile structures (F). Scale bars: A—20 μm; B—40 μm; C—80 μm; D—75 μm; E—100 μm; F—80 μm. (Photo credits: Andreas Heyland (A–C), Svetlana Maslakova (D–F).)

Here, we emphasize an additional and perhaps more fundamental critique: *the term "set-aside cell" did not adhere to an ontology consistent with cell or developmental biology*. The definition of a set-aside cell is clear enough: "We term the specific patches of cells from which the juvenile arises in maximal indirect development 'set-aside cells,' because they are in some manner withheld from the differentiation processes that in the late embryo generate the structure of the larva *per se*." (Peterson et al. 1997, p. 624). However, the cells and tissues to which this term was applied do not share a consistent set of properties. To make this point, we briefly reanalyze an example that Davidson and colleagues held up as an archetype for both maximal indirect development and the set-aside cell hypothesis itself: juvenile rudiment formation in sea urchins. Towards the end of the chapter (Section 1.5), we will reintroduce sea urchin development in the context of an exploration in parallel patterns in the evolution of complex life histories in diverse animals via deferred development that employs rudiments.

The juvenile body in most sea urchin taxa begins to form midway through the larval period when mesodermal coelomic sac cells induce a small field of oral ectoderm cells on the left side of the larva to invaginate (so-called "rudiment invagination"). Subsequent tissue interactions between these two germ layers (see Figure 1.3A) pattern these epithelial cells into juvenile ectoderm and neuroectoderm, and coelomic cells into juvenile mesoderm and germ cells (Chia and Burke 1978; Campanale et al. 2014).

Peterson et al. (1997) considered both the ectodermal invagination and the coelomic mesoderm as examples of set-aside cells. However, this characterization is not accurate for either cell type. Prior to their invagination and subsequent interaction with mesoderm, the larval epithelial cells are morphologically and functionally indistinguishable from adjacent, differentiated larval epithelial cells. They are not "set aside"; they undergo a change in function. Similarly, in the sense intended by Peterson et al., the somatic mesodermal cells are not set aside either. For example, the hydrocoel functions in larvae as a differentiated excretory organ (Ruppert and Balser 1986) and is later remodeled to produce the water vascular system of the juvenile, among other things.

Beyond the particular characteristics of these purported "set-aside cells" themselves, the term "set aside" itself is problematic for the comparison of development strategies used by different lineages of organisms, as it has two distinct connotations according to the Cambridge English Dictionary. One definition is "to save for a particular purpose." It is this connotation that Davidson and colleagues were employing, as they envisioned embryos literally segregating cells or cell potential in a certain physical portion of the embryo. However, the second definition of the term "set aside" conveys the opposite meaning: "to decide not to consider something... to state that [something] is no longer in effect." In the former definition, something is saved for future use; in the latter definition, something is removed from future consideration! To make matters more confusing still, this second connotation of "set aside" precedes the set-aside cell hypothesis in the scientific literature with respect to future cell fate.

Buss (1987) used the term to refer to germ cells, as having been set aside relative to cells having a somatic function. In an opposite usage, Bell and Koufopanou (1991) refers to the replicative potential of flagellated somatic cells in volvocalean (colonial green algal) life cycles (see Figure 1.1A) as having been set aside to ensure ongoing motility of the colony while germ cells undergo mitosis. Pehrson and Cohen (1986), reported that descendants of small micromeres of sea urchin embryos come to reside in coelomic sacs and are thus set aside. These authors used the term with some precision, in which these micromere descendant cells were kept in a mitotically quiescent state.[4] Later, Truman and Riddiford (1999) in their paper on the evolution of larvae and metamorphosis in holometabolous insects refer to cells that form the imaginal discs as having been set aside and then later refer to cells/tissues that form the juvenile as having been developmentally deferred. Thus, they used the terms interchangeably.

To conclude:

i. what were identified as set-aside cells in the hypothesis of that name do not have a consistent set of properties; and

[4] It was later shown (Yajima and Wessel 2011; Campanale et al. 2014; Wessel et al. 2014) that descendants of the small micromeres give rise to primordial germ cells; in this sense in particular, Pehrson & Cohen's usage was *post hoc* consistent with Buss (1987).

ii. opposing connotations exist in the use of "set aside" in the English language and by extension in the wider literature.

> Given the varying uses in the literature for 'set-aside' the term 'deferred development' or 'deferred-use' cells' seems more precise, and here we advocate it in place of 'set aside.'

The presumed goal of Davidson and colleagues in proposing their set-aside cell hypothesis is one that we share: to gain an understanding of broader issues in the evolution of animals and their life histories through an examination of deferred developmental programs in disparate organisms. Unfortunately, the ambiguities inherent in their analyses that we outline above undermined this worthy goal. In the next section, we set aside (second meaning!) the evolutionary theoretical dimensions of the set-aside cell hypothesis and propose an ontology of deferred development in the service of resolving such ambiguities.

1.4 ONTOLOGIES OF DEFERRED DEVELOPMENT AND DEFERRED-USE CELLS

The concept of deferred development implies that specification and terminal differentiation occur relatively later for some populations of cells compared to others. This idea is axiomatic to developmental biologists and thus may seem to need no elaboration. Nevertheless, our basic claim is that both the reasons and the phenomenology for these delays differ within and among taxa, and therefore, for the purpose of comparisons, it may be useful to devise a systematic ontology of deferred development (Table 1.1). In this section, we attempt to distinguish instances in which deferred development occurs as a by-product of something else (Category 1 in Table 1.1) from instances in which

i. ecological requirements of life history stages have resulted in heterogeneity indevelopmental rates of distinct ontogenetic processes (Categories 2 and 3); or
ii. selection has generated cells or entire tissues whose developmental program is occurring in the context of a functioning earlier stage (Categories 4–6).

Before proceeding, we note that all Categories (1–6) could occur in a single organism, and therefore, there is no implied grade of complexity or hierarchy among these categories. Furthermore, we are mindful that accelerated development of one part of an embryo could be interpreted as the deferred development of another part.

The first category of among-lineage rate heterogeneity is called "Consequential Delay" (Category 1 in Table 1.1), and it is driven by *physical constraints*. For example, the vegetal hemispheres of many frog embryos have very high yolk content, such that cleavage furrows cannot easily proceed through them. One result of this animal-vegetal disparity in yolk is that cleavage proceeds more rapidly in the animal hemisphere, and there are thus more cells in the animal hemisphere at any given time during early cleavage stages (Barresi and Gilbert 2016).

TABLE 1.1
An Ontology of Deferred Development

Category of Deferred Development	Feature	Examples	Additional Notes
1. Consequential delay	Physical constraints, leading to "basal" heterochronies.	Relative size of progenitor cells; yolk content leading to differences in cell division rates in animal versus vegetal portions of amphibian embryos.	The delays here are not selected for in and of themselves; they could be examples of spandrels (sensu Gould and Lewontin 1979).
2. Rate prioritization	Selection for rate heterogeneity among cell lineages. In some cases, this could be a functional or developmental constraint.	*Caenorhabditis elegans* heterochronic (*het*) mutation; delayed onset of oogenesis in drosophilids until after pupariation.	Relative developmental rates are determined by needs of morphogenesis (i.e., internal to the developing organism).
3. Rate modulation	Processes that can vary temporally (plasticity). Distinction with Category 2: processes are responsive to environmental signals.	Hatching plasticity; gating in insect eclosion; delayed rudiment formation in poorly fed urchin larvae; timing of terminal filament formation in response to food or temperature.	"Programmed" sensitivity to environmental input; varies with mosaicism versus regulation.
4. Deferred development—cell populations	Cell populations that have been specified (or may have partially differentiated) but are nonfunctional until signaled.	Histoblasts in *Drosophila* cuticle formation; germ cells; axillary cells in ribbon worms; all stem cells; muscle development in precocial birds; delayed ovary proliferation in *Drosophila sechellia* ovarian primordia; juvenile tissue differentiation in slime star mesogen.	"Ecological" input required, minimum state of development of other tissues; timing may be programmed, but not fixed in absolute time.

(Continued)

TABLE 1.1 (*Continued*)
An Ontology of Deferred Development

Category of Deferred Development	Feature	Examples	Additional Notes
5. Deferred development—delayed life history shift	Deferral is now at the level of entire tissues or organs and is associated with a life history transformation such as in metamorphic life histories.	Posterior growth zone in marine annelids; flight in hemimetabolous insects; rudiment formation in pencil urchins (cidaroids); juvenile tissue differentiation in sea cucumbers.	Similar to Category 4 but pertains to entire tissues or organs.
6. Deferred development—rudiments allow rapid deployment of deferred function in life histories	Initially, nonfunctional cells/organs develop in parallel with functional larval cells. Paradoxically, less “deferred” in appearance than Category 5, but we hypothesize that Category 6 is evolutionarily derived relative to Category 5 and that the function itself is still deferred.	Echinoderm rudiments; imaginal discs in ribbon worm and holometabolous insects; rudiments in oweniid polychaetes; precocious formation of juvenile bodies within compound ascidian tadpoles.	Morphogenetic aspects of metamorphosis are decoupled from the habitat transition. The deferred tissues and organs are segregated in rudiments.

The salient feature of Category 1 is that examples of "Consequential Delay" should not be assumed to be adaptations in and of themselves; they may be a result of selection for something else (i.e., spandrels sensu Gould and Lewontin 1979). In the case of frog embryos cited above, the delay of vegetal relative to animal hemisphere cleavage is likely a consequence of selection for increased maternal nutrition (i.e., selection for deferred-use molecules, Chapter 2) and therefore egg size.

Category 2 is entitled "Rate Prioritization" and can be distinguished from Category 1 in that there is presumed selection in Category 2 for among-lineage heterogeneity in rates or timing of development *to meet the needs of morphogenesis.* A clear example of Category 2 involves ciliation in developing trochophore larvae. This larval form, present in the life cycle of various spiralians (e.g., mollusks, annelids, and the less well-known entoprocts) is characterized by very rapid differentiation of a band of ciliated cells called the "troch." These trochoblast cells cease dividing and arrange themselves into a band of functional cilia when there are only 63 cells in the embryo (Kooij et al. 1998). This developmental pattern is presumably driven by the need for rapid development to a swimming stage (Staver and Strathmann 2002). The differential storage of large amounts of maternally derived tubulin mRNA or protein in presumptive trochoblasts would constitute evidence that the rate of trochoblast cell differentiation is accelerated, as opposed to the notion that development of the remainder of the embryo is deferred.

Roundworms (phylum: Nematoda) offer numerous, well-studied examples that can be placed in Category 2. Nematode embryos have stereotyped cleavage patterns, producing fixed cell lineages, where different groups of founder cells divide at different rates in a manner that reflects ontogenetic needs. Furthermore, the ontogeny of the nematode *Caenorhabditis elegans* has been subjected to decades of intense study through the characterization of myriad mutations that perturb the meticulously well-characterized cell lineage in this worm (Horvitz 1990). *C. elegans* genes have historically been named for the classes of phenotypes they produce. One example is the "heterochronic" (*het*) gene class, so named because mutations in these genes result in temporal displacements of developmental events (reviewed in Moss 2007). Therefore, in normal development, *het* genes can be thought of as orchestrators of the proper control of ontogenetic timing. *Het* mutations indicate that when such events occur out of sequence, normally functioning worms are not produced. This observation alone speaks to the relationship between careful control of the relative timing of developmental events in order to maximize an organism's fitness.

It is possible that Category 2 rate heterogeneity may be an evolutionary precondition (a preadaptation sensu Gould 1984) for occurrence of relative shifts in the timing of embryonic events in response to environmental conditions (i.e., phenotypic plasticity, or "rate modulation"; Category 3 in Table 1.1). This heterogeneity may provide the variation for selection for fixed differences in rates within and ultimately between related species (namely, intra- and interspecific variation; Category 4 in Table 1.1).

Categories 3 and 4 in our proposed ontology represent a critical distinction from Categories 1 and 2, in that *ecological inputs intervene to modulate rates of division among cell lineages or populations, or the timing of specification or differentiation of cells.* Specifically, Category 3 refers to developmental processes subject to heterochronic phenotypic plasticity: a change in relative timing of developmental events due to prevailing environmental factors. Category 4, then, represents ecologically driven selection on the relative timing of events, which contrasts with the internal needs of morphogenesis at the basis of Category 2 deferral.

An example of Category 3 deferral builds on the example we introduced in Section 1.3 of the "echinus rudiment" in the sea urchin pluteus larva: the left-side ectodermal invagination that contacts an overlying mesodermal compartment, thus initiating juvenile development within the larval body (see Figure 1.3A). If the larva of the Mediterranean urchin *Paracentrotus lividus* is well fed, its ectodermal invagination occurs when the pluteus has six larval arms. If instead the larva is poorly fed, the ectodermal invagination is deferred until the pluteus has the full complement of eight larval arms (Strathmann et al. 1992). These authors interpret this heterochronic plasticity as a differential investment in feeding larval structures (the larval arms) at the expense of rapid progression to the juvenile stage (rudiment development).

A classic example of Category 4 deferral involves divergent strategies in growth among birds: grow quickly into helpless chicks (altricial) or grow slowly into highly functioning chicks (precocial). A functional correlate of the altricial versus precocial strategies is seen in the development of skeletal muscle. In altricial species such as the European starling (*Sturnus vulgaris*), the chicks hatch with less-developed skeletal muscle. In precocial species such as the northern bobwhite quail (*Colinus virginianus*), the chicks hatch with more fully developed skeletal muscles and can thus perform better at hatching. The key to this difference seems to be deferred differentiation of the skeletal muscle in altricial taxa associated with their shorter relative incubation periods (Ricklefs et al. 1979a,b), an example of Category 4 deferral of muscle cell differentiation. Chapter 8 discusses the role of skeletal muscle satellite cells in animal muscle regeneration, cells that would fall into Category 4.

Importantly, all stem cells would also fall into Category 4 in the sense that stem cells are fate restricted to one or a few cell types, but their terminal differentiation requires signaling based upon needs of homeostasis or regeneration. However, many stem cells seem to differ from other kinds of Category 4 deferred-use cells in both their properties of self-renewal and in their residence in a niche, a biochemical and structurally discrete compartment that maintains stem cell quiescence. See Chapter 11 for a discussion of the adaptive immune system as examples of Category 4 deferred-use cells, and Chapters 13 and 14 discuss exciting new developments regarding the role of microbiota and stem cell regulation and differentiation.

An example from insect reproduction is useful for illustrating the distinction between Categories, 2, 3, and 4. The functional unit of the ovary in most insects is the ovariole (see Figure 1.4C). An ovariole is an assembly line of sorts for the production of what can be very large (millimeters in length) eggs from microscopic primordial germ cells. Ovaries in different insects contain as few as one and as many as hundreds of ovarioles, with the advantage of the latter being simultaneous, and therefore higher rates of oogenesis (see Hodin 2009 for review).

Determination of ovariole number in drosophilid fruit flies occurs in the larval stage, via a stack of mesodermal cells that cap every mature ovariole called the terminal filament (Figure 1.4). Terminal filament precursors divide throughout larval development and then form into stacks in the last larval stage (the third instar in drosophilids; Figure 1.4A). At the onset of metamorphosis (the pupariation stage), the

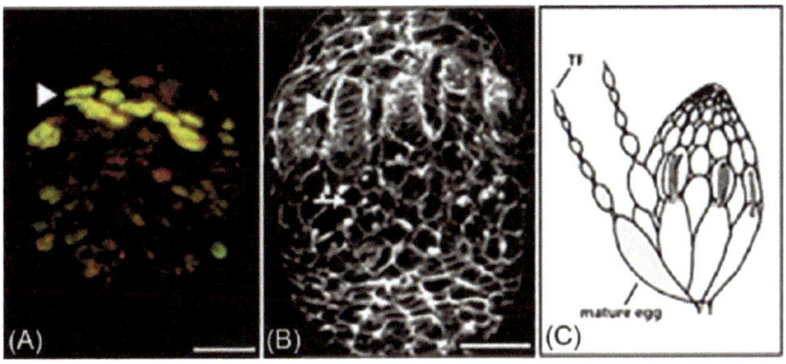

FIGURE 1.4 Ovary development in the fruit fly *Drosophila melanogaster*, anterior is up in all panels. (A) An ovary primordium late in larval development (midthird instar) beginning the process by which presumptive terminal filament (TF) cells (here labeled in yellow and indicated with the *white arrowhead*) are forming into TF stacks. Unlabeled primordial germ cells are just posterior of these cells. (B) At pupariation (the onset of metamorphosis), the TF stacks (*white arrowhead*) have formed in this phalloidin-stained ovary primordium, demarcating where each of the ovarioles will develop during the pupal stage. A primordial germ cell is indicated with the *white arrow*. (C) Schematic drawing of an adult ovary. TF stacks still cap each mature ovariole (*gray shading*): individual assembly lines for the production of numerous mature eggs throughout the adult female's life. Scale bar in (A) is 10 µm and in (B) is 20 µm. Note that the mature ovary (C) is far larger: more than 1 mm in (anterior-posterior) length.

number of terminal filaments in the adults is determined by the number of terminal filaments formed at that point (Figure 1.4B). The posterior-most cells in the terminal filament stack along with other somatic cells form a bona fide stem cell niche (Xie and Spradling 2000; Panchal et al. 2017) in response to the insect molting hormone ecdysone (Gancz et al. 2011). After pupariation, stem cell niches signal the germ line stem cells and ultimately the associated somatic follicle stem cells of the ovary to divide and then begin to differentiate, a process that continues into and throughout the adult stage. This delay of germ and follicle cell development until after pupariation is an example of Category 2 deferral: the needs of morphogenesis (namely organization of germ and somatic reproductive cells into ovarioles) dictate a delay in proliferation and differentiation until the terminal filament stacks are formed at the end of larval life.

The determination of terminal filament number is both subject to phenotypic plasticity (Category 3) and intra- and interspecific variation (Category 4). Raising *Drosophila melanogaster* larvae on low-quality food causes a reduction in terminal filament number via changes in the relative timing of terminal filament cell differentiation just before pupariation (Category 3; Hodin and Riddiford 2000a). Reduced ovariole numbers are phenocopied in *D. sechellia*, a unique relative of *D. melanogaster* that feeds exclusively on a toxic fruit. These flies have much lower fecundity than *D. melanogaster*, with correspondingly low terminal filament numbers (Hodin and Riddiford 2000a). The developmental basis for the interspecific differences in terminal filament numbers between *D. melanogaster* and *D. sechellia* is differences in cell proliferation in the ovary primordia throughout larval development, a clear example of Category 4 deferral.

Our proposed Categories 5 and 6 of deferred development in Table 1.1 refer to lineages of embryonic cells retained in an undifferentiated state. However, unlike in

Category 4, *Categories 5 and 6 of deferral pertain to cells that produce a complex structure or structures unto themselves.* Category 5 differs from Category 6 in that in the latter, the primordia of the complex structure or structures in question are segregated into rudiments: distinct compartments of the developing organism destined specifically for future use (Wilson 1932; Strathmann 2000).

In the next section, we employ the examples of three taxa—insects, echinoderms, and ribbon worms—to help distinguish among Categories 4–6 deferral and to exemplify how the independent origin of Category 6 deferral via rudiments appears to have been key to the extreme examples of radical metamorphosis seen within these three taxa.

1.5 INDEPENDENT EVOLUTION OF EXTREME PATTERNS IN DEFERRED DEVELOPMENT

Our conclusion above is that what were referred to as "set-aside cells" are more correctly (and hopefully more usefully) described as "deferred development" or "deferred-use cells" of different kinds (Section 1.3, Table 1.1). As such, the issue of the possible homology (common evolutionary origin) of disparate deferred-use cells may be most appropriately approached in terms of the nature and degree of deferred development under consideration. In other words, whereas all instances in animals of deferred development could have a common evolutionary origin at or below the base of the Metazoa (see also Sogabe et al. 2019), instances of juvenile rudiments (anlage) developing in parallel and within the larvae of oweniid polychaetes, compound ascidians, ribbon worms, echinoderms, and insects (Category 6 in Table 1.1, Figure 1.5) are almost certainly examples of homoplasy (i.e., similarity not resulting

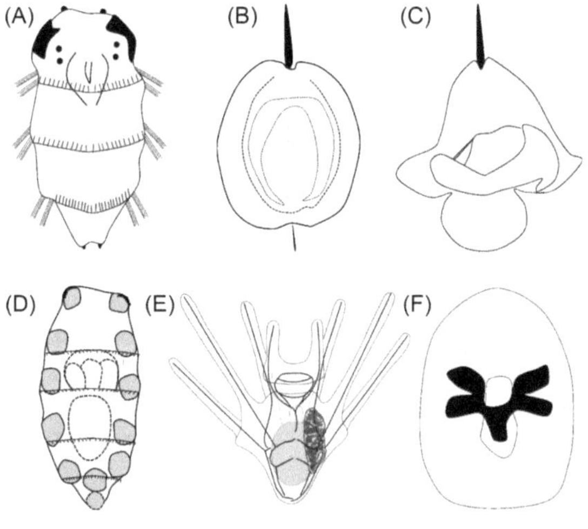

FIGURE 1.5 Development of juvenile bodies relative to larval bodies across taxa, illustrative of Categories 4–6 deferral. (A) Annelid trochophore larva: Category 4. (B) Ribbon worm decidula larva (after Maslakova 2010): a possible intermediate between Categories 5 and 6. (C) Ribbon worm pilidium larva (after Maslakova 2010): Category 6. (D) Sea cucumber doliolaria larva (after Mortensen 1921): Category 5. (E) Sea urchin pluteus larva: Category 6. (F) Mesogen of the slime seastar *Pteraster tesselatus* (after Janies and McEdward 1993): Category 4.

from common ancestry), resulting from selection for more rapid larval-to-juvenile transitions (Strathmann 2000).

Furthermore, a consideration of the numerous evolutionary origins of multicellularity across the tree of life should give pause when postulating ancient origins of particular types of deferred-use cells. It is tautological that deferred development occurs in taxa that undergo development: namely, multicellular organisms. In that sense, any commonalities seen in deferral strategies or mechanisms among independently evolved multicellular taxa—whatever Category as denoted in Table 1.1—must be examples of homoplasy. Similarly, among descendants of particular instances of the evolution of multicellularity, we encourage restraint in ascribing homology to ontogenetic processes in taxa that appear to have a shared mechanism (or Category) of deferral in a given developmental process.

With those caveats in mind, we now contrast the manner in which juvenile bodies are constructed relative to larval development in three animal lineages in which there is compelling evidence that an extreme pattern of metamorphosis has evolved independently: ribbon worms (phylum: Nemertea), echinoderms, and insects. According to Davidson et al. (1995) and Peterson et al. (1997), planktotrophic echinoderms and ribbon worms are prototypical examples of "maximal indirect development" via a feeding larval stage, which differs in fundamental ways from the adult body plan. Nevertheless, there is general agreement—based on phylogenetics as well as characters such as embryonic cleavage patterns, formation of the larval mouth, larval morphology, and larval feeding modes—that echinoderm and ribbon worm larvae are not homologous (Strathmann 1978; Lacalli 1993; Thollesson and Norenburg 2003; Dunn et al. 2014). As for insects, Davidson et al. (1995) and Peterson et al. (1997) placed insect development into a different category entirely, curiously asserting that insects were direct developers. Given their dramatic and famous larval to adult transitions, most authors consider the holometabolous insects—a monophyletic group, including butterflies, beetles, flies, bees, wasps, ants, and other lesser known orders—to be archetypal examples of indirect development with complete metamorphosis (see, e.g., the numerous definitions of metamorphosis in Bishop et al. 2006). In any case, there is no dispute that what most authors call complete metamorphosis in insects evolved independently from that in our other two example taxa.

1.5.1 Nemerteans

Ribbon worms (phylum: Nemertea) are a clade of unsegmented mostly marine worms, and like many marine animal phyla, extant ribbon worms have a diversity of developmental modes from feeding to nonfeeding. Most modern phylogenies place ribbon worms in a superphylum called the Lophotrochozoa, which includes mollusks, annelids, and lophophorates such as bryozoans and brachiopods (Dunn et al. 2014). In addition to sharing spiral cleavage with other members of this clade, several of them (annelids, mollusks, and entoprocts) have a trochophore larva, and most larval biologists consider the unusual larval forms in ribbon worms to be modified trochophores (e.g., Maslakova 2010).

One class of ribbon worms called the Pilidiophora has a unique and derived feeding larval form called the pilidium (Figure 1.5; Thollesson and Norenburg

2003; Maslakova 2010). In the pilidium larval life cycle, all major cell types in the larva have differentiated before development of the juvenile worm begins. The juvenile worm begins to form by the invagination of several imaginal disc rudiments; paired rudiments arise as invaginations of the larval ectoderm, while unpaired rudiments appear to be mesenchymal (Maslakova 2010) (Figure 1.3D). Each of these discs—cephalic pair, cerebral pair, trunk pair, and singular posterior dorsal—is a discontiguous section that forms a defined part of the juvenile body. These discs coalesce around the larval stomodaeum,[5] so the gut potentially functions as a physical lattice, if not an organizer. Invaginated paired discs pinch away from the larval ectoderm forming a bilayered rudiment, in which the thick inner layer becomes the juvenile ectoderm and the outer layer encases the entire worm in a thin membrane called the amnion (Maslakova 2010; Figures 1.3B and 1.5C). Similar to indirect development in sea urchins, development of the pilidium larva culminates in a catastrophic metamorphosis, in which the juvenile worm rapidly erupts from and consumes the larval body.

The growth of the pilidium larva occurs via a population of potentially pluripotent cells that reside in specific regions of the larval epithelium in the indentations between the larval lobes and lappets. To draw an analogy to plant ancillary meristems, Bird et al. (2014) termed these pluripotent cell–containing regions "axils." There are four such outer and inner axils in the larva that form growth zones, and the majority of larval growth is attributable to cells in these axils. Interestingly, the imaginal discs are derived (at least in part) from precisely the same axils that, postembryonically, produced the majority of larval cells. Recent cell fate mapping data on the ribbon worm *Maculaura alaskensis* (von Dassow and Maslakova 2017) indicates that the axil cells are homologous to cells that, in a typical trochophore larva, would terminally differentiate early (Category 2, Table 1.1) and form a band of cilia called the prototroch (Figure 1.5A). Instead of this ancestral program of differentiation, axil cells continue to divide throughout larval development and give rise to the ciliated band and other novel pilidium tissues. The authors suggest that selection for this change in cellular division rates and profiles may have been a key innovation in the origin of the pilidium—a maximally indirect-developing larva characteristic of ribbon worms in the class Pilidiophora. Therefore, in this example the notion of a deferred-use cell rather than a "set-aside cell" rises above semantics: *these potentially pluripotent axillary cells are proliferative during the entire life of the larva and during the formation of the juvenile body.* Moreover, as Bird et al. (2014) were aware, this result alone argues against one of the central tenets of the set-aside cell hypothesis: that larval bodies are built from cells with intrinsically limited capacity for cell division and that this limitation is reached at the end of embryonic development.

Because the pilidium larva is derived from within the Nemertea, it becomes of interest to know whether this set of four proliferative zones, or axils, are novelties (apomorphies) as well. *Pantinonemertes californiensis,* a member of the Hoplonemertea (sister taxon to Pilidiophora), has a nonfeeding planuliform larvae called a decidula (Maslakova 2010; Figure 1.5B). In the decidula, which transforms into a juvenile much more gradually than a pilidium, some of the juvenile structures

[5] A stomodaeum is a blind-end gut, an unusual configuration for a bilaterian larva.

arise from spatially segregated invaginations of larval epidermis (Hiebert et al. 2010); gene expression studies are consistent with the hypothesis of homology between ectodermal invaginations and pilidium imaginal discs (Hiebert and Maslakova 2015). The larval epidermis is shed or resorbed during development of the decidula larva (Hiebert et al. 2010), representing a minimal kind of metamorphosis. Hiebert and Maslakova (2015) suggested that the juvenile epidermal invaginations and epidermal shedding could represent a strategy to minimize or eliminate interruptions to larval ciliation patterns, which would otherwise reduce swimming performance.

Several groups within Lophotrochozoa (annelids, mollusks, and the less well-known entoprocts) have a trochophore larva, characterized by the presence of the preoral transverse ciliary band called the prototroch, which is derived from the same early-differentiating trochoblast cell lineage referred to above and serves as the primary swimming organ in the larva. Classic trochophore development is exemplified in annelids, in which the larval anterior-posterior axis is coaxial with the adult one (Figure 1.5A); larval growth involves additions of segments at the posterior, lengthening the larval into a more recognizably annelid appearance. The presumed ancestral annelid-like condition can be considered Category 5 in our ontology: deferral of posterior segments relative to formation of the feeding trochophore larva. By contrast, the pilidium with its imaginal discs would be an example of Category 6: deferral via the formation of juvenile rudiment structures (the imaginal discs) early in larval development, alongside—but at orthogonal axes to—the growing pilidium larva (Figure 1.5C). Although the polarity remains uncertain, the Hoplonemertean nonfeeding decidula larva (Figure 1.5B) may represent an intermediate between Categories 5 and 6 in that there are populations of cells that form juvenile parts interposed heterogeneously with the larval epidermis, but those rudiments are coaxial with the larva, and the metamorphosis is thus less radical.

1.5.2 Echinoderms

Echinoderms (sea urchins, sea stars, brittle stars, sea cucumbers, and sea lilies) have among the most radical metamorphoses described for animals (Figures 1.3A–C, 1.5D–F). The canonical echinoderm larva is a bilateral feeding form, which metamorphoses into a pentamerally symmetric adult (Chia and Burke, 1978). These larvae can do so in dramatic fashion: within as little as 15 min, the swimming planktonic form (Figure 1.3E) degrades, and a bottom-dwelling locomotory juvenile emerges (Figure 1.3F).

To accomplish this rapid transition, echinoderms employ a similar tactic described above for the pilidium larva: they initiate juvenile development early in the larval period through development of a segregated rudiment. In Section 1.3, we described the early stages of rudiment development in echinoids (sea urchins and sand dollars): via an invagination of a portion of larval ectoderm underlying the mesodermal hydrocoel. Together—and along with additional coelomic participants—the invaginated ectoderm and the hydrocoel form the oral field of the pentameral juvenile, with tube feet and spines developing within a functioning, feeding larva (Figures 1.3A, B, 1.5F). In the oldest living group of echinoids, the pencil urchins (order: Cidaroida), this process is less extreme. Cidaroids differ from typical echinoids in that they do not have a rudiment invagination, they do not develop definitive "adult-type" spines until the juvenile stage, and they retain much of the larval epidermis as juveniles (Emlet 1988). In this sense, they have a less radical metamorphic process, more akin to that seen in the sea cucumbers (Figure 1.5D), which likewise retains their larval epidermis into the juvenile stage (Chia and Burke 1978).

With respect to our proposed ontology, noncidaroid echinoids (subclass: Euechinoidea) such as the purple urchin *Strongylocentrotus purpuratus* exhibit Category 6 deferral of juvenile development via rudiment structures formed early in the larval period, in contrast to the more Category 5-like cidaroids and sea cucumbers. In addition to these broader patterns, echinoderms demonstrate numerous evolutionary transitions from feeding to nonfeeding larval development. And these derived life history modifications are accompanied by simpler ciliated larval forms that exhibit less extreme metamorphic transitions, which we might call a reversion to a more "Category 4-like" state. In the most extreme example known, development in the slime sea star *Pteraster tesselatus* exhibits minimal traces of a bilateral larva. In *P. tesselatus,* the embryo develops directly to the juvenile via an abbreviated nonfeeding dispersive stage referred to as a mesogen, in which pentameral symmetry emerges early in development via five circumferential outpocketings of the gut along the animal-vegetal axis of the embryo (Janies and McEdward 1993; Figure 1.5F). The manner in which the mesogen transitions into a juvenile slime star is thus analogous to the annelid example described above (Figure 1.5A). However, in annelids, the pattern of deferral is the seeming ancestral (plesiomorphic) condition, whereas in slime stars, Category 4 deferral is clearly derived (an apomorphy) from Categories 5–6 sea star ancestor (Janies and McEdward 1993).

1.5.3 Insects

Our third and final comparison is insects, a group that Davidson et al. (1995) placed in a separate category of development from the majority of animals due to their derived early nuclear division patterns without complete cleavage.[6] And although much of embryogenesis in insects is indeed unique relative to even their sister group, the Crustacea, a broader look at the evolution of insect life histories reveals a striking parallel to our descriptions of ribbon worm and echinoderm metamorphosis above.

Like all arthropods, insects grow between molts, and the stages of development are denoted by the number and type of molts having occurred. The most ancient living insect lineage is a primitively wingless group referred to as the Ametabola, which have no distinct metamorphosis (Figure 1.6). Juvenile Ametabola such as silverfish look nearly identical to adult silverfish except that they are not yet reproductively mature (Figure 1.6A). In addition, silverfish have indeterminate molting, as they continue to molt after reaching reproductive maturity. By contrast, all of the more derived, winged insects have determinate molting. With the sole exception of mayflies, insects with wings cannot molt, and the adult stage is the final stage (Figure 1.6B, C).

Among the winged insects are two main groups characterized by distinct life histories. The hemimetabolous insects (Figure 1.6B)—such as true bugs (Hemiptera),

[6] An interesting example of deferred development in its own right: deferral of cytokinesis.

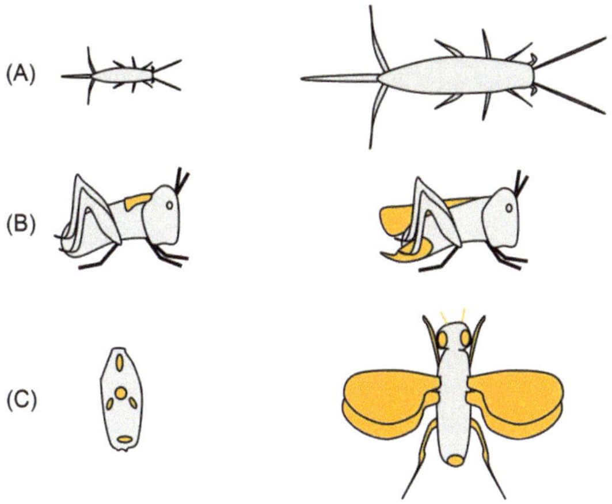

FIGURE 1.6 Life history strategies in insects. Stylized versions of (A) ametabolous, (B) hemimetabolous and (C) holometabolous life histories, indicating deferred-use structures (*in gold*) specified in immature stages (*left-side images*), and differentiated in adult stages (*right-side images*). (A) Absence of deferred structures in Ametabola. (B) Differentiation of wings and reproductive structures specified in nymphs (left), whose differentiation is deferred until the final molt to the adult stage (right) in Hemimetabola. (C) Rudiments, in the form of imaginal discs, in larvae (left) that produce most of the adult structures (right) aside from the abdomen in Holometabola. Note that holometabolous insects have a pupal stage not shown here.

grasshoppers, cockroaches, and mantids—are so-called because they have a subtle metamorphosis between their nymphal stages that lack functional wings and reproductive structures and their adult stage that has these features. The second main group of winged insects are the Holometabola (Figure 1.6C), so-called because they have a famously dramatic metamorphosis. Examples of the holometabolous insects are flies, beetles, and butterflies, each of which has a larval form (maggot, grub, and caterpillar) that passes through a metamorphic pupal stage before emerging as what is generally a strikingly different looking, winged adult.

Underlying these three different life history types in insects (Figure 1.6)—are corresponding differences in deferred development via rudiments. In the Ametabola, there is an obligatory delay of several molts until the attainment of sexual maturity, and this delay is accompanied by corresponding deferral of differentiation of the internal and external reproductive structures (Lindsay 1940; Figure 1.6A). Still, the timing of these differentiation events in silverfish and other Ametabola is quite variable and occurs gradually over a span of several molts (Lindsay 1940; Wigglesworth 1972). According to our proposed ontology, the delay in the cellular and developmental events accompanying reproductive maturity in the Ametabola is thus an example of Category 4 deferral.

In the Hemimetabola, by contrast, there is a clearer distinction between preadult ("nymph") and adult stages. Although the primordia for both the wings and the genitalia are present during hemimetabolous nymphal stages (Figure 1.6B), there is a discontinuous increase in their prominence in the final molt to the adult stage (Wigglesworth 1972). Furthermore, this nymphal-to-adult molt often corresponds to a dramatic change in habitat and corresponding physiology, such as from aquatic-to-terrestrial in some hemimetabolous insects, including stoneflies (order: Plecoptera). This metamorphic life history shift from nymph to adult in hemimetabolous insects, and the associated greater distinction between larval and juvenile structures when compared to the Ametabola, thus qualifies as an example of Category 5 deferral in our proposed ontology.

The monophyletic Holometabola share several features that correspond to a more dramatic life history shift than seen in the Hemimetabola. First, as previously mentioned, the morphology of holometabolous preadult ("larval") stages are morphologically quite distinct from the adults (Figure 1.6C). Second, there is a bona fide metamorphic life stage in holometabolous insects—the pupa or chrysalis—which intervenes between larva and adult. And third, as in echinoderms and ribbon worms, we see a trend toward early segregation of the primordia of adult structures in the form of ectodermal invaginations (Figure 1.6C). These rudiment structures, as in the ribbon worms, are called imaginal discs, and the cells that form them are specified in the embryo and then proliferate during larval stages before differentiating in the pupa.

As we observed in the trochophore-to-pilidium derivation, here we again see evidence for an evolutionary progression toward increased reliance on rudiments, with correspondingly greater portions of the adult being specified by imaginal discs in crown group Holometabola. A phylogenetic analysis by Truman and Riddiford (1999) indicates that early specification and proliferation in imaginal discs arose independently in several different holometabolous lineages. Specifically, these authors document a reiterated progression from Category 5 to Category 6 deferral in what appears to be a minimum of six different holometabolous lineages, in response to hypothesized selection for more rapid metamorphic transitions.

We want to reiterate that the categories in our proposed ontology are not hierarchical in nature, nor do they automatically imply an evolutionary series, as exemplified by adult cuticle formation in the holometabolous insects. Whereas the imaginal discs in the Holometabola generate the adult limbs, head, and dorsal thorax (Figure 1.6C), their adult abdominal cuticle is derived from small nests of histoblast cells. These cells make larval and pupal cuticle, proliferate during metamorphosis, and then produce adult abdominal cuticle before eclosion into the adult stage (Ninov et al. 2009). In this sense, the deferral of histoblast proliferation and formation of adult cuticle until the metamorphic signal is received is an example of Category 4 deferral in the Holometabola, as the histoblasts are specialized to only produce abdominal cuticle. Meanwhile, during the same deferral period when histoblast cells are either proliferating or producing larval or pupal cuticle, the majority of the adult body is forming via imaginal disc rudiments of the Category 6 variety.

And again, in insects, there are examples where Category 5 or 6 deferral is evolutionarily lost. Gall midges (Diptera: Cecidomyiidae) are typical holometabolous insects, with larval maggots that metamorphose through a pupal stage into winged adults. One group of gall midges are mushroom pests in the maggot stage. When a female finds a mushroom on which to oviposit, she lays female-determined eggs that will hatch into "paedogenetic" maggots: their ovaries mature during larval development, far earlier than in typical flies (see Section 1.4); the eggs activate parthenogenetically and hatch inside the mother maggot; and the baby maggots consume the mother and then emerge to continue feeding on the mycelium (reviewed in Went 1979). In this extreme abbreviated life cycle, metamorphosis and all signs of adult development are skipped entirely, and development of the ovaries is accelerated via early activation of ecdysone responsiveness particularly in the ovary primordium (Hodin and Riddiford 2000b). When the mushroom patch dies off, then the larvae will proceed through metamorphosis to produce male and female winged adults that will mate and disperse to find a new mushroom patch to exploit. An analogous case is seen in aphids with telescoped generations (see Dixon 1985). These examples, as we discussed above with the insect abdominal histoblasts and the slime star mesogen (Figure 1.5F), exemplify how our six proposed categories of deferred development should not be considered a unidirectional hierarchical progression.

1.6 SUMMARY AND CONCLUSION

Nearly all multicellular life invokes a cyclical need to integrate cell division with ontogenetic processes so that divergent cell functions arise in the right place at the right time. Additional demands of multicellular ontogeny are to make the bodies of later life cycle stages from those of earlier ones and to repair or replace parts. Collectively, these elements of ontogeny have converged upon the need for mechanisms to maintain and then express developmental potential in a temporally and spatially structured manner. Pluripotentiality is thus an essential commodity of multicellular life that can be retained by either physically segregating populations of cells whose terminal developmental programs are deferred to varying degrees until signaled, or initiating regulated program of re- or transdifferentiation. In metazoans, the prevailing mode of development relies on deferred-use developmental programs, rather than transdifferentiation; this may be an evolutionarily derived state. Because there are multiple ways in which animal ontogeny is time-structured, using examples of development in selected animals, we have attempted to formalize an ontology of deferred development and deferred-use cells. We hope that this exercise engenders comparisons among taxa about how different lineages employ deferred development and deferred-use cells to contend with the demands of ontogeny.

ACKNOWLEDGEMENTS

The authors thank members of Journal Club at Friday Harbor Laboratories, and L. Riddiford, R.R. Strathmann and Rachel Merz in particular, for numerous insightful comments that improved an earlier draft of this chapter. We thank Svetlana Maslakova for providing pictures of ribbon worms and for reviewing and commenting upon Section 1.5A. JH acknowledges support from a University of Washington Royalty Research Fund.

REFERENCES

Alvorado, A.S. 2000. Regeneration in the metazoans: why does it happen? *Bioessays* 22: 578–590.

Barresi, M.J.F., and S.F. Gilbert. 2016. *Developmental Biology* (11th ed.). Sinauer: New York.

Bell, G., and V. Koufopanou. 1991. The architecture of the life cycle in small organisms. *Phil. Trans. R. Soc. Lond. B* 332: 81–89.

Bird, A.M., G. von Dassow, and S.A. Maslakova. 2014. How the pilidium grows. *Evodevo* 5: 13.

Bishop, C.D., D.F. Erezyilmaz, T. Flatt, C. Georgiou, M. Hadfield, A. Heyland, J. Hodin, M.W. Jacobs, S. Maslakova, A. Pires, A.M. Reitzel, S. Santagata, K. Tanaka, and J.H. Youson. 2006. What is metamorphosis? *Int. Comp. Biol.* 46: 655–661.

Buss, L.W. 1987. *The Evolution of Individuality*. Princeton University Press: Princeton.

Campanale, J.P., T. Gökirmak, J.A. Espinoza, N. Oulhen, G.M. Wessel, and A. Hamdoun. 2014. Migration of sea urchin primordial germ cells. *Dev. Dyn.* 243: 917–927.

Chia, F.S., and R.D. Burke. 1978. Echinoderm metamorphosis: fate of larval structures. In *Settlement and Metamorphosis of Marine Invertebrate Larvae*, (eds.) F.S. Chia and M.E. Rice, pp. 219–234. Elsevier: New York.

Cohen, J. 1985. Metamorphosis: Introduction, usages, and evolution. In *Metamorphosis*, (eds.) M. Bulls and M. Brownes. Clarendon: Oxford.

Davidson, E.H., K.J. Peterson, and R.A. Cameron. 1995. Origin of bilaterian body plans: evolution of developmental regulatory mechanisms. *Science* 270: 1319–1325.

Dixon, A.F.G. 1985. *Aphid Ecology*. Chapman and Hall: New York.

Dunn, C.W., G. Giribet, G.D. Edgecombe, and A. Hejnol. 2014. Animal phylogeny and its evolutionary implications. *Annu. Rev. Ecol. Evol. Syst.* 45: 371–395.

Emlet, R.B. 1988. Larval form and metamorphosis of a “primitive” sea urchin, Eucidaris thouarsi (Echinodermata: Echinoida: Cidaroida), with implications for developmental and phylogenetic studies. *Biol. Bull.* 174: 4–19.

Extavour, C.G., and M. Akam. 2003. Mechanisms of germ cell specification across the metazoans: epigenesis and preformation. *Development* 130: 5869–5884.

Gancz, D., T. Lengil, and L. Gilboa. 2011. Coordinated regulation of niche and stem cell precursors by hormonal signaling. *PLoS Biol.* 9: e1001202.

Gould, S.J. 1984. Challenges to neo-darwinism and their meaning for a revised view of human consciousness. In *The Tanner Lectures on Human Values*, (ed.) S.M. McMurrin, Vol. 6, pp. 53–74. Cambridge University Press: Cambridge.

Gould, S.J., and R.C. Lewontin. 1979. The spandrels of San Marco and the Panglossian paradigm: a critique of the adaptationist programme. *Proc. R. Soc. Lond. B Biol. Sci.* 205: 581–598.

Grosberg, R.K., and R.R. Strathmann. 1998. One cell, two cell, red cell, blue cell: the persistence of a unicellular stage in multicellular life histories. *Trends Ecol. Evol.* 13: 112–116.

Grosberg, R.K., and R.R. Strathmann. 2007. The evolution of multicellularity: a minor major transition. *Annu. Rev. Ecol. Evol. Syst.* 38: 621–654.

Hiebert, L.S., G. Gavelis, G. von Dassow, and S.A. Maslakova. 2010. Five invaginations and shedding of the larval epidermis during development of the hoplonemertean Pantinonemertes californiensis (Nemertea: Hoplonemertea). *J. Nat. Hist.* 44(37–40): 2231–2347.

Hiebert, L.S., and S.A. Maslakova. 2015. Expression of Hox, Cdx, and Six3/6 genes in the hoplonemertean Pantinonemertes californiensis offers insight into the evolution of maximally indirect development in the phylum Nemertea. *Evodevo* 6: 26. doi:10.1186/s13227-015-0021-7.

Hodin, J. 2009. Chapter 11: She shapes events as they come: plasticity in insect reproduction. In *Phenotypic Plasticity of Insects: Mechanisms and Consequences*, (eds.) D. Whitman and T.N. Ananthakrishnan, pp. 423–521. Science Publishers, Inc. Enfield: New Hampshire.

Hodin, J., and L.M. Riddiford. 2000a. Different mechanisms underlie phenotypic plasticity and interspecific variation for a reproductive character in drosophilids (Insecta: Diptera). *Evolution* 54: 1638–1653.

Hodin, J. and L.M. Riddiford. 2000b. Parallel alterations in the timing of ovarian ecdysone receptor and ultraspiracle expression characterize the independent evolution of larval reproduction in two species of gall midges (Diptera: Cecidomyiidae). *Dev. Genes Evol.* 210: 358–372.

Horvitz, H.R. 1990. Genetic control of *Caenorhabditis elegans* cell lineage. *Harvey Lect.* 84: 65–77.

Janies, D.A., and L.R. McEdward. 1993. Highly derived coelomic and water-vascular morphogenesis in a starfish with pelagic direct development. *Biol. Bull.* 185: 56–76.
Jenner, R.A. 2000. Evolution of animal body plans: the role of metazoan phylogeny at the interface between pattern and process. *Evol. Dev.* 2: 208–221.
Juliano, C.E., S.Z. Swartz, and G.M. Wessel. 2010. A conserved germline multipotency program. *Development* 137: 4113–4126.
Kirk, D.L. 2001. Germ-soma differentiation in *Volvox*. *Dev. Biol.* 238: 213–223.
Kooij, A., C.P.W.M. van der Veraart, and A.E. van Loon. 1998. Cyclin A, cyclin B and string-like are regulated separately in cell cycle arrested trochoblasts of *Patella vulgata* embryos. *Dev. Genes Evol.* 207: 524–534.
Korotkova, G.P. 1970. Regeneration and somatic embryogenesis in sponges. In *Biology of the Porifera: Symp. Zool. Soc. London No. 25*, (ed.) W.G. Fry, pp. 423–436. Academic Press: New York.
Lacalli, T.C. 1993. Ciliary bands in echinoderm larvae: evidence for structural homologies and a common plan. *Acta Zool.* 74: 127–133.
Lai, A.G., and A. Boobaker. 2018. Time to uncover deep conservation or convergence of adult stem cell evolution and regenerative processes. *Devl. Biol.* 433: 118–131.
Lee, R.E. 2008. *Phycology* (5th ed.). Cambridge University Press: New York.
Lewontin, R.C. 2001. In the beginning was the word. *Science* 291: 1263–1264.
Lindsay, E. 1940. The biology of the silverfish, *Ctenolepisma longicaudata* Esch., with particular reference to its feeding habits. *Royal Soc. Vic.* 52: 35–83.
Maslakova, S.A. 2010. The invention of the pilidium larva in an otherwise perfectly good spiralian phylum Nemertea. *Int. Comp. Biol.* 50: 734–743.
Mortensen, T.H. 1921. *Studies of the Development and Larval Forms of Echinoderms*. G.E.C. Gad, Copenhagen, Denmark.
Moss, E.G. 2007. Heterochronic genes and the nature of developmental time. *Curr. Biol.* 17: R425–R434.
Muller, H.J. 1964. The relation of recombination to mutational advance. *Mutat. Res.* 106: 2–9.
Ninov, N., C. Manjón, and E. Martín-Blanco. 2009. Dynamic control of cell cycle and growth coupling by ecdysone, EGFR, and PI3K signaling in Drosophila histoblasts. *PLoS Biol.* 7: e1000079. doi:10.1371/journal.pbio.1000079.
Panchal, T., X. Chen, E. Alchits, Y. Oh, J. Poon, J. Kouptsova, F.A. Laski, and D. Godt. 2017. Specification and spatial arrangement of cells in the germline stem cell niche of the Drosophila ovary depend on the Maf transcription factor Traffic jam. *PLoS Genet.* 13: e1006790.
Pehrson, J.R., and L.H. Cohen. 1986. The fate of the small micromeres in sea urchin development. *Dev. Biol.* 113: 522–526.
Peterson, K.J., R.A. Cameron, and E.H. Davidson. 1997. Set-aside cells in maximal indirect development: evolutionary and developmental significance. *Bioessays* 19: 623–631.
Raff, R.A. 1996. *The Shape of Life*. University of Chicago Press: Chicago.
Raff, R. 2008. Origins of the other metazoan body plans: the evolution of larval forms. *Phil. Trans. R. Soc.* 363: 1473–1479.
Ricklefs, R.E. 1979a. Adaptation, constraint, and compromise in avian postnatal development. *Biol. Rev. Cambr. Phil. Soc.* 54: 269–290.
Ricklefs, R.E. 1979b. Patterns of growth in birds. V. A comparative study of development in the starling, common tern and Japanese quail. *Auk* 96: 10–30.
Ruppert, E.E., and E.J. Balser. 1986. Nephridia in the larvae of hemichordates and echinoderms. *Biol. Bull.* 171: 188–196.
Santelices, B., and J. Alvarado. 2006. Applying the concept of metamorphosis to the crustose-to-erect thallus transition of macroalgae. *Int. Comp. Biol.* 46: 713–718.
Slack, J.M.W., and D. Tosh. 2001. Transdifferentiation and metaplasia—switching cell types. *Curr. Opinion Genet. Dev.* 11: 581–586.

Sogabe, S., W.L. Hatleberg, K.M. Kocot, T.E. Say, D.S. Stoupin, K.E. Roper, L. Fernandez-Valverde, S.M. Degnan and B.M. Degnan. 2019. Pluripotency and the origin of animal multicellularity. *Nature* 570: 519–522. doi:10.1038/s41586-019-1290-4.

Soldner, F., and R. Jaenisch. 2018. Stem cells, genome editing, and the path to translational medicine. *Cell* 175: 615–632.

Staver, J.M., and R.R. Strathmann. 2002. Evolution of fast development of planktonic embryos to early swimming. *Biol. Bull.* 203: 58–69.

Strathmann, R.R. 1978. The evolution and loss of feeding larval stages of marine invertebrates. *Evolution* 32: 894–906.

Strathmann, R.R. 2000. Functional design in the evolution of embryos and larvae. *Semin. Cell Dev. Biol.* 11: 395–402.

Strathmann, R.R., L. Fenaux, and M.F. Strathmann. 1992. Heterochronic developmental plasticity in larval sea urchins and its implications for evolution of nonfeeding larvae. *Evolution* 46: 972–986.

Swartz, S.Z., and G.M. Wessel. 2015. Germ line versus soma in the transition from egg to embryo. In *The Maternal-to-Zygotic Transition*, (ed.) H. Lipshitz, Current Topics in Developmental Biology, Vol. 113. Elsevier Press: Amsterdam.

Takahashi, K., and S. Yamanaka. (2006). Induction of pluripotent stem cells from mouse embryonic and adult fibroblast cultures by defined factors. *Cell* 126: 663–676.

Thollesson, M., and J.L. Norenburg. 2003. Ribbon worm relationships – a phylogeny of the phylum Nemertea. *Proc. R. Soc. London, Ser. B* 270: 407–415.

Thorson, G. 1950. Reproductive and larval ecology of marine bottom invertebrates. *Biol. Rev.* 25: 1–157.

Truman, J.W., and L.M. Riddiford. 1999. The origins of insect metamorphosis. *Nature* 401: 447–452.

Truman, J.W., and L.M. Riddiford. 2002. Endocrine insights into the evolution of metamorphosis in insects. *Annu. Rev. Entomol.* 47: 467–500.

Vervoort, M. 2011. Regeneration and development in animals. *Biol. Theory* 6: 25–35.

von Dassow, G., and S.A. Maslakova. 2017. The trochoblasts in the pilidium larva break an ancient spiralian constraint to enable continuous larval growth and maximally indirect development. *EvoDevo* 8: 19. doi:10.1186/s13227-017-0079-5.

Went, D.F. 1979. Paedogenesis in the dipteran insect *Heteropeza pygmaea*: an interpretation. *Int. J. Invertebr. Reprod.* 1: 21–30.

Werner, E.E. 1988. Size, scaling and the evolution of complex life cycles. *In Size-Structured Populations*, (eds.) B. Ebenman and L. Persson, pp. 60–81. Springer-Verlag: Berlin.

Wessel, G.M., L. Brayboy, T. Fresques, E.A. Gustafson, N. Oulhen, I. Ramos, A. Reich, S.Z. Swartz, M. Yajima, and V. Zazueta. 2014. The biology of the germ line in echinoderms. *Mol. Rep. Dev.* 81: 679–711.

Wigglesworth, V.B. 1972. *The Principles of Insect Physiology* (7th ed.). Chapman & Hall: London.

Wilson, D.P. 1932. On the mitraria larva of *Owenia fusiformis* Delle Chiaje. *Philos. Trans. R. Soc. Lond. B Biol. Sci.* 221: 231–334.

Wolpert, L. 1999. From egg to adult to larva. *Evol. Devl.* 1: 3–4.

Xie, T., and A.C. Spradling. 2000. A niche maintaining germ line stem cells in the Drosophila ovary. *Science* 290: 328–330.

Yajima, M., and G.M. Wessel. 2011. Small micromeres contribute to the germline in the sea urchin. *Development* 138: 237–243.

2 Deferred-Use Molecules and Decision-Making in Development

Sally A. Moody
George Washington University

Steven L. Klein
National Science Foundation

CONTENTS

2.1 INTRODUCTION

Many organisms have evolved cell populations that are set aside during embryonic stages for use later in the life cycle. These often are kept in a quiescent state until needed to form intermediate larval structures or even adult structures that will emerge after metamorphosis. Well-recognized examples include larval insect imaginal discs, which form adult structures after metamorphosis, and primordial germ cells, which become gametes after the organism reaches sexual maturity.

Other deferred-use cells remain quiescent as a stem cell population in protected niches. For example, stem cells in amphibian larvae become activated by thyroid hormone signaling during metamorphosis to restructure various organs into the adult forms or to produce new structures such as the lungs (Buckholz and Shi, 2018). In many organisms, adult tissues retain stem cells that are called upon to replace aging and damaged cells and to repair injured tissues. *The process of deferred use also exists at the molecular level within single cells.* In particular, the oocytes of many animals have evolved mechanisms for setting aside mRNAs and proteins whose functional activities are deferred until after fertilization.

2.1.1 Deferred-Use Molecules Can Specify Deferred-Use Cells

Embryonic development begins with fertilization, followed by multiple cellular interactions that transform the zygote into a multicellular organism. In many species, three essential patterning events occur during the first few cell divisions:

1. the *body axes* are established,
2. the *germ line* is set aside as a separate population from somatic cells, and
3. the *primary embryonic germ layers* are specified.

Thereafter, the germ layers interact and differentiate into the tissues and organs of the body. Remarkably, in many animals, these first three essential events occur prior to transcription of the zygotic genome. Instead of relying on differential zygotic transcription in different cell lineages—some of which are set-aside cells, for which see Chapter 1—in these animals, these events are controlled by *deferred-use molecules*, which are RNAs and proteins that are either synthesized by maternal cells and transported into the oocyte or synthesized by the oocyte itself. Developmental processes use several strategies to prepare for the future, including deferred-use cells and deferred-use molecules. In this chapter, we concentrate on *deferred-use molecules as an alternative mechanism for cell determination.*

Deferred-use, maternal molecules are stored for use after fertilization and direct the embryonic cells that inherit them to accomplish the early embryonic specification events. These molecules have historically been referred to as "maternal determinants" because they are "maternal" (i.e., synthesized by the mother/oocyte) and they "determine" aspects of embryonic cell fate. Evidence for the existence of these molecules first came from observing the movement of cytosolic particles in transparent marine invertebrates and then from more recent gene discovery assays in a large variety of animals. Some of these molecules are kept silent during oogenesis and activated immediately upon fertilization to orchestrate such early developmental processes as establishing the embryonic axes and specifying the primary germ layers. Other molecules are kept silent during early embryogenesis and activated only after organogenesis (e.g., the germ line).

2.1.2 Deferred-Use Molecules and the First Cell Divisions of the Embryo

In animals, the developmental period regulated by maternal molecules encompasses the first few synchronous cell cycles after fertilization; the exact number of cell cycles and the duration of maternal influence depend on the length of the cell cycle in each species. For example, in mice, it lasts about 10h during the first cell cycle and, in sea urchins and the nematode *Caenorhabditis elegans*, about 1–2h during the first two cell cycles. In fruit flies (*Drosophila*), the South African clawed frog (*Xenopus laevis*), and zebrafish (*Danio rerio*), it lasts 2–6h, encompassing 6–10 cell cycles (Abrams and Mullins, 2009; Tadros and Lipshitz, 2009; Lee et al., 2014). It is thought that providing "premade" maternal molecules allows the embryonic cells to rapidly divide—i.e., replicate their genomes—without waiting for new gene transcription to occur. This mechanism seems different from stem cells whose cell division rates are prolonged compared to differentiated cells. Using premade molecules during rapid cell divisions allows the embryo to complete immediately necessary patterning steps as quickly as possible and to amass the required number of cells to undergo later morphogenetic movements and organogenesis. It is notable that the species in which these processes have been best studied in the laboratory develop quite rapidly (echinoderms, fly, nematode, *Xenopus*, zebrafish), suggesting that they have a special evolutionary requirement for speed.

The situation seems to be very different in animals that grow more slowly, such as birds, reptiles, mammals, urodele amphibians, and some fishes (e.g., Johnson et al., 2003). In fact, the maternal control period of plants ends during the first cell cycle, as discussed in Section 2.5. In all these species, it is likely that the earliest patterning events are not controlled by deferred-use molecules but rather by inductive interactions and zygotic gene products.

In many animals, the period of maternal control comes to an end when the zygotic genome becomes activated at the *maternal to zygotic transition* (MZT). The cell cycle lengthens, providing enough time for the transcription of new genes. Additionally, most maternal gene products are degraded (Tadros and Lipshitz, 2009; Lei et al., 2013; Miller, 2015; Yang et al., 2015). The differential expression of the zygotic genome begins to direct cells to differentiate into different tissues and cell types. However, aspects of this differential expression are set up by region-specific localization of maternal determinants, a lasting testament to the significant role played by deferred-use molecules. For example, the determination of the dorsal-ventral (D-V) axis by maternal Wnt signaling (discussed in Section 2.3) leads to the specification of the organizer mesoderm and its specific expression of transcription factors and secreted factors that pattern the mesoderm and induce the neural ectoderm.

In this chapter, we concentrate on maternal molecules that regulate the early patterning events that lead to the formation of the body axes and the primary germ layers. We also discuss the role of maternal molecules that regulate the determination of primordial germ cells. We describe how their role was discovered and *discuss several mechanisms by which they exert different effects in different regions, cell lineages, and deferred-use cells.*

2.2 AN HISTORICAL PERSPECTIVE

The question of how a fully formed organism arises from the egg has been considered, at least philosophically, since the time of Aristotle. Some early philosophers argued that all of the elements of the world must be "preformed," whereas others argued that they must unfold gradually in a series of steps (reviewed by Klein and Moody, 2016). In the 19th century, this philosophical discussion was revisited when the egg was recognized to be a single cell that divided into many cells to give rise to a fully formed, adult body. The burning question then was to determine whether the elements of the adult body were preformed in the egg or whether they were acquired by unfolding mechanisms that could involve interactions between parts (reviewed by Wilson, 1925 and in Dunn, 1917). This was first addressed by directly observing cell divisions from fertilization to the formation of germ layers and tissues in small, transparent invertebrates, including leech, ascidians, marine worms, mollusks, and echinoderms. These species were used because they develop external to the mother and are large enough to visualize cytoplasmic particles (even with 19th-century microscopes). Additionally, they divide in regular cleavage patterns so that cell lineages could be followed and are composed of only a small number of cells that could be tracked to a defined phenotype (reviewed by Wilson, 1925).

Major findings were made in animals whose fertilized eggs contain pigment granules that could be followed visually through each cell division into specific tissue lineages. A famous example is an ascidian egg that contains yellow pigment granules; careful tracing of the cells that inherited these granules showed that they become confined to the progenitors of somitic muscle (Conklin, 1905). This observation was not interpreted to mean that this special cytoplasm contained tiny "preformed" muscle cells, but rather that it contained some kind of cellular information that led to the specification of cells as muscle forming (Dunn, 1917). The critical role of this yellow cytoplasm was demonstrated experimentally decades later by transferring it to a non-muscle precursor cell; the recipient cell changed fate to become muscle, indicating that the cytoplasm contained "determinant" information (Whittaker, 1980). These findings indicated that *muscle development was regulated by substances locally packaged in the egg by the mother*, i.e., by maternal determinants. Later, it was discovered that maternal mRNAs and not the pigment granules themselves provided the muscle fate–determining information (Jeffrey, 1985; Swalla and Jeffrey, 1995; reviewed by Nishida, 2012). This finding led to the elucidation of the gene regulatory network required for ascidian muscle formation (Yu et al., 2019). Additionally, visually following cytoplasmic granules through early cleavage stages of several species revealed cytoplasmic inclusions that are inherited specifically by the germ cell lineage and contain maternal molecules required for gamete formation (see Section 2.3).

Dozens of additional studies using careful fate mapping of each embryonic cell indicated that the body axes (anterior-posterior [A–P] and D-V) and primary embryonic germ layers (endoderm, mesoderm, ectoderm) are at least partially defined during the maternal control period (reviewed by Klein and Moody, 2016). Oogenesis, which can take several months, thus involves processes that synthesize, transport, localize, and store many RNAs and proteins that direct the initial patterning events. We now recognize that sequential gene activation/silencing, differential signaling, inductive events, and instructive cell-cell interactions—i.e., "unfolding" processes—regulate tissue and organ development. Nonetheless, although the egg may not contain a "preformed" animal as envisioned by some early philosophers and embryologists, it does go to extraordinary lengths to construct a prepattern of deferred-use molecules upon which the embryo will be assembled.

2.3 DEFERRED-USE MOLECULES THAT REGULATE EMBRYONIC DEVELOPMENT

Since the discovery of "special cytoplasm" over a century ago, many studies have shown that *numerous RNAs and proteins are synthesized during oogenesis and sequestered to different cytoplasmic domains of the unfertilized egg* (reviewed by Davidson, 1990; Bowerman, 1998; Sullivan et al., 1999; King et al., 2005; Heasman, 2006; White and Heasman, 2008; Abrams and Mullins, 2009; Yang et al., 2015; Ma et al., 2016; Escobar-Aguirre et al., 2017). Many eggs contain maternal molecules that specify the A-P and D-V axes and the germ cell lineage. There also is evidence that molecules localized to different regions of the egg bias the formation of endoderm and ectoderm.

2.3.1 Factors That Specify the Body Axes

Each *Drosophila* ovary is oriented along the A-P axis of the mother's body and consists of 16–18 *ovarioles*. The ovariole is a niche containing an anteriorly located stem cell population (anterior germarium) and a series of egg chambers each consisting of 16 germ line–derived sister cells that are in direct communication via cytoplasmic bridges. These germ line cells are surrounded by a layer of follicle cells that are derived from somatic tissue. One of the 16 cells becomes the oocyte and takes a posterior position in the egg chamber; the remaining 15 cells become nurse cells (Figure 2.1A). During egg maturation, the nurse cells synthesize a large number of proteins and mRNAs that are transported into the oocyte via the cytoplasmic bridges. These molecules are then utilized to pattern the body axes of the future embryo and to establish the germ line (Section 2.3B). In this system, the ovariole niche is specified and set aside during larval stages for adult use (reviewed by Lehmann, 2012; Allbee et al., 2018); it is intriguing that this set-aside population of cells is an important source of deferred-use molecules for early embryonic patterning.

Several of these maternal molecules set up the embryonic body axes in the oocyte prior to fertilization (reviewed by Kenyon, 2007; Ratnaparkhi and Courey, 2014; Ma et al., 2016). An initial cellular indicator of the A-P axis occurs at the midpoint of oogenesis when the oocyte takes a posterior position in the egg chamber and its nucleus takes a posterior position within the ooplasm. The maternal transcripts synthesized by the nurse cells in the ovariole niche and transported into the oocyte subsequently are localized to different A-P positions with relation to the nucleus (Figure 2.1A). For example, *oskar* and *nanos* mRNAs are localized to the posterior pole of the oocyte, where they contribute to specifying the posterior axis of the embryo and the germ line. *bicoid* mRNA, whose encoded protein is required for the specification of anterior cell fates, is transported from the anteriorly located nurse cells into the anterior pole of the oocyte where it becomes tethered to the cytoskeleton. Sequestering each of these transcripts results in localized translation of proteins that then diffuse through the cytoplasm, setting up a gradient that is high at the site of the localized transcripts and low at the opposite pole. *These opposing gradients are critical for regulating the differential expression of the zygotic genes required to form the embryonic A-P axis.*

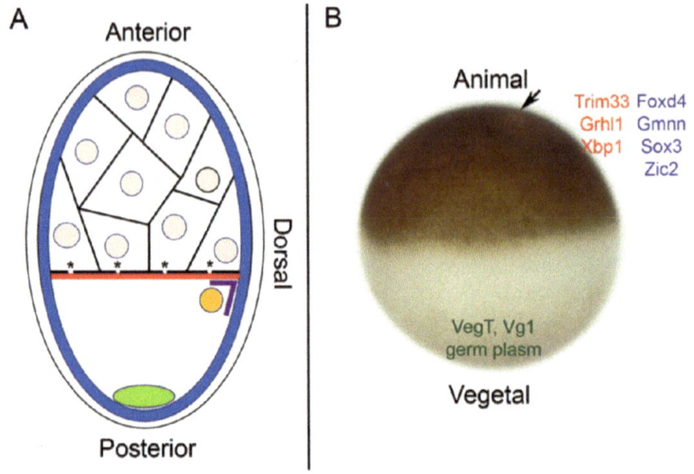

FIGURE 2.1 Molecular asymmetry in oocytes. (A) The *Drosophila* egg case (blue), which is surrounded by somatic follicle cells (white), consists of 15 anteriorly located nurse cells that are connected to the posterior-located oocyte by cytoplasmic bridges (*). Nurse cells provide mRNAs that are transported to the posterior pole (green; e.g., *nanos*, *oskar*) or tethered to the anterior pole (red; e.g., *bicoid*). Localization of these maternal molecules initiates formation of both the A-P axis and the germ line. During mid-oogenesis, the oocyte nucleus (orange) migrates to one side of the anterior pole of the oocyte. Nurse cells supply *gurken* mRNA (purple) that localizes to the cytoplasm between the nucleus and the cell membrane, initiating formation of the D-V axis. (B) The *Xenopus* oocyte is visually polarized along the animal-vegetal axis, with pigment granules present in the animal half but not in the vegetal half. The maternal pronucleus is located in the animal pole (arrow indicates its faint white appearance). Listed are known vegetally localized mRNAs and germ plasm (green), known animal-localized mRNAs required for ectoderm formation (red), and known animal-localized mRNAs required for neural ectoderm formation (blue). VegT, vegetal T-box containing; Vg1, gdf1, growth differentiation factor 1; Trim33, tripartite motif containing 33; Grhl1, grainyhead-like 1; Xbp1, X-box binding protein 1; Foxd4, forkhead box D4-like1; Gmnn, geminin; Sox3, SRY-box 3; Zic2, zic family member 2.

Determination of the D-V axis in *Drosophila* eggs also relies on localized maternal molecules. At mid-gestation, rearrangements of microtubules direct the plus end of each microtubule to the posterior pole and the minus end to the anterior pole. Concomitantly, the oocyte nucleus moves from the posterior pole to an anterior "corner" location (Figure 2.1A). *gurken* mRNA, supplied by the nurse cells, is preferentially localized to the cytoplasm between the nucleus and the "corner" cell membrane. As a consequence of this localization, its epidermal growth factor–like protein is locally translated and secreted. The adjacent follicle cells respond by repressing *pip*, whose protein product is required to activate ventral fate. Thus, *local translation of maternally supplied gurken mRNA leads to the specification of the dorsal side of the fly embryo by repressing a ventralizing factor.*

In *Xenopus*, the asymmetry of the mature egg is visually obvious at the macroscopic level because it possesses distinct "animal" and "vegetal" poles (Figure 2.1B).

The animal hemisphere contains dark melanin pigment granules embedded in an actin network located just below the plasma membrane. The animal cytoplasm contains the maternal pronucleus at the pole, a central region that is relatively free of yolk platelets, and a peripheral region that contains small yolk platelets (Danilchik and Gerhart, 1987). In contrast, the vegetal hemisphere does not contain pigment granules and appears yellowish-white in color due to the high content of medium- and large-size yolk platelets.

The very large size of the *Xenopus* egg (~1.5mm diameter) enabled it to be physically dissected into different regions from which mRNAs can be extracted to identify region-specific transcripts. Initially, this was accomplished by constructing cDNA libraries and performing expression screening assays (Rebagliati et al., 1985; Weeks and Melton, 1987; Mosquera et al.,1993; Elinson et al., 1993; reviewed by King, 2017), more recently by microarray and RNA-Seq assays (Cuykendall and Houston, 2010a; Grant et al., 2014; De Domenico et al., 2015). These approaches discovered localized mRNAs that are required for the formation of the germ cell lineage (Section 2.3B) and the primary embryonic germ layers (Section 2.3C).

The maternally supplied molecules that regulate A-P axis formation are likely those that are restricted to the animal rather than to the vegetal region (Figure 2.1B) and that participate in germ layer formation (Section 2.3C). In contrast, those that are known to be involved in establishing the D-V axis are vegetally localized. Some are entrapped in the vegetal cortex, whereas others are constituents of the vegetal cytoplasm itself (reviewed by Houston, 2012). These factors are moved from a vegetal location in the egg to a dorsal location in the embryo immediately after fertilization (Section 2.4B). This results in local, dorsal activation of Wnt signaling via several maternal components of the pathway (White and Heasman, 2008). Maternally supplied β-catenin protein, which is the nuclear effector of Wnt signaling, is ubiquitously distributed in the egg but becomes nucleus associated only on the dorsal side after fertilization.

Two upstream components of the Wnt pathway also are localized to the vegetal region. Maternal *Wnt11* mRNA, entrapped in the vegetal cortex during oogenesis (Ku and Melton, 1993), is released from the vegetal cytoskeleton upon egg maturation and relocated to the dorsal region after fertilization (Schroeder et al., 1999; Tao et al., 2005). Embryos depleted of maternal *Wnt11* mRNA are ventralized and express diminished levels of zygotic genes required for the formation of the dorsal organizer, including two direct β-catenin targets: *Siamois* and *Xnr3* (Brannon et al., 1997; McKendry et al., 1997). Although the molecules that establish the D-V axis in *Xenopus* appear to exert their effects only after fertilization and before the first cell division, there is evidence that Wnt11 and other components of the pathway may act during late oogenesis to influence regional levels of β-catenin protein prior to fertilization (Houston, 2012). In this case, factors in the unfertilized oocyte would be examples of the earliest regulators of deferred-use molecules.

2.3.2 Germ Line Determinants

The unfertilized eggs of many non-mammalian species contain a specialized cytoplasm often referred to as "germ plasm." This region of the oocyte cytoplasm is distinctive due to its electron dense appearance; it contains granules composed of mRNAs, RNA-binding proteins, mitochondria, and ribosomes. Watching the movements of these granules showed that after fertilization, germ plasm is inherited by a subset of embryonic cells that later in development will differentiate into primordial germ cells and ultimately into the gametes. For example, in *C. elegans*, fluorescently labeled cytoplasmic inclusions, called P granules, were visually tracked in the transparent embryos exclusively into the germ line precursors (Strome and Wood, 1982); genetically depleting P granules typically leads to infertility (Strome and Updike, 2015). Manually depleting the germ plasm in frogs and flies also results in loss of primordial germ cells and infertility (Buehr and Blackler, 1970; Illmensee and Mahowald, 1974).

Numerous studies have identified a large number of mRNAs and proteins in the germ plasm that are required for proper gamete formation, including *oskar, nanos, tudor*, and *vasa* (Houston and King, 2000; Kloc et al., 2001; Zhou and King, 2004; Haston and Reijo Pera, 2007; Yang et al., 2015; Strome and Updike, 2015; Lehmann, 2016; Aguero et al., 2017b; Escobar-Aguirre et al., 2017). Many of the germ line–specific transcription factors required for gametogenesis are silenced during oogenesis and early embryogenesis and are only activated upon the later differentiation of primordial germ cells. Thus, *the use of these gamete-specifying maternal molecules is deferred for days, weeks, or months*, and deferred-use molecules can function after cell specification.

It is important to note that there is *no evidence for a similar germ line–specific cytoplasm in the eggs of some nonfly insects, mammals, or urodele amphibians* (Extavour and Akam, 2003; Johnson et al., 2003; Johnson and Alberio, 2015). In many of these animals, induction of primordial germ cells takes place during mesoderm formation and morphogenesis. In amphibians, later induction of the germ line in urodeles is thought to be the ancestral state, whereas sequestration of germ plasm in germ cells in anurans is considered a derived state (Johnson et al., 2003). Nonetheless, the genes encoding many of the maternally provided mRNAs are required in the germ cell lineage of those animals in which primordial germ cells are induced in order for gametes to form (Haston and Reijo Pera, 2007; Strome and Updike, 2015). Thus, although protection of the germ cell lineage by segregating a subset of cells that inherit germ plasm is not universal, *the genes that regulate gamete differentiation are mostly evolutionarily conserved*, whether used immediately after fertilization or later in already specified cell lineages.

2.3.3 Factors That Specify *Xenopus* Primary Embryonic Germ Layers

Endoderm and mesoderm: The technique of screening cDNA libraries made it possible to identify localized mRNAs in *Xenopus*. Three animal and one vegetally localized mRNAs were identified by screening a cDNA library constructed from mature, polarized oocytes with probes representing mRNAs extracted from dissected animal or vegetal regions (Rebagliati et al., 1985). The maternal mRNA localized to the vegetal pole (*Vg1*) encodes a TGF-β growth factor that is required for endoderm formation and proper mesoderm induction (Weeks and Melton, 1987; Dale et al., 1993; Thomsen and Melton, 1993; Birsoy et al., 2006). A similar approach revealed a maternally supplied T-box transcription factor, VegT (Lustig et al., 1996; Zhang and King, 1996; Stennard et al., 1996; Horb and Thomsen, 1997). VegT-depleted embryos do not form endoderm; maternal VegT also is required for initiating mesoderm induction by activating zygotic *nodal*-related genes (Zhang et al., 1998; Clements et al., 1999; Kofron et al., 1999; Xanthos et al., 2001; White and Heasman, 2008).

Ectoderm and neural fate: Animal-enriched mRNAs also have been identified by these assays (Figure 2.1B). Fate mapping studies showed that cells derived from this region of the fertilized egg later gave to the embryonic ectoderm and the nervous system (Dale and Slack, 1987; Moody, 1987a,b; Moody and Kline, 1990). *Do these molecules specify ectodermal and/or neural fate?* Most experimental evidence has been interpreted to suggest that the ectoderm forms by default in regions where vegetal determinants are absent. This interpretation is supported by numerous experiments showing that the animal cap ectoderm of the *Xenopus* blastula can form nearly any tissue if exposed to the right signaling molecules at the correct developmental time (e.g., Asashima et al., 2009; Kurisaki et al., 2010). Recent studies also show that these cells are the equivalent of mammalian embryonic stem cells in that they maintain the gene expression pattern of pluripotency (Buitrago-Delgado et al., 2015).

Conversely, many studies suggest that *Xenopus* animal pole blastomeres and their maternal cytoplasmic content are required for formation of the ectoderm, particularly neural ectoderm. Deletion of single dorsal-animal blastomeres causes significant deficits in the neural plate and neural tube (Kageura and Yamana, 1984; Gallagher et al., 1991; Kageura, 1995). Dorsal-animal blastomeres also retain their original neural fate when transplanted to sites that normally give rise to epidermis (Kageura and Yamana, 1986; Kageura, 1990; Gallagher et al., 1991; Moody et al., 2000). Thus, these cells appear to have an autonomous bias toward neural fate. In fact, they express neural markers even when they are removed during cleavage stages—when the zygotic genome is silent—and cultured as explants in the absence of growth factors (Gallagher et al., 1991; Hainski and Moody, 1992; Grant et al., 2014; Gaur et al., 2016). Similar to the transfer of yellow crescent material in ascidians described in Section 2.2, microinjecting whole cytoplasm or total RNA isolated from dorsal-animal neural precursors into ventral-animal epidermis precursors caused these cells to form ectopic neural tissue (Hainski and Moody, 1992). These examples support the idea that the autonomous ability of dorsal-animal blastomeres to become neural cells is likely influenced by maternal molecules inherited from the polarized egg.

What is the evidence that specific maternal mRNAs are responsible for the specification of the ectodermal fate of *Xenopus* animal blastomeres? Several screens have identified a large number of maternal transcripts that are enriched in the animal hemisphere (reviewed by Sullivan et al., 1999; King et al., 2005; Grant et al., 2014) (Figure 2.1B). Knock-down and ectopic expression assays indicate that some of these promote ectoderm and/or repress mesoderm formation (Zhang et al., 2003, 2004; Dupont et al., 2005; Zhang and Klymkowsky, 2007; Sasai et al., 2008; Xu et al., 2012). For example, the maternal transcript for Foxi2 is required to activate zygotic Foxi1e, which is essential for activating numerous ectodermal genes (Cha et al., 2012). Animal-located maternal transcripts encoding Coco, a secreted antagonist of TGF-β signaling, are required to prevent mesoderm-inducing signals diffusing from the vegetal pole, thus preserving the ectodermal domain (Bates et al., 2013).

Some of the animal-enriched maternal transcripts promote a neural fate. When any of Foxd4, Geminin, Zic2, or Sox11 is ectopically expressed in isolated epidermis progenitor blastomeres, the explants initiate zygotic neural gene expression; i.e., they *change fate from epidermal to neural* (Gaur et al., 2016). Interestingly, these neural-promoting maternal transcripts are found in both ventral-animal epidermal blastomeres-progenitors and dorsal-animal neural blastomeres-progenitors, suggesting that their dorsal localized activity may be translationally regulated. In fact, both dorsal-specific polyadenylation and ventral-specific inhibitory Wnt signaling have been implicated (Pandur et al., 2002, and see Section 2.4C).

2.4 PROCESSES THAT LOCALIZE DEFERRED-USE MOLECULES

Localization of maternal molecules is key for these critical early embryonic events. How is their localization accomplished? There is evidence for several mechanisms, including (i) transport to specific subcellular domains in the oocyte, (ii) transport by cytoplasmic movements in response to fertilization, and (iii) local activation or silencing of protein translation/function in a specific lineage or during a specific developmental time period.

2.4.1 Localization during Oogenesis

During *Drosophila* oogenesis, maternal transcripts exert their local effects via all three processes. In one case, *nanos* mRNA appears to diffuse freely throughout the oocyte cytoplasm but becomes entrapped in the actin cytoskeleton in the posterior pole cytoplasm by several localized proteins that bind to sequences in the *nanos* 3' UTR (Forrest and Gavis, 2003). In contrast, kinesin actively transports *oskar* mRNA along the A-P oriented microtubular array whose plus ends are at the posterior pole. There, *oskar* mRNA is tethered to microfilaments in the cortex of the posterior oocyte cytoplasm. *bicoid* mRNA, on the other hand, complexes to a number of proteins that transport it along microtubular arrays from the nurse cells into the anterior pole of the oocyte. Once in the oocyte, *bicoid* mRNA is transported by dynein to the minus ends of the A-P oriented microtubular array where it becomes anchored. In general, fly maternal mRNAs appear to be localized by a combination of association with proteins that facilitate transport along oriented microtubular arrays and entrapment within a local filamentous cytoskeleton.

In *Xenopus*, germ line determinants are transported en masse from the nucleus to the vegetal pole in a complex called the *mitochondrial cloud* (Heasman et al., 1984; Houston, 2012). This complex contains a dense network of mitochondria, endoplasmic reticulum, and electron-dense "germinal granules" that are composed of RNA-binding proteins and several RNAs that play a role in germ line specification (Heasman et al., 1984; King et al., 2005; Yang et al., 2015).

During early stages of oogenesis when the cell is still relatively small, the mitochondrial cloud assembles adjacent to the nucleus. It subsequently disassociates and moves to the nearby cortex of the cell. This "transport" is over a very short distance and does not require microtubules. Rather the entire complex appears to diffuse and then become entrapped in the cortical cytoskeleton throughout later phases of oogenesis and after fertilization (Chang et al., 2004; Houston, 2012). Recent work indicates that specific RNA-binding proteins, such as Hermes/Rbpms, associate with germ line mRNAs in the nucleus to transport them specifically to granules within the mitochondrial cloud (Aguero et al., 2016).

In addition, some maternal molecules are actively transported from the animal-located egg nucleus to the vegetal pole. For example, *Vg1*, *VegT*, and *Wnt11* mRNAs are transported along microtubules via protein complexes that recognize sequences in their 3' UTRs (reviewed by Kindler et al., 2005; Houston, 2012). Another mechanism is entrapment by the vegetal cytoskeleton, much as *nanos* mRNA is entrapped at the posterior pole of fly. In fact, the *Xenopus* vegetal cortical cytoskeleton has a rich array of localized mRNAs associated with it (Elinson et al., 1993). Interestingly, studies in which VegT was depleted demonstrate a nontranscriptional role for this protein: it disrupts the vegetal intermediate filament array, resulting in the release and mislocalization of vegetally localized *Wnt11* and *Vg1* mRNAs (Heasman et al., 2001; Kloc et al., 2005). In contrast, although there are many mRNAs enriched in the animal hemisphere of the *Xenopus* egg (Figure 2.1B), there is no evidence that they are actively transported or sequestered. It is possible that animal-enriched transcripts are simply limited in their diffusion from the nucleus through a viscous cytoplasm that is jam-packed with yolk platelets.

2.4.2 Localization after Fertilization

When the sperm fertilizes the mature egg, it causes an asymmetry by introducing the male pronucleus and its associated centrosomes. The paternal centrosomes assemble bundles of microtubules that will form the mitotic spindle of the zygote, as well as nucleate microtubular arrays that can cause cytoplasmic reorganizations.

This was discovered in *C. elegans* by the ability to observe the movement of the P granules, which flow to the posterior pole of the zygote and are then inherited by the P_1 blastomere, the precursor of the germ line (Strome and Wood, 1982; reviewed by Strome and Updike, 2015). This cytoplasmic flow moves the sperm pronucleus and centrosomes to one end of the zygote, which then is specified as the posterior pole of the embryo (Bowerman, 1998; Lyczak et al., 2002) (Figure 2.2A). The localization of P granules to the posterior-most germ line precursor cell at each postfertilization mitotic division requires microfilaments, microtubules, and Mes-1, a cell membrane protein that influences asymmetrical protein localization and the orientation of the mitotic spindle. Interestingly, germ line specification involves both active transport of P granules into the posterior germ line daughter cell and degradation of P granules left behind in the cytoplasm of the anterior somatic daughter cell (Strome and Wood, 1983; Hird et al., 1996).

Genetic screens for mutations that affect these movements demonstrate that cytoplasmic flow also triggers an asymmetric distribution of various PAR proteins that are associated with the zygote's cortical domain (Figure 2.2A). PAR-1 and PAR-2 become localized to the posterior cortex, whereas PAR-3 and PAR-6 become localized to the anterior cortex. These proteins are essential for setting up the A-P axis of the nematode embryo, including the movement of P granules to the posterior blastomere at the first cell division. Both cortical microfilaments and sperm astral

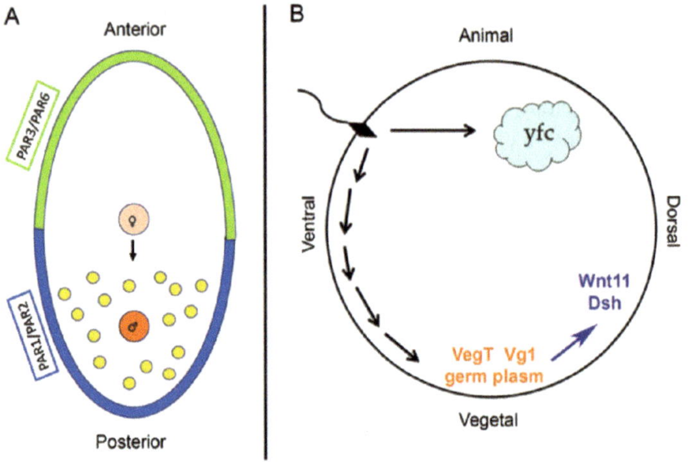

FIGURE 2.2 Cytoplasmic movements resulting from fertilization cause molecular asymmetry in embryos. (A) In *Caenorhabditis elegans*, the paternal pronucleus (dark orange, male symbol) determines the posterior pole of the zygote, causing cytoplasmic movements that (1) pull the female pronucleus (light orange, female symbol) toward that pole for eventual fusion and asymmetric cell division and (2) relocate the P granules (yellow) to the posterior cytoplasm. PAR proteins also become segregated to different cell membrane regions. PAR3/PAR6 defines the anterior region (green), and PAR1/PAR2 defines the posterior region (blue). Localization of these maternal molecules initiates formation of both the A-P axis and the germ line. (B) In *Xenopus*, the head of the sperm (black diamond) penetrates the egg randomly in the animal hemisphere. Centrosomes associated with the paternal pronucleus organize microtubular arrays (black arrows) that (1) extend into the animal pole region and shift yolk-free cytoplasm (yfc) toward the dorsal side and (2) extend into the vegetal pole. The latter causes a rotation of the cortical cytoplasm toward the vegetal pole and shifts some molecules, such as Wnt11 and Dsh toward the dorsal, vegetal region. The maternal mRNAs for *VegT*, *Vg1*, and the germ line, however, are tethered to the vegetal cytoskeleton and therefore do not shift. Movement of these maternal molecules initiates formation of the A-P and D-V axes and the germ line. VegT, vegetal T-box containing; Vg1, gdf1, growth differentiation factor 1; Wnt11, wnt family member 11; Dsh, disheveled segment polarity.

microtubules are implicated in setting up this asymmetry. It is thought that the differential localization of the PAR proteins influences the posterior positioning of the first mitotic spindle, leading to an *asymmetric cell division that gives rise to two daughter cells with very different fates: the anterior daughter as the precursor of the somatic lineages and the posterior daughter as the precursor of the germ cell lineage*. PAR proteins also are required for the posterior localization of maternal transcription factors, such as SKN-1 and PAL-1, as well as germ line determinants, and for the anterior localization of the maternal GLP-1 transcript. Although these events occur during the nematode maternal control phase, some experiments suggest that the paternal pronucleus may contribute unknown factors or provide a substrate to which maternal factors bind and are thus transported to the posterior pole (Lyczak et al., 2002; Miller, 2015).

In *Xenopus*, the unfertilized egg is already molecularly distinct at the animal and vegetal poles (Figure 2.1A), contributing to establishing the germ line and the primary germ layers (Section 2.3). However, as in nematodes, sperm entry introduces a new asymmetry (Figure 2.2B). The sperm can penetrate the egg at any position in the animal hemisphere, the site of which is aptly called the "sperm entry point" (SEP). This triggers the elaboration of two microtubular arrays from the sperm centrosomes. One radiates into the animal hemisphere, a component of which appears to shift the central cytoplasm to the prospective dorsal marginal region (Figure 2.2B). This yolk-free cytoplasm is rich in RNAs and mitochondria. Although its movement is implicated in contributing to D-V axis formation (Denegre and Danilchik, 1993; Brown et al., 1993; Yost et al., 1995), to our knowledge, the maternal molecules within this distinctive cytoplasm have not yet been identified.

A second microtubular array extends from the SEP in the animal pole to the vegetal pole. It forms a few microns beneath the cell membrane in the actin-rich cytoplasmic cortex and causes a thin layer of yolk-free cortical cytoplasm to rotate toward the vegetal pole during the first cell cycle (Elinson and Rowning, 1988; Houliston and Elinson, 1991, 1992). Numerous experiments have demonstrated that this cytoplasmic movement establishes the D-V axis of the embryo (reviewed by White and Heasman, 2008; Houston, 2012). For example, a localized needle prick or electric shock in the animal hemisphere of a mature egg can substitute for sperm entry and induce a cortical rotation that predicts the orientation of the D-V axis.

What is the molecular basis for cytoplasmic movement initiating the D-V axis? *Microtubule assembly appears to be key.* Treatments that disrupt microtubule polymerization, such as UV irradiation, cold, pressure, and microtubule disrupting drugs, can prevent cortical rotation and the formation of D-V polarity. Interestingly, these treatments can be overridden by causing rotation with gravitational force. For example, UV irradiation of the vegetal half of fertilized eggs that blocks the formation of the microtubular arrays and suppresses dorsal development can be reversed; tilting embryos with respect to gravity during the first cell cycle mechanically results in cortical rotation and rescues dorsal axis formation (reviewed by Houston, 2012).

Another cytoplasmic movement, referred to as a *fast transport system*, localizes some intracellular components of the Wnt pathway to the D-V region (Houston, 2012). This movement, which not only is independent of cortical rotation, also relies on vegetal microtubule assembly, was detected by tagging proteins with GFP. Both disheveled and GSK-binding protein were shown to associate with small vesicular organelles that are rapidly transported along microtubular arrays to the D-V region. However, the molecular motors responsible for this fast transport have yet to be identified. It has been proposed that relocation of these vegetally located Wnt pathway components to the dorsal quadrant after fertilization is responsible for the nuclear accumulation of β-catenin in cells that later in development, at the onset of gastrulation, will form the dorsal-most region of the mesoderm. It is this mesodermal subset of cells, called the organizer in amphibians and the node in mammals and birds, which will turn on a zygotic program that will pattern the mesoderm in the D-V axis and induce the adjacent ectoderm to form the neural plate.

Recently, one mechanism for directing cortical rotation to the vegetal pole has been elucidated. Mei et al. (2013) reported that an RNA-binding protein, Dnd1, originally identified for its critical role in germ line development, also is essential for D-V axis formation. Knockdown of Dnd1 interferes with the formation of vegetal cortical microtubules, resulting in failed relocalization of Wnt signaling components to the dorsal side. These authors show that Dnd1 binds to the 3'UTR of *trim36* mRNA, another vegetally localized maternal molecule, and anchors it to the vegetal cortex. This results in a high concentration of Trim36 protein at the vegetal pole. This entrapment is critically important because Trim36, an E3 ubiquitin-protein ligase, is required for microtubule assembly (Cuykendall and Houston, 2010b).

2.4.3 Local Activation and Silencing

mRNAs are the most abundant type of maternal deferred-use molecules deposited in the egg. They can be *localized* to produce an enriched protein source upon activation, for example, as described in Section 2.4A for the fly A-P and D-V determinants. Alternatively, the *local translation* of ubiquitously distributed mRNAs can be regulated spatially and temporally to provide a local effect. In *C. elegans,* for example, the majority of maternal transcripts are not localized. Instead, localized protein activity appears to be regulated predominantly by local translation via MEX-3 and SPN-4, local protein degradation via MEX-5 and MEX-6, and localized posttranslational modifications (reviewed by Lyczak et al., 2002). In *Drosophila*, localized protein translation can be accomplished even if the transcript is available throughout the egg. For example, caudal protein is required for activating zygotic genes in the posterior region of the embryo. Although maternally supplied *caudal* mRNA is distributed evenly throughout the egg, it is translationally repressed in the anterior region by high levels of bicoid protein (Ma et al., 2016). Thus, caudal protein only has a posterior activity.

Another mechanism for activating local protein synthesis is by *deregulation of repression*. For example, *oskar* mRNA is only translated when located in the posterior pole cytoplasm; this seems to be regulated by factors that bind to the 5' UTR of its transcript that thereby reverses its 3' UTR-mediated translational repression (Lehmann, 2016). In *Xenopus*, Dnd1 protein binds to eIF3f, a repressive component in the 43S preinitiation complex, thus relieving *nanos1* mRNA from its silenced state (Aguero et al., 2017a).

A fourth mechanism is *regulation of the adenylation state* of the transcript. For example, maternal gene products localized to the dorso-anterior corner of the *Drosophila* egg render *gurken* mRNA in a polyadenylated, translatable state. In contrast, *gurken* mRNAs that are not successfully localized to that position are not fully adenylated and therefore are poorly translated (Derrick and Weil, 2017).

Local silencing of maternal transcripts is another common mechanism. For example, the function of many germ line determinant molecules is to silence other mRNAs in the germ plasm that are required for gamete differentiation much later in development. In fly, some germ line mRNAs are kept in a silent state by maternal microRNAs that bind to 3' UTR sequences and prevent translation, or by maternal RNA-binding proteins, such as Nanos and Pumillio, which bind to 3' UTR and thereby recruit a deadenylase complex to the transcript (Slaidina and Lehmann, 2014).

Another mechanism for local silencing is lineage-specific silencing of transcription. In *C. elegans*, the presence of P granules prevents somatic mRNAs from accumulating in embryonic and adult germ cells (Updike et al., 2014; Knutson et al., 2017). In both fly and nematode, overall transcription is silenced in the primordial germ cells by inhibition of positive transcription elongation factor b (P-TEFb) (Strome and Updike, 2015). In the worm, PIE-1 binds P-TEFb and prevents it from interacting with RNA polymerase II (Pol II). Depletion of PIE-1 results in germ line blastomeres expressing somatic fates. In the fly, a different protein, polar granule component, binds to P-TEFb with the same effect. In many species, Nanos silences transcription in primordial germ cells by preventing the phosphorylation of Pol II; this may be an indirect effect since Nanos is an RNA-binding protein. In *Xenopus*, zygotic transcription of endoderm target genes of maternal VegT is prevented in primordial germ cells by a lack of an activating phosphorylation of RNA polymerase II (Venkatarama et al., 2010). Ascidians and mouse employ other molecules to repress somatic gene expression in the germ line. These studies demonstrate *transcriptional silencing to be an evolutionary conserved mechanism for protecting the primordial germ cells from expressing inappropriate genes.*

Localized activation/silencing of maternal transcripts also has been reported as an important mechanism for germ layer decisions in ascidian embryos. For example, three maternal transcripts, *CsEndo-1, CsEndo-2,* and *CsEndo-3*, are ubiquitously distributed up to the eight-cell stage and then found only in the endoderm lineage (Imai et al., 1999), suggesting localized degradation in the non-endodermal blastomeres. In other cases, the molecular mechanisms of silencing are better understood. For example, Gata.a activity is required in the animal blastomeres to activate ectoderm-specific gene expression, even though its maternal transcripts are ubiquitously distributed. Recent work demonstrated that a physical interaction of Gata.a protein with β-catenin and Tcf7 inhibits its activity in the vegetal endoderm lineage (Oda-Ishii et al., 2016, 2018). In the absence of Gata.a activity in vegetal cells, the β-catenin/Tcf7 complex directly promotes the transcription of endomesoderm genes and indirectly inhibits transcription of ectoderm genes. Conversely, ectoderm forms in the presence of Gata.a activity in the animal hemisphere, which is free from maternal β-catenin and Tcf7. This Gata.a/β-catenin-Tcf7 interaction continues in the marginal zone at the next cell cycle to segregate the mesoderm from the endoderm lineages (Imai et al., 2016).

Local transcript degradation is another way to control local protein activity. At the MTZ transition, many maternal transcripts are degraded as zygotic transcription begins. In many animals, this occurs by deadenylation of the transcript or ubiquitin-mediated degradation (Tadros and Lipshitz, 2009). Interestingly, some transcripts are protected from MTZ degradation, particularly in the germ line in which it is important to preserve the maternal germ plasm-associated transcripts (Strome and Updike, 2015). There also is lineage-specific degradation. For example, germ line determinants that are mislocalized to somatic cells often are specifically degraded, thus avoiding ectopic production of gametes. In *C. elegans*, the first zinc finger of the PIE-1 germ line determinant protein is required for its degradation in somatic cells, whereas its second zinc finger is required for its localization to the germ line (reviewed by Spike and Strome, 2003). Many of the molecules responsible for nematode A-P axis, such as PAR proteins, also are involved in anterior degradation and posterior protection of germ line molecules.

Many of the studies described above that identified local protein translation or posttranslational modifications were made possible by the availability of specific antibodies. However, in species in which similar antibodies are not available, it has been technically challenging to determine whether maternally provided proteins or their local translation contribute to early embryonic events. Recent advances in the sensitivity of mass spectrometry are beginning to reveal local differences in protein abundance and activity. For example, using single-cell capillary electrophoresis-electrospray ionization mass spectrometry, Onjiko et al. (2015) showed that *Xenopus* ventral-animal blastomeres (epidermal progenitors) and dorsal-animal blastomeres (neural progenitors) have distinctive metabolite profiles that drive their respective gastrulation movements into epidermal versus neural fates. A similar proteomic analysis showed that these two blastomeres have distinct protein profiles (Lombard-Banek et al., 2016). It is likely that with continued technological improvements in sensitivity (e.g., Lombard-Banek et al., 2019), the issue of *whether the maternal mRNAs commonly inherited by these cells are differentially translated to direct later fate choices will be resolved.*

2.5 DO DEFERRED-USE MOLECULES REGULATE PLANT DEVELOPMENT?

The long history of studying plant development also included the concept of maternal determinants (Dunn, 1917). However, early evidence for maternal determinants was more difficult to acquire in plants. This is undoubtedly because plant eggs and embryos are more difficult to observe directly than are the marine invertebrates favored by early animal embryologists. These difficulties include the fact that the egg and embryo are surrounded by an embryo sac that includes a nutritive endosperm consisting of starch, oils, and proteins. Accordingly, the role for maternal determinants in plants has been directly examined only recently (reviewed by Zhao and Sun, 2015).

RNA sequencing analysis has made it possible to examine whether, and when, plant embryos also undergo an MZT. Recent examinations of corn and rice development compared the transcriptional profiles of embryonic stages to those of egg and sperm. These studies revealed that the zygotic genome becomes activated during the first cell cycle, which lasts for about 35 h in corn and about 11.5 h in rice. In corn, about 3,600 zygotic genes are activated during the S phase of the first cell cycle, beginning at about 4 h post-fertilization. This represents about 9% of the maize transcriptome and includes genes known to be involved in establishing polarity and cell fate (Chen et al., 2017). In rice, about 500 zygotic genes are activated soon after the sperm and egg pronuclei fuse, beginning at about 2 h post-fertilization. An additional 5,000 zygotic genes are activated by 4.5 h post-fertilization (Anderson et al., 2017). The Gene Ontology (GO) terms of these genes indicate that they are involved in embryogenesis and patterning. Thus, axis specification and cell fate determination may be controlled by the plant's zygotic genome, rather than by deferred-use maternal components. Interestingly, unlike animals, plant maternal transcripts do not seem to be degraded at the MZT (Zhao and Sun, 2015). This difference may reflect the fact that, unlike animals, plants experience explosive growth during the first few cell cycles, which would rapidly dilute maternal transcripts, obviating the need for their degradation.

These observations suggest that *maternal determinants would have minor, if any, roles in the early developmental events of plants.* Perhaps, deferred-use molecules are less necessary in plants because of the long duration of first cell cycle, compared to flies, worms, fish, and amphibians. A long cell cycle is likely to allow both zygotic gene transcription and DNA replication. Of course, it remains possible that the egg and zygote contain maternal RNAs that direct some critical early events that are yet to be detected. If present, they might be discovered by looking for maternal RNAs and proteins that are differentially located in specific regions of the egg or early zygote, as has been done extensively in the much larger animal embryos.

2.6 CONCLUSIONS

Many organisms have evolved deferred-use cell populations that reside in specialized niches to allow formation of adult structures during intermediate developmental stages and to replace damaged tissue, as described in detail in this volume. But single cells, particularly oocytes, have mechanisms for setting aside molecules whose functional activities are deferred for future developmental processes. In a sense, the fertilized egg is the ultimate stem cell and its cytoplasm is the niche, i.e., it facilitates activation of molecules (cells in other niches) to initiate their function and maintains the molecules by keeping them in place.

An important mechanism that influences cell fate prior to gastrulation in a number of animals is the region-specific inheritance of maternally synthesized mRNAs and proteins. Characterization of the spatial and temporal expression of these molecules during embryonic development has been critical for understanding how the descendants of the zygote, which are genomically equivalent, become different and give rise to all the different types of tissues and organs. The examples we have presented for specifying the body axes, the germ line, and the primary germ layers likely only scratch the surface of the remarkable richness in the number of deferred-use maternal molecules and the processes used to localize them. In fact, nearly 2,200 maternal transcripts have been detected during early fly embryogenesis, >70% of which show subcellular localization (Lécuyer et al., 2007).

This work also may reveal cellular mechanisms that can be utilized in the adult as well, such as local activation of protein synthesis and region-specific degradation of proteins and mRNAs. Work across a wide variety of animals show many commonalities; however, plants appear to have evolved different strategies. Although it has been relatively straightforward to identify regionally sequestered transcripts, recent technological advances are beginning to make it possible to measure proteins and small molecules in a region-specific fashion, including in single embryonic cells. These approaches are likely to enhance our molecular understanding of the activities of deferred-use molecules in a wide range of developmental processes.

ACKNOWLEDGEMENT

Dr. Klein's efforts were supported by the National Science Foundation while working at the Foundation. Any opinions, findings, and conclusions or recommendations expressed in this material are those of the authors and do not necessarily reflect the views of the National Science Foundation.

REFERENCES

Abrams, E. W. and Mullins, M. C. 2009. Early zebrafish development: it's in the maternal genes. *Curr. Opin. Genet. Dev.* 19: 396–403.

Aguero, T., Jin, Z., Chorghade, S., Kalsotro, A., King, M. L. and Yang, J. 2017a. Maternal dead-end 1 promotes translation of nanos1 by binding the eIF3 complex. *Development* 144: 3755–3765.

Aguero, T., Kassmer, S., Alberio, R., Johnson, A. and King, M. L. 2017b. Mechanisms of vertebrate germ cell determination. *Adv. Exp. Med. Biol.* 953: 383–440.

Aguero, T., Zhou, Y., Kloc, M., Chang, P., Houliston, E. and King, M. L. 2016. Hermes (Rbpms) is a critical component of RNP complexes that sequester germline RNAs during oogenesis. *J. Dev. Biol.* 4: 2.

Allbee, A. W., Rincon-Limas, D. E. and Biteau, B. 2018. Lmx1 is required for the development of the ovarian stem cell niche in Drosophila. *Development* 145. doi:10.1242/dev.163394.

Anderson, S. N., Johnson, C. S., Chesnut, J., Jones, D. S., Khanday, I., Woodhouse, M., Li, C., Conrad, L. J., Russell, S. D. and Sundaresan, V. 2017. The zygotic transition is initiated in unicellular plant zygotes with asymmetric activation of parental genomes. *Dev. Cell* 43: 349–358.

Asashima, M., Ito, Y., Chan, T., Michiue, T., Nakanishi, M., Suzuki, K., Hitachi, K., Okabayashi, K., Kondow, A. and Ariizumi, T. 2009. *In vitro* organogenesis from undifferentiated cells in *Xenopus*. *Dev. Dyn.* 238: 1309–1320.

Bates, T., Vonica, A., Heasman, J., Brivanlou, A. H. and Bell, E. 2013. Coco regulates dorsoventral specification of germ layers via inhibition of TGFβ signalling. *Development* 140: 4177–4181.

Birsoy, B., Kofron, M., Schaible, K., Wylie, C. and Heasman, J. 2006. Vg1 is an essential signaling molecule in *Xenopus* development. *Development* 133: 15–20.

Bowerman, B. 1998. Maternal control of pattern formation in early *Caenorhabditis elegans* embryos. *Curr. Top. Dev. Biol.* 39: 73–117.

Brannon, M., Gomperts, M., Sumoy, L., Moon, R. T. and Kimelman, D. 1997. A beta-catenin/XTcf-3 complex binds to the *siamois* promoter to regulate dorsal axis specification in *Xenopus*. *Genes Dev.* 11: 2359–2370.

Brown, E. E., Denegre, J. M. and Danilchik, M. V. 1993. Deep cytoplasmic rearrangements in ventralized *Xenopus* embryos. *Dev. Biol.* 160: 148–156.

Buckholz, D. R. and Shi, Y. B. 2018. Dual function model revised by thyroid hormone receptor alpha knock out frogs. *Gen. Comp. Endocrinol.* 265: 214–218.

Buehr, M. L. and Blackler, A. W. 1970. Sterility and partial sterility in the South African clawed toad following the pricking of the egg. *J. Embryol. Exp. Morphol.* 23: 375–384.

Buitrago-Delgado, E., Nordin, K., Rao, A., Geary, L. and LaBonne, C. 2015. Shared regulatory programs suggest retention of blastula-stage potential in neural crest cells. *Science* 348: 1332–1335.

Cha, S. W., McAdams, M., Kormish, J., Wylie, C. and Kofron, M. 2012. *Foxi2* is an animally localized maternal mRNA in *Xenopus*, and an activator of the zygotic ectoderm activator Foxi1e. *PLoS One* 7: e41782.

Chang, P., Torres, J., Lewis, R. A., Mowry, K. L., Houliston, E. and King, M. L. 2004. Localization of RNAs to the mitochondrial cloud in *Xenopus* oocytes through entrapment and association with endoplasmic reticulum. *Mol. Biol. Cell* 15: 4669–4681.

Chen, Y., Strieder, N., Krohn, N. G., Cyprys, P., Sprunck, S., Engelmann, J. C. and Dresselhous, T. 2017. Zygotic genome activation occurs shortly after fertilization in maize. *Plant Cell* 29: 2106–2125.

Clements, D., Friday, R. V. and Woodland, H. R. 1999. Mode of action of VegT in mesoderm and endoderm formation. *Development* 126: 4903–4911.

Conklin, E. G. 1905. Organization and cell-lineage of the ascidian egg. *J. Acad. Natl. Sci. Philadelphia* 13: 1–119.

Cuykendall, T. N. and Houston, D. W. 2010a. Identification of germ plasm-associated transcripts by microarray analysis of *Xenopus* vegetal cortex RNA. *Dev. Dyn.* 239: 1838–1848.

Cuykendall, T. N. and Houston, D. W. 2010b. Vegetally localized *Xenopus* trim36 regulates cortical rotation and dorsal axis formation. *Development* 136: 3057–3065.

Dale, L., Matthews, G. and Colman, A. 1993. Secretion and mesoderm-inducing activity of the TGF-beta-related domain of *Xenopus* Vg1. *EMBO J.* 12: 4471–4480.

Dale, L. and Slack, J. M. 1987. Fate map for the 32-cell stage of *Xenopus laevis. Development* 99: 527–551.

Danilchik, M. V. and Gerhart, J. C. 1987. Differentiation of the animal-vegetal axis in *Xenopus laevis* oocytes. I. Polarized intracellular translocation of platelets establishes the yolk gradient. *Dev. Biol.* 122: 101–112.

Davidson, E. H. 1990. How embryos work: a comparative view of diverse modes of cell fate specification. *Development* 108: 365–389.

De Domenico, E., Owens, N. D., Grant, I. M., Gomes-Faria, R. and Gilchrist, M. J. 2015. Molecular asymmetry in the 8-cell stage *Xenopus tropicalis* embryo described by single blastomere transcript sequencing. *Dev. Biol.* 408: 252–268.

Denegre, J. M. and Danilchik, M. V. 1993. Deep cytoplasmic rearrangements in axis-respecified *Xenopus* embryos. *Dev. Biol.* 160: 157–164.

Derrick, C. J. and Weil, T. T. 2017. Translational control of *gurken* mRNA in *Drosophila* development. *Cell Cycle* 16: 23–32.

Dunn, L. C. 1917. Nucleus and cytoplasm as vehicles of heredity. *Amer. Nat.* 51: 286–300.

Dupont, S., Zacchigna, L., Cordenonsi, M., Soligo, S., Adorno, M., Rugge, M. and Piccolo, S. 2005. Germ-layer specification and control of cell growth by Ectodermin, a Smad4 ubiquitin ligase. *Cell* 121: 87–99.

Elinson, R. P., King, M. L. and Forristall, C. 1993. Isolated vegetal cortex from *Xenopus* oocytes selectively retains localized mRNAs. *Dev. Biol.* 160: 554–562.

Elinson, R. P. and Rowning, B. 1988. A transient array of parallel microtubules in frog eggs: potential tracks for a cytoplasmic rotation that specifies the dorso-ventral axis. *Dev. Biol.* 128: 185–197.

Elinson, R. P., King, M. L. and Forristall, C. 1993. Isolated vegetal cortex from *Xenopus* oocytes selectively retains localized mRNAs. *Dev Biol* 160: 554–562.

Escobar-Aguirre, M., Elkouby, Y. M. and Mullins, M. C. 2017. Localization in oogenesis of maternal regulators of embryonic development. *Adv. Exp. Med. Biol.* 953: 173–207.

Extavour, C. G. and Akam, M. 2003. Mechanisms of germ cell specification across the metazoans: epigenesis and preformation. *Development* 130: 5869–5884.

Forrest, K. M. and Gavis, E. R. 2003. Live imaging of endogenous RNA reveals a diffusion and entrapment mechanism for *nanos* mRNA localization in *Drosophila. Curr. Biol.* 13: 1159–1168.

Gallagher, B. C., Hainski, A. M. and Moody, S. A. 1991. Autonomous differentiation of dorsal axial structures from an animal cap cleavage stage blastomere in *Xenopus. Development* 112: 1103–1114.

Gaur, S., Mandelbaum, M., Herold, M., Majumdar, H. D., Neilson, K. M., Maynard, T. M., Mood, K., Daar, I. O. and Moody, S. A. 2016. Neural transcription factors bias cleavage stage blastomeres to give rise to neural ectoderm. *Genesis* 54: 334–349.

Grant, P. A., Yan, B., Johnson, M. A., Johnson, D. L. E. and Moody, S. A. 2014. Novel animal pole-enriched maternal mRNAs are preferentially expressed in neural ectoderm. *Dev. Dyn.* 243: 478–496.

Hainski, A. M. and Moody, S. A. 1992. *Xenopus* maternal RNAs from a dorsal animal blastomere induce a secondary axis in host embryos. *Development* 116: 347–355.

Haston, K. M. and Reijo-Pera, R. A. 2007. Germ line determinants and oogenesis. In *Principles of Developmental Genetics* (1st edition), ed. S. A. Moody, pp. 150–172. New York: Elsevier.

Heasman, J. 2006. Maternal determinants of embryonic cell fate. *Sem. Cell Dev. Biol.* 17: 93–98.

Heasman, J., Quarmby, J. and Wylie, C. C. 1984. The mitochondrial cloud of *Xenopus* oocytes: the source of germinal granule material. *Dev. Biol.* 105: 458–469.

Heasman, J., Wessely, O., Langland, R., Craig, E. J. and Kessler, D. S. 2001. Vegetal localization of maternal mRNAs is disrupted by VegT depletion. *Dev. Biol.* 240: 377–386.

Hird, S. N., Paulsen, J. E. and Strome, S. 1996. Segregation of germ granules in living *Caenorhabditis elegans* embryos: cell-type-specific mechanisms for cytoplasmic localisation. *Development* 122: 1303–1312.

Horb, M. E. and Thomsen, G. H. 1997. A vegetally localized T-box transcription factor in *Xenopus* eggs specifies mesoderm and endoderm and is essential for embryonic mesoderm formation. *Development* 124: 1689–1698.

Houliston, E. and Elinson, R. P. 1991. Evidence for the involvement of microtubules, EER, and kinesin in the cortical rotation of fertilized frog eggs. J. Cell Biol. 114: 1017–1028.

Houliston, E. and Elinson, R. P. 1992. Microtubules and cytoplasmic reorganization in the frog egg. *Curr. Top. Dev. Biol.* 26: 53–70.

Houston, D. W. 2012. Cortical rotation and messenger RNA localization in *Xenopus* axis formation. *WIREs Dev. Biol.* 1: 371–388.

Houston, D. W. and King, M. L. 2000. Germ plasm and molecular determinants of germ cell fate. *Curr. Top. Dev. Biol.* 50: 155–181.

Illmensee, K. and Mahowald, A. P. 1974. Transplantation of posterior polar plasm in *Drosophila*. Induction of germ cells at the anterior pole of the egg. *Proc. Nat. Acad. Sci. USA* 71: 1016–1020.

Imai, K., Hudson, C., Oda-Ishin, I., Yasuo, H. and Satou, Y. 2016. Antagonism between β-catenin and Gata.a sequentially segregates the germ layers of ascidian embryos. *Development* 143: 4167–4172.

Imai, K., Satoh, N. and Satou, Y. 1999. Identification and characterization of maternally expressed genes with mRNAs that are segregated with the endoplasm of early ascidian embryos. *Int. J. Dev. Biol.* 43: 125–133.

Imai, K., Hudson, C., Oda-Ishin, I., Yasuo, H. and Satou, Y. 2016. Antagonism between β-catenin and Gata.a sequentially segregates the germ layers of ascidian embryos. *Development* 143: 4167–4172.

Jeffrey, W. R. 1985. Identification of proteins and mRNAs in isolated yellow crescents of ascidian eggs. *J. Embryol. Exp. Morphol.* 89: 275–287.

Johnson, A. D. and Alberio, R. 2015. Primordial germ cells: the first cell lineage or the last cells standing? *Development* 142: 2730–2739.

Johnson, A. D., Drum, M., Bachvarova, R. F., Masi, T., White, M. E. and Crother, B. I. 2003. Evolution of predetermined germ cells in vertebrate embryos: implications for macroevolution. *Evol. Dev.* 5: 414–431.

Kageura, H. 1990. Spatial distribution of the capacity to initiate a secondary embryo in the 32-cell embryo of *Xenopus laevis*. *Dev. Biol.* 142: 432–438.

Kageura, H. 1995. Three regions of the 32-cell embryo of *Xenopus laevis* essential for formation of a complete tadpole. *Dev. Biol.* 170: 376–386.

Kageura, H. and Yamana, K. 1984. Pattern regulation in defect embryos of *Xenopus laevis. Dev. Biol.* 101: 410–415.

Kageura, H. and Yamana, K. 1986. Pattern formation in 8-cell composite embryos of *Xenopus laevis. J. Embryol. Exp. Morphol.* 91: 79–100.

Kenyon, K. L. 2007. Patterning the anterior-posterior axis during Drosophila embryogenesis. In *Principles of Developmental Genetics* (1st edition), ed. S. A. Moody, pp. 173–200. New York: Elsevier.

Kindler, S., Wang, H., Richter, D. and Tiedge, H. 2005. RNA transport and local control of translation. *Annu. Rev. Cell Dev. Biol.* 21: 223–245.

King, M. L. 2017. Maternal messages to live by: a personal historical perspective. *Genesis* 55: e23007. doi:10.1002/dvg.23007.

King, M. L., Messitt, T. J. and Mowry, K. L. 2005. Putting RNAs in the right place at the right time: RNA localization in the frog oocyte. *Biol. Cell* 97: 19–33.

King, M. L. 2017. Maternal messages to live by: a personal historical perspective. *genesis* 55. doi: 10.1002/dvg.23007.

Klein, S. L. and Moody, S. A. 2016. When family history matters: the importance of lineage analyses and fate maps for explaining animal development. *Curr. Top. Dev. Biol.* 117: 93–112.

Kloc, M., Bilinski, S., Chan, A. P., Allen, L. H., Zearfoss, N. R. and Etkin, L. D. 2001. RNA localization and germ cell determination in *Xenopus. Int. Rev. Cytol.* 203: 63–91.

Kloc, M., Wilk, K., Vargas, D., Shirato, Y., Bilinski, S. and Etkin, L. D. 2005. Potential structural role of non-coding and coding RNAs in the organization of the cytoskeleton at the vegetal cortex of *Xenopus* oocytes. *Development* 132: 3445–3457.

Knutson, A. K., Egelhofer, T., Rechtsteiner, A. and Strome, S. 2017. Germ granules prevent accumulation of somatic transcripts in the adult *Caenorhabditis elegans* germline. *Genetics* 206: 163–178.

Kofron, M., Demel, T., Xanthos, J., Lohr, J., Sun, B., Sive, H., Osada, S., Wright, C., Wylie, C. and Heasman, J. 1999. Mesoderm induction in *Xenopus* is a zygotic event regulated by maternal VegT via TGFbeta growth factors. *Development* 126: 5759–5770.

Ku, M. and Melton, D. A. 1993. Xwnt-11: a maternally expressed *Xenopus* wnt gene. *Development* 119: 1161–1173.

Kurisaki, A., Ito, Y., Onuma, Y., Intoh, A. and Asashima, M. 2010. *In vitro* organogenesis using multipotent cells. *Hum. Cell* 23: 1–14.

Lécuyer, E., Yoshida, H., Parthasarathy, N., Alm, C., Babak, T., Cerovina, T., Hughes, T. R., Tomancak, P. and Krause, H. M. 2007. Global analysis of mRNA localization reveals a prominent role in organizing cellular architecture and function. *Cell* 131: 174–187.

Lee, M. T., Bonneau, A. R. and Giraldez, A. J. 2014. Zygotic genome activation during the maternal-to-zygotic transition. *Annu. Rev. Cell Dev. Biol.* 30: 581–613.

Lehmann, R. 2012. Germ line stem cells: origin and destiny. *Cell Stem Cell* 10: 729–739.

Lehmann, R. 2016. Germ plasm biogenesis – an Oskar-centric perspective. *Curr. Top. Dev. Biol.* 116: 679–707.

Lei, L., Xukun, L. and Dean, J. 2013. The maternal to zygotic transition in mammals. *Mol. Aspects Med.* 34: 919–938

Lombard-Banek, C., Moody, S. A., Manzini, M. C. and Nemes, P. March 3, 2019. Microsampling capillary electrophoresis mass spectrometry enables single-cell proteomics in complex tissues: developing cell clones in live Xenopus laevis and zebrafish embryos. *Anal. Chem.* 91: 4797–4805. doi:10.1021/acs.analchem.9b00345.

Lombard-Banek, C., Reddi, S., Moody, S. A. and Nemes, P. 2016. Label-free quantification of proteins in single embryonic cells by capillary electrophoresis high-resolution mass spectrometry. *Mol. Cell. Proteomics* 15: 2756–2768.

Lustig, K. D., Kroll, K. L., Sun, E. E. and Kirschner, M. W. 1996. Expression cloning of a *Xenopus* T–related gene (*Xombi*) involved in mesodermal patterning and blastopore lip formation. *Development* 122: 4001–4012.

Lyczak, R., Gomes, J. E. and Bowerman, B. 2002. Heads or tails: cell polarity and axis formation in the early *Caenorhabditis elegans* embryo. *Dev. Cell* 3: 157–166.

Ma, J., He, F., Xie, G. and Deng, W. M. 2016. Maternal AP determinants in the *Drosophila* oocyte and embryo. *WIREs Dev. Biol.* 5: 562–581.

McKendry, R., Hsu, S. C., Harland, R. M. and Grosschedl, R. 1997. LEF-1/TCF proteins mediate wnt-inducible transcription from the *Xenopus nodal-related 3* promoter. *Dev. Biol.* 192: 420–431.

Mei, W., Jin, Z., Lai, F., Schwend, T., Houston, D. W., King, M. L. and Yang, J. 2013. Maternal dead-end1 is required for vegetal cortical microtubule assembly during *Xenopus* axis specification. *Development* 140: 2334–2344.

Miller, D. 2015. Confrontation, consolidation, and recognition: the oocyte's perspective on the incoming sperm. *Cold Spring Harbor Perspect. Med.* 5: a023408.

Moody, S. A. 1987a. Fates of the blastomeres of the 16-cell stage *Xenopus* embryo. *Dev. Biol.* 119: 560–578.

Moody, S. A. 1987b. Fates of the blastomeres of the 32-cell stage *Xenopus* embryo. *Dev. Biol.* 122: 300–319.

Moody, S. A., Chow, I. and Huang, S. 2000. Intrinsic bias and lineage restriction in the phenotype determination of dopamine and neuropeptide Y amacrine cells. *J. Neurosci.* 20: 3244–3253.

Moody, S. A. and Kline, M. J. 1990. Segregation of fate during cleavage of frog (*Xenopus laevis*) blastomeres. *Anat. Embryol.* 182: 347–362

Mosquera, L., Forrestall, C., Zhou, Y. and King, M. L. 1993. A mRNA localized to the vegetal cortex of *Xenopus* oocytes encodes a protein with a nanos-like zinc finger domain. *Development* 117: 377–386.

Nishida, H. 2012. The maternal muscle determinant in the ascidian egg. *Wiley Interdiscip. Rev. Dev. Biol.* 1: 425–433.

Oda-Ishii, I., Abe, T. and Satou, Y. 2018. Dynamics of two key maternal factors that initiate zygotic regulatory programs in ascidian embryos. *Dev. Biol.* 437: 50–59.

Oda-Ishii, I., Kubo, A., Kari, W., Suzuki, N., Rothbächer, U. and Satou, Y. 2016. A maternal system initiating the zygotic developmental program through combinatorial repression in the ascidian embryo. *PLoS Gen.* 12: e1006045.

Onjiko, R. M., Moody, S. A. and Nemes, P. 2015. Single-cell mass spectrometry reveals small molecules that affect cell fates in the 16-cell embryo. *Proc. Natl. Acad. Sci. USA* 112: 6545–6550.

Pandur, P. D., Sullivan, S. A. and Moody, S. A. 2002. Multiple maternal influences on dorsal-ventral fate of *Xenopus* animal blastomeres. *Dev. Dyn.* 225: 581–587.

Ratnaparkhi, G. S. and Courey, A. J. 2014. Signaling cascades, gradients, and gene networks in dorsal/ventral patterning. In *Principles of Developmental Genetics* (2nd edition), ed. S. A. Moody, pp. 131–151. New York: Elsevier.

Rebagliati, M. R., Weeks, D. L., Harvey, R. P. and Melton, D. A. 1985. Identification and cloning of localized maternal RNAs from *Xenopus* eggs. *Cell* 42: 769–777.

Sasai, N., Yakura, R., Kamiya, D., Nakazawa, Y. and Sasai, Y. 2008. Ectodermal factor restricts mesoderm differentiation by inhibiting p53. *Cell* 133: 878–890.

Schroeder, K. E., Condic, M. L., Eisenberg, L. M. and Yost, H. J. 1999. Spatially regulated translation in embryos: asymmetric expression of maternal Wnt-11 along the dorsal-ventral axis in *Xenopus*. *Dev. Biol.* 214: 288–297.

Slaidina, M. and Lehmann, R. 2014. Translational control in germline stem cell development. *J. Cell. Biol.* 207: 13–21.

Spike, C. A. and Strome, S. 2003. Germ plasm: protein degradation in the soma. *Curr. Biol.* 13: R837–R839.

Stennard, F., Carnac, G. and Gurdon, J. B. 1996. The *Xenopus* T-box gene, *Antipodean*, encodes a vegetally localised maternal mRNA and can trigger mesoderm formation. *Development* 122: 4179–4188.

Strome, S. and Updike, D. 2015. Specifying and protecting germ cell fate. *Nat. Rev. Mol. Cell Biol.* 16: 406–416.

Strome, S. and Wood, W. B. 1982. Immunofluorescence visualization of germ-line-specific cytoplasmic granules in embryos, larvae, and adults of *Caenorhabditis elegans*. *Proc. Natl. Acad. Sci. USA* 79: 1558–1562.

Strome, S. and Wood, W. B. 1983. Generation of asymmetry and segregation of germ-line granules in early *C. elegans* embryos. *Cell* 35: 15–25.

Sullivan, S. A., Moore, K. B. and Moody, S. A. 1999. Early events in blastomere fate determination. In *Cell Lineage and Cell Fate Determination*, ed. S. A. Moody, pp. 207–321. New York: Academic Press.

Swalla, B. J. and Jeffrey, W. R. 1995. A maternal RNA localized in the yellow crescent is segregated to the larval muscle cells during ascidian development. *Dev. Biol.* 170: 353–364.

Tadros, W. and Lipshitz, H. D. 2009. The maternal-to-zygotic transition: a play in two acts. *Development* 136: 3033–3042.

Tao, Q., Yokota, C., Puck, H., Kofron, M., Birsoy, B., Yan, D., Asashima, M., Wylie, C. C., Lin, X. and Heasman, J. 2005. Maternal *wnt11* activates the canonical wnt signaling pathway required for axis formation in *Xenopus* embryos. *Cell* 120: 857–871.

Thomsen, G. H. and Melton, D. A. 1993. Processed Vg1 protein is an axial mesoderm inducer in *Xenopus*. *Cell* 74: 433–441.

Updike, D. L., Knutson, A. K., Egelhofer, T. A., Campbell, A. C. and Strome, S. 2014. Germ-granule components prevent somatic development in the *C. elegans* germline. *Curr. Biol.* 24: 970–975.

Venkatarama, T., Lai, F., Luo, X., Zhou, Y., Newman, K. and King, M. L. 2010. Repression of zygotic gene expression in the *Xenopus* germ line. *Development* 137: 651–660.

Weeks, D. L. and Melton, D. A. 1987. A maternal mRNA localized to the vegetal hemisphere in *Xenopus* eggs codes for a growth factor related to TGF–beta. *Cell* 51: 861–867.

White, J. A. and Heasman, J. 2008. Maternal control of pattern formation in *Xenopus laevis*. *J. Exp. Zool., Part B* 310: 73–84.

Whittaker, J. R. 1980. Acetylcholinesterase development in extra cells caused by changing the distribution of myoplasm in ascidian embryos. *J. Embryol. Exp. Morphol.* 55: 343–354.

Wilson, E. B. 1925. *The Cell in Development and Inheritance*. New York: MacMillan.

Xanthos, J. B., Kofron, M., Wylie, C. and Heasman, J. 2001. Maternal VegT is the initiator of a molecular network specifying endoderm in *Xenopus laevis*. *Development* 128: 167–180.

Xu, S., Cheng, F., Liang, J., Wu, W. and Zhang, J. 2012. Maternal xNorrin, a canonical Wnt signaling agonist and TGF-β antagonist, controls early neuroectoderm specification in *Xenopus*. *PLoS Biol.* 10: e1001286.

Yang, J., Aguero, T. and King, M. L. 2015. The *Xenopus* maternal-to-zygotic transition from the perspective of the germline. *Curr. Top. Dev. Biol.* 113: 271–303.

Yost, H. J., Phillips, C. R., Boore, J. L., Bertman, J., Whalon, B. and Danilchik, M. V. 1995. Relocation of mitochondria to the prospective dorsal marginal zone during *Xenopus* embryogenesis. *Dev. Biol.* 170: 83–90.

Yu, D., Oda-Ishii, I., Kubo, A. and Satou, Y. 2019. The regulatory pathway from genes directly activated by maternal factors to muscle structural genes in ascidian embryos. *Development* 146(3). doi:10.1242/dev.173104.

Zhang, C., Basta, T., Hernandez-Lagunas, L., Simpson, P., Stemple, D. L., Artinger, K. B. and Klymkowsky, M. W. 2004. Repression of nodal expression by maternal B1-type SOXs regulates germ layer formation in *Xenopus* and zebrafish. *Dev. Biol.* 273: 23–37.

Zhang, C., Basta, T., Jensen, E. D. and Klymkowsky, M. W. 2003. The beta-catenin/VegT-regulated early zygotic gene Xnr5 is a direct target of SOX3 regulation. *Development* 130: 5609–5624.

Zhang, J., Houston, D. W., King, M. L., Payne, C., Wylie, C. and Heasman, J. 1998. The role of maternal VegT in establishing the primary germ layers in *Xenopus* embryos. *Cell* 94: 515–524.

Zhang, J. and King, M. L. 1996. *Xenopus* VegT RNA is localized to the vegetal cortex during oogenesis and encodes a novel T-box transcription factor involved in mesodermal patterning. *Development* 122: 4119–4129.

Zhang, C. and Klymkowsky, M. W. 2007. The sox axis, nodal signaling, and germ layer specification. *Differentiation* 75: 536–545.

Zhao, P. and Sun, M.-X. 2015. The maternal-to-zygotic transition in higher plants: available approaches, critical limitations, and technical requirements. *Curr. Top. Dev. Biol.* 113: 373–398.

Zhou, Y. and King, M. L. 2004. Sending RNAs into the future: RNA localization and germ cell fate. *IUBMB Life* 56: 19–27.

3 Coevolution of the Cell Cycle and Deferred-Use Cells

Siim Pauklin
University of Oxford

CONTENTS

3.1 THE CONNECTION OF CELL CYCLE AND CELL FATE DECISIONS IN STEM CELLS

The cell cycle and cell fate decisions are interlinked in a broad range of developmental contexts in many organisms. Coordination of stem cell proliferation and differentiation is essential for normal development, organ homeostasis, and tissue repair through a direct interplay between cell cycle progression and differentiation in somatic stem cells in the skin, brain, gut, and hematopoietic system (Fuchs 2009; Lange and Calegari 2010; Li and Clevers 2010). Particularly important insight to the processes coordinating cell fate and the cell cycle has been derived from *pluripotent stem cells*. Research in the past decades has shown that different cell cycle phases and mitosis provide intricate layers for regulating stem cell self-renewal, exit from the stem cell state, and guide stem cell differentiation to various cellular identities (Soufi and Dalton 2016). These processes involve coupling of the cell cycle with distinct gene expression programs via molecular processes mediated by transcription factors, cell type-specific patterns of epigenetic modifications, and chromatin organization. Figure 3.1 depicts the coordination of cell cycle and cell fate decisions.

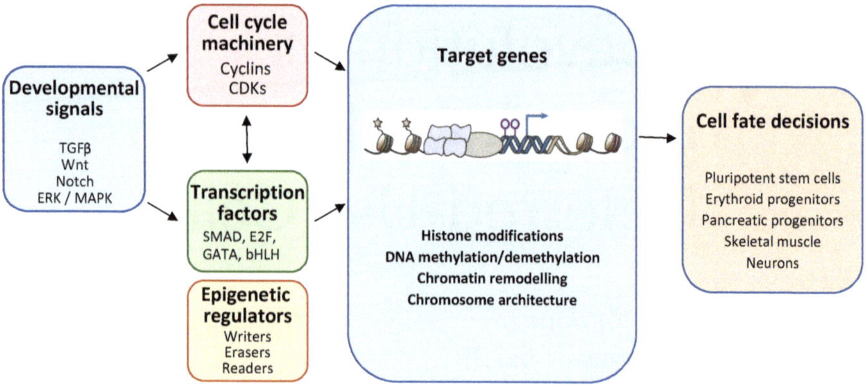

FIGURE 3.1 The cell cycle regulates developmental processes. Crosstalk between cell cycle control and cell fate specification involves developmental signals and cyclin-dependent protein kinases (CDKs) that target transcription factors that control expression of developmental genes via epigenetic mechanisms. In addition, CDKs modulate the epigenetic landscape and chromosome architecture around developmental genes. Induction of target genes determines important cell fate decisions and subsequent lineage commitment. Abbreviations: bHLH, basic helix-loop-helix; ERK/MAPK, mitogen-activated protein kinases/extracellular signal–regulated kinases; SMAD, mothers against decapentaplegic homolog; TGFβ, transforming growth factor beta.

The cell cycle represents a sequence of events in which chromosomes are replicated during DNA synthesis in S phase and segregated to the arising daughter cells via mitosis in M phase. These key events are separated by gap phases, G1 and G2, which control the correct timing and order of cellular events. The gap phase known as G1 starts the cell cycle after cell division and marks the stage when the arising cell either continues proliferation and prepares for DNA replication, or exits from the cell cycle and enters a nonproliferative stage known as G0 phase. The gap phase known as G2 follows DNA synthesis in S phase and represents a stage when the cell prepares for the subsequent cell division into two daughter cells.

Thus, the cell cycle of a continuously proliferating cell can be divided into G1, S, G2, and M phases, which will be repeated upon cell division. These successive cell cycle stages are regulated by the activity of cyclins and cyclin-dependent kinases (CDKs) that phosphorylate intracellular proteins essential for the different cycle phases and their ordered progression. A central decision stage for the exit or continuation for the cell cycle, known as the restriction (R-) point, is made in the G1 phase and regulated by CDK-mediated phosphorylation of retinoblastoma (RB) family proteins (Blagosklonny and Pardee 2002). This mechanism, which governs the growth and proliferation of most cell types, has an important role in stem cell self-renewal, differentiation, and quiescence in a broad range of developmental contexts (Tetteh et al. 2015; Li and Clevers 2010).

A cell needs to reach sufficient volume in G1 phase before progressing to S phase and committing to another round of cell division (Ginzberg et al. 2015).

This cell size–associated mechanism is widely conserved in evolution, for instance, budding yeast have a size-regulated restriction point in G1 phase called "Start" that resembles the R-point in mammals (Jorgensen and Tyers 2004). In addition to cell size, the R-point connects cell cycle decisions to cell fate decisions via external growth-promoting signals. In the presence of such mitogenic signals, normal cells progress through the restriction point and enter S phase due to the activation of genes necessary for cell cycle progression. On the other hand, when mitogenic signals are absent, the cells will neither enter S phase nor replicate their DNA. In pluripotent stem cells, exit from the cell cycle due to MYC proto-oncogene depletion leads to a developmental state that resembles the diapause stage (Scognamiglio et al. 2016). In addition to regulating stem cells, deregulation of the G1 checkpoint mechanism plays an important role in tumorigenesis, since the cells will not be able to block cell cycle progression upon environmental stress or the absence of mitogenic signals.

3.2 COORDINATION OF CELL CYCLE AND CELL FATE DECISIONS IS PRESENT IN MANY SPECIES

The connection between cell cycle and cell fate decisions is present across the whole evolutionary tree. For instance, haploid budding yeast switch between two alleles of the mating-type locus via the HO endonuclease mediated recombination event in G1 phase, which leads to cells that can mate and form diploid cells (Thon et al. 2018). The amoeba *Dictyostelium* enters a prespore identity upon nutrient starvation conditions in G1 phase while undertaking a prestalk identity when the signal is received in S and G2 phases (Gomer and Firtel 1987).

Cell cycle and cell fate decisions are also interconnected in multicellular organisms, spanning the nematode *Caenorhabditis elegans* (Ambros 1999), zebrafish (Bouldin and Kimelman 2014), frogs, and sea squirts (Ogura et al. 2011). This interconnection also plays an important role in different tissues during mammalian embryogenesis. During neocortical development, murine cortical progenitor cells initiate different cell fate decisions when cells undergo cycling compared to when they are not cycling (McConnell and Kaznowski 1991). Murine fetal erythropoiesis requires S phase to induce the erythroid differentiation program via a GATA1-dependent transcriptional regulation (Pop et al. 2010). Endocrine progenitors during pancreatic development initiate distinct cell fate decision depending on stage of the cell cycle when the cells receive differentiation signals: early G1 phase cells undergo an asymmetric cell division, whereas cells in late G1 phase progress through the cell cycle and give rise to two differentiated endocrine daughter cells (Kim et al. 2015).

While many events in multipotential stem cells are linked to their G1 phase, other cell cycle phases also can impact cell fate choice. For instance, a G2 phase transitioning time of *Drosophila* cells during bristle patterning is linked to cell fate decisions between a microchaete or a neural fate: a process known as lateral inhibition causes cells with elevated *Notch* signaling to divide first, compared to cells with lower *Notch* signaling, which extend their G2 phase and delay daughter cell production (Hunter et al. 2016).

3.3 THE MOLECULAR MECHANISMS COORDINATING THE CELL CYCLE WITH STEM CELL SELF-RENEWAL AND DIFFERENTIATION

The correct balance in proliferation and differentiation of stem cells is essential for embryogenesis as well as for the formation and repair of adult tissues (Oshimori and Fuchs 2012; Coronado et al. 2013; Sela et al. 2012; Calder et al. 2012; Singh et al. 2013, 2015; Chazaud et al. 2006; Pauklin and Vallier 2015).

Human embryonic stem cells (hESCs) have an interconnection between cell cycle, self-renewal, and differentiation (Pauklin and Vallier 2015), while they exert a metastable state with heterogeneity at the single cell level. Metastability is a central feature of many types of stem cells and seems to be generally connected to cell cycle progression (Pauklin and Vallier 2013). Accordingly, hESCs are characterized by a specific cell cycle profile with a short G1 phase, which restricts entry into lineage commitment and maintains stem cell self-renewal (Pauklin and Vallier 2013; Singh et al. 2012, 2015); differentiation is associated with an increase in G1 phase length (Calder et al. 2012; Coronado et al. 2013).

Pluripotent stem cells at the preimplantation developmental stage and cells maintained in vitro proliferate rapidly due to a truncated G1 phase and a shortened delay in S phase initiation. This shortened G1 phase supports the maintenance of pluripotency because cells reside for a briefer time period in the G1 cell cycle phase where cells are more prone to differentiation in response to external stimuli. When differentiation is initiated, elongation of the G1 phase makes cells more susceptible to irreversible germ layer commitment (Neganova et al. 2009; Pauklin and Vallier 2013; Boward et al. 2016; Ruiz et al. 2011). This cell cycle–mediated process results in a cell cycle–dependent capacity of differentiation in which hESCs initiate differentiation into the three germ layers during the transition from early to late G1 phase, while the S phase and G2/M primarily support pluripotency over differentiation (Gonzales et al. 2015; Pauklin and Vallier 2013). These mechanisms are governed by the cell cycle regulators cyclin D1–D3 that control differentiation signals TGFβ/Activin/Nodal-Smad2/3 pathway. hESCs in early G1 phase can only initiate differentiation into endoderm, whereas hESCs in late G1 are limited to neuroectoderm differentiation. The activity of Activin/Nodal signaling during cell cycle progression is controlled by cyclin D proteins that activate CDK4/6 and lead to the phosphorylation of Smad2 and Smad3 in their linker region. This mechanism blocks Smad2/3 shuttling in the nucleus in late G1, thereby preventing endoderm and allowing neuroectoderm specification (Pauklin and Vallier 2013). The G1 phase master regulators cyclin D1–D3 proteins also control mammalian stem cell differentiation and lineage specification on chromatin as transcription factors by blocking the expression of endoderm and mesoderm genes through transcriptional corepressors and inducing the transcription of neuroectoderm genes through transcriptional coactivators (Pauklin et al. 2016).

Bivalent domains consisting of H3K4me3 and H3K27me3 modifications are enriched on developmental genes in G1 phase in pluripotent stem cells, and there is a particular increase in H3K4me3 marks in G1 phase (Singh et al. 2012, 2015). On the other hand, the H3K27me3 mark remains relatively stable on developmental genes throughout the cell cycle (Singh et al. 2012, 2015). This is consistent with the observation that developmental genes are primed for activation every time they go through G1 phase but are not activated unless the appropriate signaling networks are also active. G1-specific epigenetic changes at developmental genes also coincide with the establishment of DNA loops that bridge distal enhancers with proximal promoters. *Altogether, this indicates that developmental genes in G1 phase are in a lineage-primed state due to changes in epigenetic landscapes and chromosome architectures, likely to be mediated via a crosstalk with transcription factors and CDKs.*

S and G2 phases attenuate exit from a pluripotent state, because hESCs in these cell cycle phases possess an intrinsic propensity toward maintaining their pluripotent state. Among the components regulating this process are DNA damage checkpoint factors ATM/ATR-CHEK2-p53 and cyclin B1 pathways (Gonzales et al. 2015). ATR/ATM enhances the activity of the TGFβ/Activin/Nodal pathway through p53 during S and G2. The upregulation of TGFβ then delays the decrease in the expression of pluripotency markers. This suggests that induction of hESC differentiation occurs during the G1 and M phases, while loss of pluripotency is achieved subsequently in S and G2 phases. Figure 3.2 depicts the connection of distinct phases of the cell cycle and pathways involved in regulating cell fate decisions and the exit from pluripotency in hESCs.

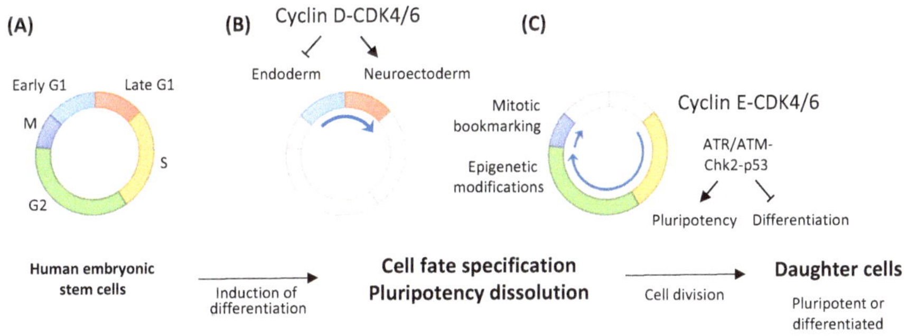

FIGURE 3.2 Cell fate specification and the cell cycle are interlinked in stem cells. (A) Human embryonic stem cells have a short G1 phase that promotes the pluripotent state and allows rapid proliferation. (B) Dissolution of pluripotency and lineage priming work in concert to orchestrate exit from pluripotency and initiate cell fate decisions in G1 phase. Cell fate specification of human embryonic stem cells (hESCs) starts in the G1 phase upon receiving differentiation signals and is regulated by a crosstalk between cyclin D-CDK4/6 and TGFβ-Smad2/3 pathway. (C) Cell fate commitment is further achieved in G2/M when exit from pluripotency is mediated through cell cycle–dependent mechanisms that are governed by cyclin E and ATR/ATM-Chk2-p53 pathway, as well as mitotic bookmarking. The daughter cells that arise after cell division either maintain their pluripotent state upon signals that support stem cell self-renewal or undergo differentiation to a developmentally restricted cell identity. Abbreviations: ATR/ATM, ataxia telangiectasia and Rad3-related protein/ataxia telangiectasia mutated; CDK4/6, cyclin-dependent kinase 4/6; Chk2, checkpoint kinase 2; p53, tumor protein p53.

3.4 CELL DIVISION, EPIGENETIC MEMORY, AND MITOTIC BOOKMARKING

Mitosis is the process of cell division that gives rise to two daughter cells from a single cell. Among the main characteristics of mitosis are (i) CDK-controlled processes of chromosome condensation and (ii) nuclear envelope breakdown (Spencer et al. 2000). During mitosis, the majority of factors regulating transcription—including developmental transcription factors, epigenetic "writers," and "readers" that deposit and interact with histone modifications—dissociate from chromatin, and cell type–specific transcription programs stop temporarily (Egli et al. 2008; Kellum et al. 1995; Minc et al. 1999; Voncken et al. 1999; McManus et al. 2006; Singh et al. 2015). This is accompanied by the removal of histone modifications (such as by histone acetylation) responsible for promoting gene expression; residual acetylations on distinct gene promoters are retained for "bookmarking" these loci throughout mitosis (Zhao et al. 2011). However, some repressive marks such as H3K27me3 and H3K9me3 remain on various loci. While mitosis is accompanied by a widespread erasure of the epigenome, bookmarking of some loci through cell division allows the arising daughter cells to restore previous gene expression programs (Zaidi et al. 2010). This process, which involves the reassembly of transcription complexes on enhancer and promoter regions, avoids the heterochromatinization of previously activated DNA loci, allowing for rapid reactivation of genes in G1 phase. *Hence, mitotic bookmarking is important for maintaining the epigenetic memory throughout cell divisions and for interconnecting cell fate decisions with cell cycle-dependent epigenetic processes.*

3.5 CELL CYCLE REGULATION AND TERMINAL DIFFERENTIATION

3.5.1 Cyclin-Dependent Kinases and Cyclin-Dependent Kinase Inhibitors

The cell cycle plays an important role in cell fate decisions by allowing the self-renewal of stem cells or progenitors while restricting further cell divisions during terminal differentiation. Differentiation is often associated with a change in cell cycle length (Coronado et al. 2013; Savatier et al. 1996), suggesting that mechanisms controlling cell cycle length and progression could be involved in cell fate decisions. Proliferation of stem cells or progenitors maintains their developmentally plastic and immature cell identity, while terminal differentiation of cells during developmental processes usually requires the exit of cells from cell cycle in G1 phase.

The progression of cell cycle in mammalian cells is primarily controlled by cyclins and CDKs, which affect key transcriptional regulators of the RB tumor suppressor protein family. The activity of the cyclin D/CDK complexes and thus cell proliferation is constrained by cyclin-dependent kinase inhibitors (CDKIs), which are subdivided into two families. The INK4 proteins (p16INK4a, p15INK4b, p18INK4c, and p19INK4d) bind to CDK4 and CDK6, and inhibit their kinase activities by interfering with their association with cyclin D proteins, while the Kip/Cip proteins (p21Cip1, p27Kip1, and p57Kip2) inhibit cyclin E-CDK2 (Sherr and Roberts 1999). Upregulation of CDKIs leads to the inhibition of CDKs during G1 phase and the hypophosphorylation of the RB protein family. This, in turn, allows RBs to bind to E2F transcription factors and thereby repress E2F target genes that are required for transitioning to S phase. The cyclin D-CDK4/6 complex can also bind to Kip/Cip CDKIs. However, this interaction enhances cyclin D-CDK4/6 activity since proteins such as p27 appear to limit INK4 CKIs' capacity to bind this complex. *Hence, a complex combination of INK4 and KIP/CIP protein regulations determines cell cycle progression.*

It was long believed that the cell cycle has only a passive role in cell fate decisions as a consequence of differentiation rather than being an active determinant guiding cell identity. Studies in various cell types have clarified this notion; for instance, neuronal differentiation is regulated by cell cycle stage–specific phosphorylation of the master transcription factors Ascl1 and Ngn2, while terminal differentiation of myoblasts is blocked by forced expression of cell proliferation–promoting factors such as cyclin D1 (Rao and Kohtz 1995) and by regulating the activity of MyoD1 during the cell cycle (Kitzmann and Fernandez 2001); terminal differentiation of myoblasts involves an upregulation of cell cycle inhibitor p21 by MyoD1 (Ruijtenberg and van den Heuvel 2015; Halevy et al. 1995). *This suggests that cell cycle regulators have an active function in early events of differentiation by controlling the activity of master transcription factors and extracellular signals.*

3.5.2 Retinoblastoma Family Proteins

The RB family of tumor suppressors (RBs: pRb, Rbl1 and Rbl2) are pivotal for cell cycle progression in mammalian cells. They control the G1 to S phase transition by reducing the transcriptional activity of E2F proteins (E2Fs), thereby leading to transcriptional repression of target genes necessary for proliferation. In turn, phosphorylation of RBs by cyclin D/CDK4–6 blocks interactions with E2Fs, permitting the induction of E2F-mediated transcription (Fiorentino et al. 2009). Besides their tumor suppressive function, RBs also are central for early mammalian development. Genetic studies in mice have shown that absence of pRb is embryonic lethal between E13 and E15 of gestation due to abnormal hematopoietic, neuronal, and eye lens development provoked by defects in cellular differentiation (Jacks et al. 1992; Korenjak and Brehm 2005).

pRb seems to direct differentiation of various cell types by controlling the activity of master regulators of differentiation. In contrast, the function of the other members of the RB family in differentiation is less well established. Rbl1$^{-/-}$ mice are viable and fertile but have impaired growth and exert myeloid hyperplasia (LeCouter et al. 1998a). More strikingly, Rbl2$^{-/-}$ embryos die between E11 and E13 due to disorganization in neural and dermatomyotomal structures (LeCouter et al. 1998b). Thus, Rbl1/2 are likely to have a function in cell fate decision, which might not be directly overlapping with pRb.

Targeted disruption of the three Rb-related genes in mESCs strongly alters their capacity of differentiation, while absence of RB protein function in hESCs induces cell death (Conklin et al. 2012). Moreover, mouse embryonic fibroblasts with a knockout for the three RB genes display a loss of G1 control and cellular immortalization (Sage et al. 2000; Dannenberg et al. 2000) that can be regarded as pathological self-renewal and that resembles the physiological self-renewal processes observed in embryonic or adult tissue-specific stem cells (Goding et al. 2014). In line with its function in controlling the self-renewal properties of stem cells or progenitors, pRb can restrict reprogramming and tumorigenesis by inhibiting pluripotency networks (Kareta et al. 2015). Considered together, these studies suggest a central function for RBs in cell fate decisions of stem cells and progenitors during embryonic development and in adult organs.

3.6 CONCLUSIONS

The cell cycle is tightly intertwined with cell fate decisions in diverse species ranging from yeast to human. Pluripotent stem cells in human have a cell cycle phase–specific responsiveness to differentiation signals, which changes from proneness to specification toward endoderm in early G1 phase and, in contrast, toward neuroectoderm in late G1 phase. The rest of the cell cycle is directed toward retaining the pluripotent identity of the stem cell. Together with lineage priming in G1 phase, the pluripotency network is prone to dissolution at this particular cell cycle phase. The rapid cycling of pluripotent stem cells with a truncated G1 phase and spending most of the time in S and G2 phases reduces the possibility for pluripotent stem cells to commit to lineage specification. During cell differentiation, there are large changes in the cell cycle profiles, which coordinate molecular complexes involving transcription factors, epigenetic modifying enzymes, epigenetic landscapes, and chromatic architecture. These cell cycle–orchestrated events ultimately govern gene expression programs unique to differentiated cell types.

REFERENCES

Ambros, V. 1999. Cell cycle-dependent sequencing of cell fate decisions in *Caenorhabditis elegans* vulva precursor cells. *Development* 126:1947–1956.

Blagosklonny, M. V., and Pardee, A. B. 2002. The restriction point of the cell cycle. *Cell Cycle* 1:103–110.

Bouldin, C. M., and Kimelman, D. 2014. Cdc25 and the importance of G2 control: insights from developmental biology. *Cell Cycle* 13:2165–2171. doi:10.4161/cc.29537.

Boward, B., Wu, T., and Dalton, S. 2016. Concise review: control of cell fate through cell cycle and pluripotency networks. *Stem Cells* 34:1427–1436. doi:10.1002/stem.2345.

Calder, A., Roth-Albin, I., Bhatia, S., Pilquil, C., Lee, J. H., Bhatia, M., Levadoux-Martin, M., McNicol, J., Russell, J., Collins, T., and Draper, J. S. 2012. Lengthened G1 phase indicates differentiation status in human embryonic stem cells. *Stem Cells Dev.* 22:279–295. doi:10.1089/scd.2012.0168.

Chazaud, C., Yamanaka, Y., Pawson, T., and Rossant, J. 2006. Early lineage segregation between epiblast and primitive endoderm in mouse blastocysts through the Grb2–MAPK pathway. *Dev. Cell* 10:615–624. doi:10.1016/j.devcel.2006.02.020.

Conklin, J. F., Baker, J., and Sage, J. 2012. The RB family is required for the self-renewal and survival of human embryonic stem cells. *Nat. Commun.* 3:1244. doi:10.1038/ncomms2254.

Coronado, D., Godet, M., Bourillot, P. Y., Tapponnier, Y., Bernat, A., Petit, M., Fanassieff, M., Markossian, S., Malashicheva, A., Iacone, R., Anastassiadis, K., and Savatier, P. 2013. A short G1 phase is an intrinsic determinant of naive embryonic stem cell pluripotency. *Stem Cell Res.* 10:118–131. doi:10.1016/j.scr.2012.10.004.

Dannenberg, J. H., van Rossum, A., Schuijff, L., and te Riele, H. 2000. Ablation of the retinoblastoma gene family deregulates G(1) control causing immortalization and increased cell turnover under growth-restricting conditions. *Genes Dev.* 14:3051–3064.

Egli, D., Birkhoff, G., and Eggan, K. 2008. Mediators of reprogramming: transcription factors and transitions through mitosis. *Nat. Rev. Mol. Cell Biol.* 9:505–516. doi:10.1038/nrm2439.

Fiorentino, F. P., Symonds, C. E., Macaluso, M., and Giordano, A. 2009. Senescence and p130/Rbl2: a new beginning to the end. *Cell Res.* 19:1044–1051. doi:10.1038/cr.2009.96.

Fuchs, E. 2009. The tortoise and the hair: slow-cycling cells in the stem cell race. *Cell* 137:811–819. doi:10.1016/j.cell.2009.05.002.

Ginzberg, M. B., Kafri, R., and Kirschner, M. 2015. Cell biology. On being the right (cell) size. *Science* 348:1245075. doi:10.1126/science.1245075.

Goding, C. R., Pei, D., and Lu, X. 2014. Cancer: pathological nuclear reprogramming? *Nat. Rev. Cancer* 14:568–573. doi:10.1038/nrc3781.

Gomer, R. H., and Firtel, R. A. 1987. Cell-autonomous determination of cell-type choice in Dictyostelium development by cell-cycle phase. *Science* 237:758–762.

Gonzales, K. A., Liang, H., Lim, Y. S., Chan, Y. S., Yeo, J. C., Tan, C. P., Gao, B., Le, B., Tan, Z. Y., Low, K. Y., Liou, Y. C., Bard, F., and Ng, H. H. 2015. Deterministic restriction on pluripotent state dissolution by cell-cycle pathways. *Cell* 162:564–579. doi:10.1016/j.cell.2015.07.001.

Halevy, O., Novitch, B. G., Spicer, D. B., Skapek, S. X., Rhee, J., Hannon, G. J., Beach, D., and Lassar, A. B. 1995. Correlation of terminal cell cycle arrest of skeletal muscle with induction of p21 by MyoD. *Science* 267:1018–1021.

Hunter, G. L., Hadjivasiliou, Z., Bonin, H., He, L., Perrimon, N., Charras, G., and Baum, B. 2016. Coordinated control of Notch/Delta signalling and cell cycle progression drives lateral inhibition–mediated tissue patterning. *Development* 143:2305–2310. doi:10.1242/dev.134213.

Jacks, T., Fazeli, A., Schmitt, E. M., Bronson, R. T., Goodell, M. A., and Weinberg, R. A. 1992. Effects of an Rb mutation in the mouse. *Nature* 359:295–300. doi:10.1038/359295a0.

Jorgensen, P., and Tyers, M. 2004. How cells coordinate growth and division. *Curr. Biol.* 14:R1014–R1027. doi:10.1016/j.cub.2004.11.027.

Kareta, M. S., Gorges, L. L., Hafeez, S., Benayoun, B. A., Marro, S., Zmoos, A. F., Cecchini, M. J., Spacek, D., Batista, L. F., O'Brien, M., Ng, Y. H., Ang, C. E., Vaka, D., Artandi, S. E., Dick, F. A., Brunet, A., Sage, J., and Wernig, M. 2015. Inhibition of pluripotency networks by the Rb tumor suppressor restricts reprogramming and tumorigenesis. *Cell Stem Cell* 16:39–50. doi:10.1016/j.stem.2014.10.019.

Kellum, R., Raff, J. W., and Alberts, B. M. 1995. Heterochromatin protein 1 distribution during development and during the cell cycle in Drosophila embryos. *J. Cell Sci.* 108:1407–1418.

Kim, Y. H., Larsen, H. L., Ru, P., Lemaire, L. A., Ferrer, J., and Grapin-Botton, A. 2015. Cell cycle-dependent differentiation dynamics balances growth and endocrine differentiation in the pancreas. *PLoS Biol.* 13:e1002111. doi:10.1371/journal.pbio.1002111.

Kitzmann, M., and Fernandez, A. 2001. Crosstalk between cell cycle regulators and the myogenic factor MyoD in skeletal myoblasts. *Cell. Mol. Life Sci.* 58:571–579.

Korenjak, M., and Brehm, A. 2005. E2F-Rb complexes regulating transcription of genes important for differentiation and development. *Curr. Opin. Genet. Dev.* 15:520–527. doi:10.1016/j.gde.2005.07.001.

Lange, C., and Calegari, F. 2010. Cdks and cyclins link G1 length and differentiation of embryonic, neural and hematopoietic stem cells. *Cell Cycle* 9:1893–1900.

LeCouter, J. E., Kablar, B., Hardy, W. R., Ying, C., Megeney, L. A., May, L. L., and Rudnicki, M. A. 1998a. Strain-dependent myeloid hyperplasia, growth deficiency, and accelerated cell cycle in mice lacking the Rb-related p107 gene. *Mol. Cell. Biol.* 18:7455–7465.

LeCouter, J. E., Kablar, B., Whyte, P. F., Ying, C., and Rudnicki, M. A. 1998b. Strain-dependent embryonic lethality in mice lacking the retinoblastoma-related p130 gene. *Development* 125:4669–4679.

Li, L., and Clevers, H. 2010. Coexistence of quiescent and active adult stem cells in mammals. *Science* 327:542–545. doi:10.1126/science.1180794.

McConnell, S. K., and Kaznowski, C. E. 1991. Cell cycle dependence of laminar determination in developing neocortex. *Science* 254:282–285.

McManus, K. J., Biron, V. L., Heit, R., Underhill, D. A., and Hendzel, M. J. 2006. Dynamic changes in histone H3 lysine 9 methylations: identification of a mitosis-specific function for dynamic methylation in chromosome congression and segregation. *J. Biol. Chem.* 281:8888–8897. doi:10.1074/jbc.M505323200.

Minc, E., Allory, Y., Worman, H. J., Courvalin, J. C., and Buendia, B. 1999. Localization and phosphorylation of HP1 proteins during the cell cycle in mammalian cells. *Chromosoma* 108:220–234.

Neganova, I., Zhang, X., Atkinson, S., and Lako, M. 2009. Expression and functional analysis of G1 to S regulatory components reveals an important role for CDK2 in cell cycle regulation in human embryonic stem cells. *Oncogene* 28:20–30. doi:10.1038/onc.2008.358.

Ogura, Y., Sakaue-Sawano, A., Nakagawa, M., Satoh, N., Miyawaki, A., and Sasakura, Y. 2011. Coordination of mitosis and morphogenesis: role of a prolonged G2 phase during chordate neurulation. *Development* 138:577–587. doi:10.1242/dev.053132.

Oshimori, N., and Fuchs, E. 2012. The harmonies played by TGF-beta in stem cell biology. *Cell Stem Cell* 11:751–764. doi:10.1016/j.stem.2012.11.001.

Pauklin, S., Madrigal, P., Bertero, A., and Vallier, L. 2016. Initiation of stem cell differentiation involves cell cycle-dependent regulation of developmental genes by Cyclin D. *Genes Dev.* 30:421–433. doi:10.1101/gad.271452.115.

Pauklin, S., and Vallier, L. 2013. The cell-cycle state of stem cells determines cell fate propensity. *Cell* 155:135–47. doi:10.1016/j.cell.2013.08.031.

Pauklin, S., and Vallier, L. 2015. Activin/nodal signalling in stem cells. *Development* 142:607–619. doi:10.1242/dev.091769.

Pop, R., Shearstone, J. R., Shen, Q., Liu, Y., Hallstrom, K., Koulnis, M., Gribnau, J., and Socolovsky, M. 2010. A key commitment step in erythropoiesis is synchronized with the cell cycle clock through mutual inhibition between PU.1 and S-phase progression. *PLoS Biol.* 8:e1000484. doi:10.1371/journal.pbio.1000484.

Rao, S. S., and Kohtz, D. D. 1995. Positive and negative regulation of D-type cyclin expression in skeletal myoblasts by basic fibroblast growth factor and transforming growth factor beta. A role for cyclin D1 in control of myoblast differentiation. *J. Biol. Chem.* 270:4093–4100.

Ruijtenberg, S., and van den Heuvel, S. 2015. G1/S inhibitors and the SWI/SNF complex control cell-cycle exit during muscle differentiation. *Cell* 162:300–313. doi:10.1016/j.cell.2015.06.013.

Ruiz, S., Panopoulos, A. D., Herrerias, A., Bissig, K. D., Lutz, M., Berggren, W. T., Verma, I. M., and Izpisua Belmonte, J. C. 2011. A high proliferation rate is required for cell reprogramming and maintenance of human embryonic stem cell identity. *Curr. Biol.* 21:45–52. doi:10.1016/j.cub.2010.11.049.

Sage, J., Mulligan, G. J., Attardi, L. D., Miller, A., Chen, S., Williams, B., Theodorou, E., and Jacks, T. 2000. Targeted disruption of the three Rb-related genes leads to loss of G(1) control and immortalization. *Genes Dev.* 14:3037–3050.

Savatier, P., Lapillonne, H., van Grunsven, L. A., Rudkin, B. B., and Samarut, J. 1996. Withdrawal of differentiation inhibitory activity/leukemia inhibitory factor up-regulates D-type cyclins and cyclin-dependent kinase inhibitors in mouse embryonic stem cells. *Oncogene* 12:309–322.

Scognamiglio, R., Cabezas-Wallscheid, N., Thier, M. C., Altamura, S., Reyes, A., Prendergast, A. M., Baumgartner, D., Carnevalli, L S., Atzberger, A., Haas, S., von Paleske, L., Boroviak, T., Worsdorfer, P., Essers, M. A., Kloz, U., Eisenman, R. N., Edenhofer, F., Bertone, P., Huber, W., van der Hoeven, F., Smith, A., and Trumpp, A. 2016. Myc Depletion induces a pluripotent dormant state mimicking diapause. *Cell* 164:668–680. doi:10.1016/j.cell.2015.12.033.

Sela, Y., Molotski, N., Golan, S., Itskovitz-Eldor, J., and Soen, Y. 2012. Human embryonic stem cells exhibit increased propensity to differentiate during the G1 phase prior to phosphorylation of retinoblastoma protein. *Stem Cells* 30:1097–1108. doi:10.1002/stem.1078.

Sherr, C. J., and Roberts, J. M. 1999. CDK inhibitors: positive and negative regulators of G1–phase progression. *Genes Dev.* 13:1501–1512.

Singh, A. M., Chappell, J., Trost, R., Lin, L., Wang, T., Tang, J., Matlock, B. K., Weller, K. P., Wu, H., Zhao, S., Jin, P., and Dalton, S. 2013. Cell–cycle control of developmentally regulated transcription factors accounts for heterogeneity in human pluripotent cells. *Stem Cell Rep.* 1:532–544. doi:10.1016/j.stemcr.2013.10.009.

Singh, A. M., Reynolds, D., Cliff, T., Ohtsuka, S., Mattheyses, A. L., Sun, Y., Menendez, L., Kulik, M., and Dalton, S. 2012. Signaling network crosstalk in human pluripotent cells: a Smad2/3-regulated switch that controls the balance between self-renewal and differentiation. *Cell Stem Cell* 10:312–326. doi:10.1016/j.stem.2012.01.014.

Singh, A. M., Sun, Y., Li, L., Zhang, W., Wu, T., Zhao, S., Qin, Z., and Dalton, S. 2015. Cell-cycle control of bivalent epigenetic domains regulates the exit from pluripotency. *Stem Cell Rep.* 5:323–336. doi:10.1016/j.stemcr.2015.07.005.

Soufi, A., and Dalton, S. 2016. Cycling through developmental decisions: how cell cycle dynamics control pluripotency, differentiation and reprogramming. *Development* 143:4301–4311. doi:10.1242/dev.142075.

Spencer, C. A., Kruhlak, M. J., Jenkins, H. L., Sun, X., and Bazett-Jones, D. P. 2000. Mitotic transcription repression in vivo in the absence of nucleosomal chromatin condensation. *J. Cell Biol.* 150:13–26.

Tetteh, P. W., Farin, H. F., and Clevers, H. 2015. Plasticity within stem cell hierarchies in mammalian epithelia. *Trends Cell Biol.* 25:100–108. doi:10.1016/j.tcb.2014.09.003.

Thon, G., Maki, T., Haber, J. E., and Iwasaki, H. 2018. Mating-type switching by homology-directed recombinational repair: a matter of choice. *Curr. Genet.* 65:351–362. doi:10.1007/s00294-018-0900-2.

Voncken, J. W., Schweizer, D., Aagaard, L., Sattler, L., Jantsch, M. F., and van Lohuizen, M. 1999. Chromatin-association of the Polycomb group protein BMI1 is cell cycle-regulated and correlates with its phosphorylation status. *J. Cell Sci.* 112:4627–4639.

Zaidi, S. K., Young, D. W., Montecino, M. A., Lian, J. B., van Wijnen, A. J., Stein, J. L., and Stein, G. S. 2010. Mitotic bookmarking of genes: a novel dimension to epigenetic control. *Nat. Rev. Genet.* 11:583–589. doi:10.1038/nrg2827.

Zhao, R., Nakamura, T., Fu, Y., Lazar, Z., and Spector, D. L. 2011. Gene bookmarking accelerates the kinetics of post-mitotic transcriptional re-activation. *Nat. Cell. Biol.* 13:1295–1304. doi:10.1038/ncb2341.

Section II

Origin of Deferred-Use Cells and Their Niches

Phylogenetic Approaches

4 The Early Evolution of Cellular Reprogramming in Animals

Nagayasu Nakanishi
University of Arkansas

David K. Jacobs
University of California Los Angeles

CONTENTS

4.1 INTRODUCTION

One of the hallmarks of animal development is progressive determination of cell fate, whereby cellular states become increasingly specialized as development proceeds, for example, cell-layer specification during gastrulation followed by cell-type differentiation during organogenesis. Developmental potential of a cell is channeled by the action of cytoplasmic determinants such as maternal factors that are asymmetrically distributed during cell division and/or by inductive interactions with other cells. Pools of stem cells, defined as undifferentiated cells with the capacity to self-renew and generate more specialized cells, often segregate from differentiating cells and tissues during animal development.

Segregated stem cells can be important not only for tissue homeostasis and sexual maturation of animals but also for replenishing damaged or lost cells during regeneration and for generating new postembryonic cell types during life cycle transition. In sea urchins, for instance, most adult tissues as well as germ line cells derive from coelomic sac cells of the larva referred to as "set-aside" cells (Davidson et al. 1995; Peterson et al. 1997), which are pluripotent stem cells segregated during embryogenesis and that remain mitotically quiescent until metamorphosis (Pehrson and Cohen 1986). We use the term *pluripotent stem cells* to refer to stem cells capable of generating somatic cells and germ cells.

Other examples of segregated stem cells include the migratory pluripotent stem cells of planarians referred to as neoblasts (reviewed by Reddien and Alvarado 2004) and those of hydrozoan cnidarians known as *interstitial stem cells* (i-cells) (reviewed by Gahan et al. 2016). Similarly, a pluripotent stem cell type—archeocytes—occurs in some sponges (poriferans) and is thought to be a major source of differentiated cells during development/metamorphosis, regeneration, reproduction, and tissue homeostasis (reviewed by Funayama 2008). Interestingly, a conserved set of germ line determinants (*piwi, vasa, bruno*, and *pl-10*) is expressed in these pluripotent stem cell populations—set-aside cells of a sea urchin (Juliano et al. 2006); i-cells in hydrozoans (Leclere et al. 2012; Rebscher et al. 2008; Seipel et al. 2004); and archeocytes of a sponge (Funayama et al. 2010). These comparative gene expression data led to the proposal that stem cells that are segregated early in development and have both somatic and germ potential—referred to as "primordial stem cells"—are a fundamentally conserved cell type of animals (Solana 2013). Yet, i-cells and archeocytes appear to be lineage-specific cell types within Cnidaria (Gold and Jacobs 2013) and Porifera (Ereskovsky 2010), respectively, casting doubt on whether early animal ancestors indeed generated pluripotent stem cells.

In addition to questions of common ancestry of pluripotent stem cells across basally branching metazoan groups, stem cells are not the only source of cells during regeneration and metamorphosis. Although it is often assumed that development generates irreversible, terminally differentiated cell types such as neurons, "terminally" differentiated cells frequently change their cellular states via reprogramming—referred to as transdifferentiation—during development and regeneration in animals (reviewed by Sanchez Alvarado and Yamanaka 2014 and by Okada 1991). For instance, eye lens cells of adult newts can regenerate from epithelial cells of the dorsal iris (Eguchi and Shingai 1971). In *Caenorhabditis elegans*, a rectal epithelial cell Y transdifferentiates into a motor neuron during larval development (Borisenko et al. 2015). Likewise, in zebrafish larvae, transdifferentiation of dorsal root ganglia sensory neurons into sympathetic neurons has been reported (Wright et al. 2010). In vitro, the striated muscle cells of hydrozoan jellyfish can transform into a variety of somatic cell types such as smooth muscle cells and neurons (Schmid et al. 1988). Given that some differentiated somatic cells are capable of altering cellular states via reprogramming, it seems reasonable to consider the contribution of differentiated somatic cells, along with stem cells, as potential sources of postembryonic cells.

What, then, is the ancestral cellular mechanism of generating postembryonic cell types in the context of development and regeneration in animals? Is it differentiation of resident pluripotent stem cells, reprogramming of differentiated somatic cells, or both? Resolving this problem requires an understanding of the processes of postembryonic cell differentiation and regeneration in early-branching lineages of animals (Figure 4.1). To this end, we review (i) developmental origins of cells that are produced at life cycle transition or during regeneration in the early-branching animal groups Cnidaria, Ctenophora, and Porifera and (ii) evidence for cell differentiation in the closest relative of animals, the choanoflagellates. We begin by examining

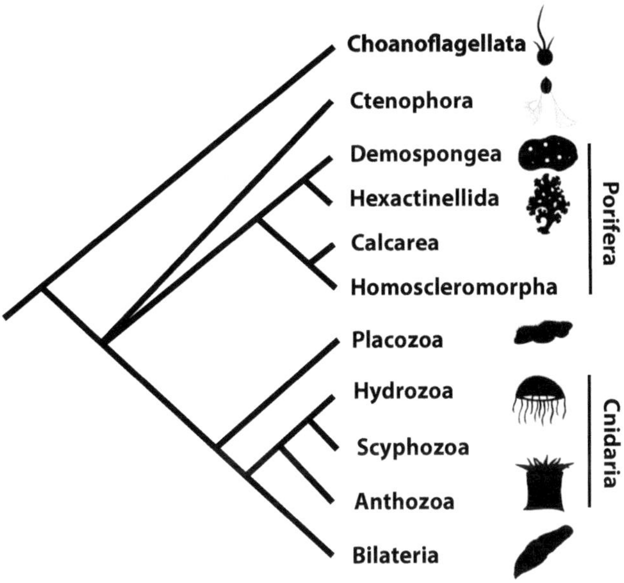

FIGURE 4.1 A consensus animal phylogeny based on current phylogenetic/phylogenomic evidence, rooted by the sister group of animals, Choanoflagellata. Note polytomy at the base of the animal tree, representing uncertainties about the branching order of Porifera and Ctenophora. Silhouette images are from phylopic.org and are under the Public Domain. Based on data from numerous sources, notably Dunn et al. (2008), Moroz et al. (2014), Ryan et al. (2013), Simion et al. (2017), Carr et al. (2008), Collins et al. (2006), Medina et al. (2001), Putnam et al. (2007), Zapata et al. (2015), Erpenbeck and Worheide (2007), and Gazave et al. (2012).

Cnidaria, a group of animals that include jellyfish, sea anemones, and corals, all with complex life cycles. We then explore Porifera (sponges), which have a biphasic life cycle, and Ctenophora (comb jellies), which undergo direct development. Finally, we briefly consider Choanoflagellata.

4.2 CNIDARIA

Cnidarians constitute a diverse group of animals that thrive in marine and freshwater environments. They include sea anemones, corals, and a variety of jellyfishes such as sea wasps and Lion's mane. Cnidaria is sister group to Bilateria and consists of Anthozoa (sea anemones and corals) and Medusozoa (jellyfishes; Staurozoa, Hydrozoa, Scyphozoa, and Cubozoa) (Collins et al. 2006; Medina et al. 2001; Putnam et al. 2007; Zapata et al. 2015). Cnidarians are generally conceived of as diploblastic, having the outer ectodermal layer and the inner endodermal layer separated by an extracellular matrix, the mesoglea, although it should be noted that cnidarian epithelia can be complex and are not always simple single-cell layers.

A cnidarian life cycle typically begins with a free-swimming planula larva that metamorphoses into a sessile polyp, which sexually matures in anthozoans and some medusozoans such as the hydrozoan *Hydra*. In most medusozoans, the polyp undergoes another round of metamorphosis to form free-swimming medusae either through lateral budding (in Hydrozoa) or transverse fission/strobilation (in Scyphozoa and Cubozoa). Each life cycle stage is characterized not only by distinct body plans but also by sets of stage-specific cell types (e.g., motor nerve net neurons of scyphozoan medusa or hair cells of anthozoan polyps).

The transition between life cycle phases appears to involve either cellular reprogramming or differentiation of segregated stem cells, depending on the taxa. In the scyphozoan *Aurelia*, metamorphosis of a planula larva into a polyp entails renewal of the endoderm, a process that has been referred to as secondary gastrulation (Yuan et al. 2008). During this process, the planula endoderm appears to undergo apoptosis (Yuan et al. 2008), and ectodermal epithelial cells of the planula larva transform into endodermal epithelial cells of a polyp (Gold et al. 2016). These data indicate that metamorphosis of a planula into a polyp in Scyphozoa involves transdifferentiation.

On the other hand, at the planula-polyp transition in hydrozoans, ectoderm undergoes apoptosis, as seen in the colonial marine hydroid *Hydractinia echinata* (Seipp et al. 2001), and endodermally derived migratory stem cells—referred to as interstitial stem cells (i-cells)—move into the ectoderm as demonstrated in the feather hydroid *Pennaria tiarella* by Summers and Haynes (1969) and in *H. echinata by* Weis and Buss (1987). Such cells presumably contribute to the formation of the polyp ectoderm, as i-cells are known to give rise to a variety of cell types, including neurons, stinging cells (cnidocytes), gland cells, and gametes in *Hydra* and *Hydractinia* (reviewed by (Bode 1996 and by Gahan et al. 2016). We note, however, that i-cells are a hydrozoan-specific cell type and that homologous cell types have not been demonstrated in nonhydrozoan cnidarians (reviewed by Gold and Jacobs 2013). Thus, while in hydrozoans it is likely that i-cells generate new cells at the planula to polyp transition, there is currently no evidence that segregated stem cells contribute to morphogenesis during life cycle transition in nonhydrozoan cnidarians.

The planula to polyp transition in the anthozoan starlet sea anemone *Nematostella vectensis* is not nearly as drastic; there is no evidence that it involves reorganization of cell layers. Nonetheless, cell shape changes occur in the body column and tentacle primordia during polyp formation (Fritz et al. 2013; Nakanishi et al. 2012). Moreover, a new polyp-tentacle-specific somatic cell type, the hair cell, develops (Nakanishi et al. 2012). The developmental origin of tentacular hair cells has not been resolved.

Limited data indicate that transformation of polyps into medusae may involve both differentiation of segregated stem cells and transformation of differentiated somatic cells. In the hydrozoan *Podocoryne carnea*, an electron microscopic study suggests that somatic cells of a polyp, namely ectodermal epithelial cells and endodermal digestive cells, transdifferentiate into exumbrellar epithelial cells and manubrial digestive cells of a medusa, while i-cells generate other cell types such as muscle cells, cnidocytes, gland cells, and gametes (Boelsterli 1977).

Helm et al. (2015) examined the development of medusa muscles in two closely related scyphozoan cnidarians, *Chrysaora quinquecirrha*, with a complete life cycle, and *Pelagia noctiluca*, which lost the polyp stage. They found that polyp muscles did not directly transform into medusa muscles in *Chrysaora* and did not transiently appear during the development of medusa muscles in direct-developing *Pelagia*. Thus, "remodeling" by which preexisting larval structures are modified to generate adult structures was ruled out in favor of "compartmentalization" by which adult structures develop de novo from segregated stem cells. However, the identity of these stem cells remains unknown. In addition, the possibility of transdifferentiation of polyp somatic cells into medusa muscle cells has not been examined.

Cnidarians generally have high regenerative potential; regeneration appears to result from either cellular reprogramming or differentiation of segregated stem cells. In *Hydra*, three stem cell lineages—tissue-specific epithelial stem cells of ectoderm and endoderm, and pluripotent interstitial stem cells—contribute to regeneration of a head or foot, without requiring cell proliferation (Cummings and Bode 1984; Hicklin and Wolpert 1973). This process of regeneration that involves repatterning of the existing tissues without growth is referred to as *morphallaxis*. i-cells can generate neurons, secretory cells, cnidocytes, and gametes but do not generate ectodermal and endodermal epithelial cells in *Hydra* (reviewed by Bode 1996). In another hydrozoan *Hydractinia*, the developmental potential of i-cells is less restricted; i-cells generate all somatic and germ cell types, including ectodermal and endodermal epithelial cells (Muller et al. 2004). During head regeneration in *Hydractinia* polyps, i-cells migrate to the decapitated site and proliferate to form a blastema from which a head is regenerated (Bradshaw et al. 2015). Thus, i-cells appear to be the major source of tissues during head regeneration in *Hydractinia*. In the anthozoan *N. vectensis*, oral structures of the polyp regenerate upon amputation, which, as in *Hydractinia*, requires cell proliferation (Passamaneck and Martindale 2012); however, cellular sources of regenerated tissues are not known.

Transdifferentiation-mediated regeneration also has been observed in Cnidaria. In the scyphozoan *Aurelia*, ectodermal fragments of a polyp—devoid of i-cells—can reconstitute the entire polyp (Steinberg 1963). Light microscopic evidence indicates that during this process of regeneration, ectodermal epithelial cells proliferate and generate endodermal cells via a dedifferentiated intermediate stage of amoeboid cells with little mitotic activity. Therefore, in *Aurelia*, ectodermal epithelial cells can transform into endodermal cells, not only during metamorphosis as discussed above but also during regeneration.

Another example of transdifferentiation-mediated regeneration comes from green hydra (*Hydra viridis*), in which the gastrodermal digestive cells contain symbiotic algae, and i-cells and cnidocytes occur exclusively in the epidermis and not in the gastrodermis (Haynes and Burnett 1963). In this hydra species, the isolated gastrodermis (lacking i-cells), but not epidermis (having i-cells), can regenerate a complete polyp, and histological observations indicated that gastrodermal cells (gland cells and/or digestive cells) directly gave rise to epidermal epitheliomuscular cells during regeneration (Haynes and Burnett 1963). Subsequent electron microscopy studies of regeneration from the isolated gastrodermis in *H. viridis* provided evidence that endodermal digestive cells directly transdifferentiated into epidermal epitheliomuscular cells (Davis et al. 1966), while gland cells transformed into cnidoblasts (cnidocyte precursor cells; Davis et al. 1966) and germ cells (Burnett et al. 1966) via an i-cell intermediate.

Similarly, a whole medusa can regenerate from i-cell-free umbrellar fragments in the hydrozoan *Clytia hemisphaerica* (formerly *Campanularia johnstoni;* Schmid and Tardent 1971). Moreover, in another hydrozoan jellyfish *Podocoryne carnea*, striated muscle cells can be induced in vitro to transform into a variety of somatic cell types, including ciliated smooth muscle cells and anti-FMRFamide-immunoreactive neurons, to form manubria or tentacles, but not the whole medusa (Schmid and Alder 1984; Schmid et al. 1988). Transdifferentiation is triggered by digesting the mesoglea adhered to mechanically isolated striated muscle cells using extracellular matrix–specific enzymes such as collagenase; without destabilization of extracellular matrix, isolated striated muscle cells remain differentiated (Schmid 1978). As expected for reprogramming, transcription and translation are required for transdifferentiation of striated muscle cells (Weber et al. 1987). Taken together, these observations suggest that differentiated somatic cells can contribute to regeneration via reprogramming in medusozoan cnidarians.

Interestingly, transformation of an advanced life cycle phase such as polyps and medusae back into earlier life cycle phases, referred to as *reverse development*, has been reported in some cnidarians (reviewed by (Piraino et al. 2004). For example, in the hydrozoan *Turritopsis,* medusae can transform into the colonial polyps connected by tube-like stolons regardless of the status of sexual maturity when exposed to environmental stress such as a sudden increase or decrease in water temperature (Piraino et al. 1996). Similarly, in the scyphozoan *Aurelia*, a strobilating polyp can generate a stack of polyps, instead of ephyrae, in response to heat stress (Kakinuma 1975), and juvenile and sexually mature medusae, as well as their tissue fragments, have been reported to transform back into a polyp (He et al. 2015). Moreover, newly settled polyps of the scleractinian cauliflower or lace coral *Pocillopora damicornis* can revert to a planula-like free-swimming form under unfavorable environmental conditions (Richmond 1985).

A few lines of evidence suggest that reverse development in *Turritopsis* primarily involves transdifferentiation (Piraino et al. 1996). First, i-cells do not appear sufficient for reverse development; isolated manubria containing a large number of replicating i-cells cannot generate polyps. Second, tissues from the exumbrellar epidermis and the gastrovascular system, which are poor in i-cells, are required for reverse development. Third, the exumbrellar epidermis lacks i-cells but is the only tissue capable of generating the secretory epidermis of stolons. Hence, differentiated ectodermal cells of the exumbrellar epidermis of the medusa must transform into those of the stolons during reverse development. The possibility that i-cells contribute to reverse development cannot be ruled out, however. The cellular bases of reverse development in *Aurelia* and *Pocillopora* remain unexplored.

In summary, cnidarians appear capable of generating postembryonic cells by either differentiation of segregated stem cells or transformation of differentiated somatic cells. Hydrozoans use a pluripotent stem cell type, the i-cells, to generate new cells at the transition from the planula to polyp, and at the transition from the polyp to medusa, and to replenish cells during head regeneration in polyps. There is also evidence that transdifferentiation of somatic cells contributes to regeneration and reverse development in hydrozoans. In nonhydrozoan cnidarians, there is currently no evidence for the presence of segregated pluripotent stem cells, but a cell lineage tracing study provides evidence for transdifferentiation in a scyphozoan cnidarian, where ectodermal cells of planulae transform into endodermal cells of polyps at metamorphosis.

4.3 PORIFERA

Sponges (Porifera) are marine and freshwater benthic animals, characterized by internal epithelial structures known as the choanocyte chambers that consist of ciliated epithelial cells (choanocytes) to filter-feed and generate a water current through the body (Bergquist 1978).

Sponges represent one of the earliest-evolving metazoan lineages composed of four diverse clades: Demospongiae, Calcareae, Homoscleromorphae, and Hexactinellidae with the phylogenetic interrelationship ([Demospongiae, Hexactinellidae], [Calcareae, Homoscleromorphae]; (Erpenbeck and Worheide 2007; Gazave et al. 2012). Fossil and biomarker evidence indicates that sponges have thrived on earth since at least 635 million years ago (Maloof et al. 2010; Love et al. 2009; Gold et al. 2016).

The sponge internal epithelial layer is typically made up of choanocytes and endopinacocytes, while the outer epithelial layer is composed of exopinacocytes that are exposed to the outer environment, and basopinacocytes that are in contact with the underlying substrate. Sandwiched between the internal and external layers is the mesohyl enriched with mesenchymal cells and collagen fibers. Following fertilization, embryogenesis typically generates elongated, radially symmetrical ciliated swimming larvae that settle onto a substrate by attaching their anterior region. The settled larvae metamorphose and grow into a sexually mature adult. Choanocyte chambers typically develop during metamorphosis, but they have been observed in larvae in some taxa, for example in the trichimella larvae of Hexactinellidae.

Two mechanisms for generating choanocytes—cellular reprogramming and differentiation from segregated stem cells—have been proposed. In some demosponges with parenchymella larvae, morphological evidence indicates that choanocyte chambers form directly from internally localized mesenchymal stem cells referred to as *archeocytes* (Brien and Meewis 1938, 1973; Meewis 1939; Bergquist and Green 1977), akin to hydrozoan i-cells. Here, all the larval epithelial cells are phagocytized by archeocytes during metamorphosis, as demonstrated in *Ephydatia fluviatilis* (Brien and Meewis 1938; Meewis 1939), *Microciona prolifera* (Meewis 1939; Misevic and Burger 1982; Misevic et al. 1990), or shed to the external media as in *Halichondria moorei, Ulosa* sp., and *Microciona rubens* (Bergquist and Green 1977). Thus, larval epithelial cells seem to play no role in the development of choanocytes. More recently, a cell lineage tracing study demonstrated that the larval archeocytes can generate choanocytes in the demosponge *Amphimedon queenslandica* (Nakanishi et al. 2014). However, development of choanocytes from larval archeocytes is unlikely to be an ancestral mechanism in sponges; larval archeocytes are not observed outside Demospongiae (e.g. Calcinea (Amano and Hori 2001); Hexactinellida (Boury-Esnault et al. 1999)).

In contrast, ciliated epithelial cells of larvae transform into choanocytes during metamorphosis across sponges. In Demospongiae, electron microscopy and cell lineage tracing studies support transformation of larval ciliated epithelial cells into choanocytes during metamorphosis in diversely represented taxa with different larval types. They include

1. parenchymella larvae (*A. queenslandica,* Nakanishi et al. 2014; Leys and Degnan 2002; Sogabe et al. 2016), the freshwater sponge *Spongilla lacustris* (Evans 1899), *Hamigera hamigera*, (Boury-Esnault 1976), *Mycale contarenii* (Borojevic and Lévi 1965), and the purple encrusting sponge *Haliclona permollis* (Amano and Hori 1996);
2. coeloblastula larvae lacking internal cells as in the melted chocolate sponge *Chondrilla australiensis* (Usher and Ereskovskyz 2005), and
3. dispherula larvae with a transient internal epithelial layer (Ereskovsky et al. 2007).

Likewise, among calcareous sponges, the amphiblastula larvae in Calcaronea and the coeloblastula/calciblastula larvae in Calcinea contain few internalized cells. Morphological data in these groups indicate that larval flagellated cells dedifferentiate into amoeboid cells via a loss of cilia and then differentiate into choanocytes (Minchin 1896; Amano and Hori 2001, 1993), similar to the pattern observed in the demosponge *A. queenslandica* (Nakanishi et al. 2014; Sogabe et al. 2016). Moreover, in the cinctoblastula larvae in Homoscleromorpha, internalized cells are rare or absent (Boury-Esnault et al. 2003; de Caralt et al. 2007), and electron microscopic evidence suggests that the juvenile internal epithelial layer is generated by tissue movement via invagination and involution of either anterior or posterior epithelium of the larva and transdifferentiation of the larval ciliated cells during metamorphosis (Ereskovsky et al. 2007). It is not known whether the outer layer epithelial cells of Hexactinellida trichimella larvae can generate choanocytes, if the larval choanocyte chambers simply grow or whether a combination of both occurs. Taken together, it is most parsimonious to assume that the last common ancestor of sponges developed choanocytes by transdifferentiation of larval ciliated epithelial cells during metamorphosis.

Similar to cnidarians, sponges have strong regenerative potential (reviewed by Simpson 1984). Morphological evidence indicates that cellular sources of regenerated tissues in demosponges are typically archeocytes (reviewed by Simpson 1984; Funayama 2008), but it remains unclear whether archeocytes contribute to regeneration in hexactinellids (Leys et al. 2007). Archeocytes are absent in homoscleromorphs and calcareans, despite their ability to regenerate (Ereskovsky et al. 2015; Korotkova 1963, 1970; Tuzet and Paris 1963). Thus, archeocytes are not a requirement for regeneration across sponges. We also note that, importantly, the lack of archeocytes in homoscleromorphs and calcareans makes it ambiguous whether the last common ancestor of sponges developed archeocytes, despite this being often assumed to be the case (e.g., Funayama 2008; Solana 2013).

Choanocytes, on the other hand, occur across all sponge clades and appear to maintain pluripotency, enabling contribution to tissue regeneration via transdifferentiation. For instance, it has been reported that choanocytes can dedifferentiate into archeocytes upon tissue damage in the demosponge *Suberites massa* (Diaz 1977), which, in turn, presumably replenish lost cell types. Moreover, electron microscopic evidence suggests that choanocytes transdifferentiate into exopinacocytes during regeneration in another demosponge *Halisarca dujardini* (Borisenko et al. 2015) and in the encrusting homoscleromorph *Oscarella lobularis* (Ereskovsky et al. 2015). These comparative data are consistent with the last common ancestor of sponges having used choanocytes to regenerate lost tissues by transdifferentiation.

In summary, while generation of postembryonic cells during metamorphosis and regeneration from the segregated stem cell type, archeocytes, seems common among demosponges, abundant morphological and cell lineage tracing data support transdifferentiation of somatic cells as a general mechanism of generating postembryonic cells in sponges. In particular, evidence for the capacity for ciliated epithelial cells of larvae and choanocytes of juvenile/adult sponges to transdifferentiate is found across major sponge clades. This supports an argument for deep ancestry of cell-type switching within sponges.

4.4 CTENOPHORA

Ctenophores, commonly known as comb jellies, are a group of gelatinous marine carnivores whose phylogenetic position relative to sponges is currently debated; recent phylogenomic studies place them as the earliest or the second earliest diverging animal lineage (Dunn et al. 2008; Moroz et al. 2014; Ryan et al. 2013; Simion et al. 2017).

Ctenophora is traditionally classified into six orders: Platyctenida, Lobata, Thalassocalycida, Cestida, Beroida, and polyphyletic Cydippida (Mertensiidae, Pleurobrachiidae, and Haeckeliidae). Phylogeny reconstruction based on 18S rRNA sequence data indicates that Mertensiidae is the earliest branching group, followed by Platyctenida, which is sister to a group consisting of [Pleurobrachiidae, [Beroida, Haeckeliidae], [Lobata, Thalassocalycida, Cestida]] (Podar et al. 2001). Platyctenes are the only benthic ctenophores; the rest are pelagic.

Ctenophores are characterized by a rotational (biradial) symmetry along the oral-aboral axis, eight longitudinal rows of ciliary comb plates (or ctene plates) used for locomotion, a pair of tentacles bearing sticky cells, referred to as colloblasts, that are employed to capture prey. An aboral apical organ composed of a gravity-sensitive statocyst is the only identifiable sensory structure. Ctenophores are diploblastic, consisting of an outer ectodermal epithelium and an inner endodermal epithelium separated by an extracellular matrix, the mesoglea. The mesoglea contains various cell types, including muscle cells. Ctenophore development is direct; a fertilized egg undergoes a stereotyped cleavage pattern that generates a free-swimming cydippid form with features of adult body plan (Martindale and Henry 1999). New cell types do not appear to develop postembryonically, but "postregeneration" can generate missing structures postembryonically (see below).

Most ctenophores (except for Beroida) can regenerate lost body parts. In the lobate warty comb jelly *Mnemiopsis leidyi* and the platyctene *Vallicula multiformis*, this includes the apical organ, comb plates, and tentacles (Coonfield 1936; Martindale 1986; Freeman 1967). Thus, somatic stem cells and/or differentiated cells capable of transdifferentiation must exist in ctenophores. The ctenophore phylogeny described above suggests that the last common ancestor of ctenophores was capable of regeneration and that regenerative potential was lost in Beroid ctenophores (Martindale 2016).

The mechanism of tissue regeneration in ctenophores is enigmatic. In *M. leidyi*, comb plates normally develop from e_1 and m_1 micromeres of the 16-cell stage embryo (Farfaglio 1963; Reverberi and Ortolani 1963; Martindale and Henry 1997a). Each quadrant of the 16-cell stage embryo contains two M cell descendants—a small m_1 micromere and a large 1M macromere—and two E cell descendants—an e_1 micromere and a 1E macromere.

When e_1 micromeres are experimentally removed, comb plates fail to develop during embryogenesis but develop from m_1 micromeres after several days (Martindale 1986; Martindale and Henry 1996). The process of generating a structure that was never present during the course of development is referred to as "*postregeneration*." When both of the cell lineages (e_1 and m_1) that normally generate comb plates are deleted, however, postregeneration does not occur (Henry and Martindale 2000). These manipulations argue that cell lineages that are not e_1 or m_1 do not generate pluripotent stem cells or somatic cells capable of forming comb plates via transdifferentiation. Instead, developmental potential to generate comb plates appears restricted to the e_1 and m_1 cell lineages. Thus, during postregeneration of comb plates upon removal of e_1 micromeres, the m_1 cell lineage must give rise to comb plate progenitor cells, transdifferentiation-competent somatic cells, or both, which enable the formation of comb plates.

Not all cell lineages are as restricted in fate as e_1 and m_1 cell lineages, however. In *M. leidyi*, 2M blastomeres at the 32-cell stage embryo normally give rise to the muscular core of the tentacles, but when they are deleted, tentacles form normally without defects in contractility (Martindale and Henry 1997a,b). Thus, non-2M cell lineages must be able to generate the muscular core in the absence of the 2M cell lineage, presumably through transdifferentiation of somatic cells or differentiation of pluri- or multipotent stem cells. The source of regenerated muscular core tissues remains unresolved.

Adult ctenophores appear to have tissue-specific (i.e. non-toti- or pluripotent) somatic stem cells that contribute to tissue homeostasis, however. Cell labeling experiments in adult sea gooseberries *Pleurobrachia pileus* indicate that fate-restricted somatic stem cells occur in localized regions at the tentacle bases, the comb rows, and aboral apical organ (Alie et al. 2011). In particular, proliferative cells in tentacle bases and comb rows were shown to differentiate into somatic cell types, colloblasts and ciliated polster cells, respectively; the fate of proliferative cells associated with the apical organ remains unclear. These data are consistent with a role for the ctenophore stem cell system in regulating tissue homeostasis. Thus, contribution of these tissue-specific somatic stem cells to replenishing missing cells during regeneration and postregeneration appears likely, although this remains to be confirmed.

In contrast to Cnidaria and Porifera, there is currently no clear evidence for transdifferentiation of somatic cells during development or regeneration in Ctenophora. Neither is there any report of i-cell- or archeocyte-like pluripotent stem cell types that are segregated early in development. However, lineage-restricted somatic stem cells seem to regulate tissue homeostasis in tentacles, comb rows, and the aboral apical organ.

4.5 CHOANOFLAGELLATA

Choanoflagellates, marine and freshwater protists, are the closest relative to animals (Steenkamp et al. 2006; Carr et al. 2008; Ruiz-Trillo et al. 2008; King et al. 2008). Comparative studies of choanoflagellates and animals therefore may provide insights into the biology of early animals (King 2004). Molecular phylogenetic analyses suggest that choanoflagellates can be divided into three major clades: Clade 1, Clade 2, and Clade 3, with the relationship [Clade 3, [Clade 1, Clade 2]] (Carr et al. 2008). Choanoflagellates are characterized by an apical flagellum surrounded by a collar of microvilli, referred to as a collar complex. These morphologically resemble choanocytes of sponges. Choanoflagellates beat flagella to generate water flow, which propels the cell and allows the collar of microvilli to capture and phagocytose bacteria.

Interestingly, some choanoflagellate taxa such as *Salpingoeca rosetta* display temporal cell differentiation depending on environmental conditions (Fairclough et al. 2010; Dayel et al. 2011). For instance, *S. rosetta* can transdifferentiate from a free-swimming form to a sessile thecate form attached to a substrate via theca, or vice versa, and can form multicellular colonies of different morphological types (rosettes and chains) via mitosis (Fairclough et al. 2010; Dayel et al. 2011). Transcriptome data from *S. rosetta* show that distinct solitary and colonial forms are characterized by differential gene expression (Fairclough et al. 2013). Thus, transformation of cellular states is under genetic control. Also, in *S. rosetta*, haploid solitary cells can directly transdifferentiate into gametes, both small and large flagellated cells (Levin and King 2013; Woznica et al. 2017). Ancestral character state reconstruction based on the molecular phylogeny of choanoflagellates suggests that multicellular colony development predated the divergence of Clade 1 and 2 or evolved independently in different choanoflagellate lineages multiple times (Carr et al. 2008). The former scenario leaves open the possibility that colony formation, and thus temporal differentiation of cellular states, is an ancestral trait of choanoflagellates.

4.6 EARLY ANIMALS WERE CAPABLE OF REPROGRAMMING SOMATIC CELLS

The data summarized above can be mapped onto metazoan phylogeny. This allows preliminary inference of the evolutionary history of postembryonic mechanisms that generate new or lost cell types (Figure 4.2).

As mentioned above, it is currently debated whether Ctenophora or Porifera is sister to the rest of the animals (e.g., Dunn et al. 2008; Moroz et al. 2014; Ryan et al. 2013; Simion et al. 2017), and thus, we assume polytomy at the base of the animal phylogeny (Figures 4.1 and 4.2A). Phylogenetically widespread instances of cellular reprogramming across early-evolving animal groups—Porifera and Cnidaria, in particular—are consistent with the hypothesis that early animals were capable of reprogramming somatic cells during postembryonic development or regeneration (Figure 4.2B). Furthermore, although it remains to be addressed whether transdifferentiation is an ancestral trait in Choanoflagellata or a derived trait of *S. rosetta* (and other colony-forming choanoflagellates), the evidence of alteration of differentiated cellular states in a choanoflagellate raises the possibility that cellular

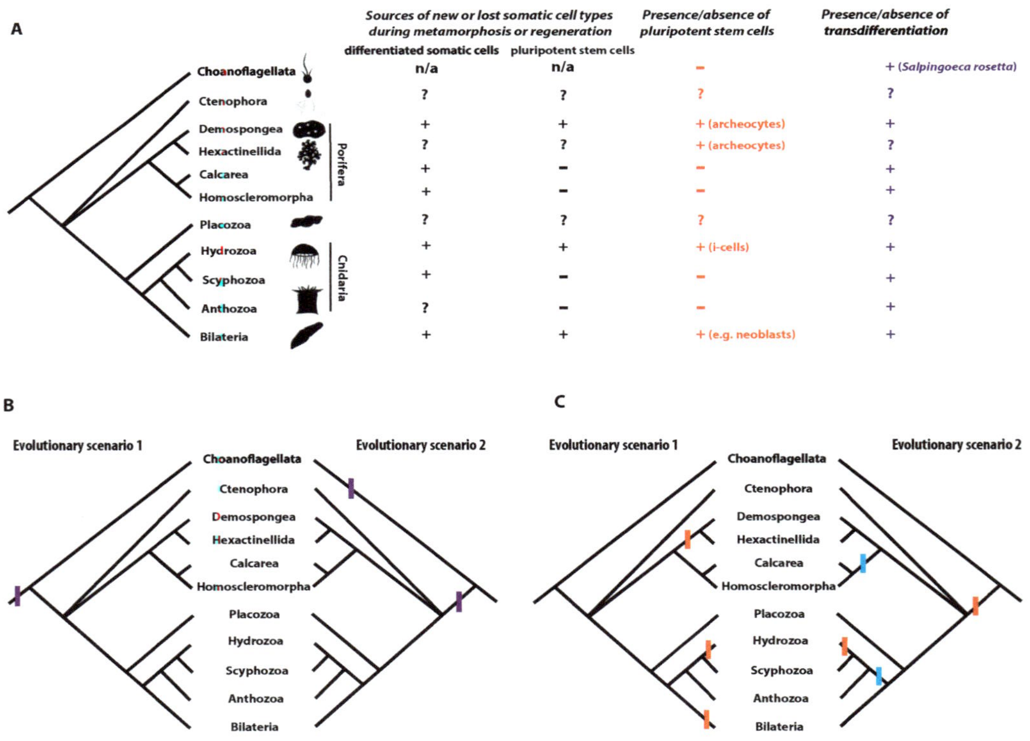

FIGURE 4.2 Possible evolutionary histories of transdifferentiation and pluripotent stem cells in animals. (A) Animal phylogeny (left) and a character matrix (right) used for reconstructing ancestral character states. Characters were scored for taxa where information was available (see text for details). (B) Alternative evolutionary scenarios for the origin of transdifferentiation. Purple lines indicate emergence. (C) Alternative evolutionary scenarios for the origins of pluripotent stem cells. Orange lines indicate emergence; blue lines indicate evolutionary losses. Silhouette images in (A) are from phylopic.org and are under the Public Domain.

reprogramming mechanisms may even predate animal origin (evolutionary scenario 1 in Figure 4.2B). Alternatively, the ability to change differentiated cellular states may have evolved independently in *S. rosetta* and the metazoan stem lineage (evolutionary scenario 2 in Figure 4.2B).

In contrast, segregated populations of pluripotent stem cells are found sporadically in divergent taxa, indicative of more complex evolutionary histories than previously assumed. This suggests multiple gains or losses of pluripotent stem cells consistent with the unique attributes of stem cells in different groups.

As discussed above, archeocytes are found only in demosponges and its sister group hexactinellids within Porifera, and interstitial stem cells (i-cells) are restricted to hydrozoans within Cnidaria. Although these stem cell populations express an orthologous set of germ line determinants (*piwi, vasa, bruno*, and *pl-10*) as discussed above, the limited occurrence across metazoan phylogeny, combined with the lack of segregated pluripotent stem cells in the outgroup taxa (e.g., choanoflagellates), leaves the ancestral state ambiguous. One possibility is that pluripotent stem cell populations evolved independently in divergent lineages (evolutionary scenario 1 in Figure 4.2C). This scenario implies that gametes were generated by transdifferentiation of somatic cells in early animals, similar to the condition encountered in sponges where a differentiated somatic cell type, the choanocyte, is thought to give rise to gametes—both eggs and sperm, but sperm only in some demosponges—reviewed by Harrison and De Vos (1991). Consistent with this hypothesis, it has been reported that striated muscle cells of a hydrozoan cnidarian have the potential to generate gametes via transdifferentiation (Schmid et al. 1988). Interestingly, both sponge choanocytes and hydrozoan striated muscle cells express *piwi* (Funayama et al. 2010; Seipel et al. 2004), indicating that "germ line" determinants can function in differentiated somatic cells, possibly, to maintain genome integrity and cellular potency (van Wolfswinkel 2014). Also noteworthy is that under this evolutionary scenario, it must be assumed that the Weismann barrier, in which genetic information flows from germ line cells to somatic cells but not vice versa, was absent in early animals, in disagreement with the primordial stem cell hypothesis that assumes otherwise (Solana 2013). Alternatively, the last common ancestor of animals may have developed pluripotent stem cells during embryogenesis, which were subsequently lost in Homoscleromorpha/Calcarea and Cnidaria independently, followed by a reversal in Hydrozoa (evolutionary scenario 2 in Figure 4.2C).

4.7 FUTURE DIRECTIONS

A number of problems remain unresolved.

First, sources of cells during metamorphosis or regeneration are unknown in some phylogenetically informative taxa. For instance, in ctenophores, although tissue-specific somatic stem cells do seem to exist, it is unclear whether and how they contribute to tissue regeneration and whether transdifferentiation has any role in the process. In anthozoan cnidarians, segregated somatic stem cells in the form of i-cells appear absent, but this does not necessitate that transdifferentiation generates postembryonic cell types during metamorphosis or regeneration; postembryonic cells could come from reserve somatic stem cells that have yet to be discovered.

Second, the molecular basis of transdifferentiation is poorly understood in early-diverging animal groups. Relevant data are currently limited to the hydrozoan cnidarian *Podocoryne*, where *bmp2/4*, *bmp5/8*, *msx*, and *piwi* have been found to be differentially expressed during transdifferentiation of striated muscle cells, indicative of their role in regulating transdifferentiation (Seipel et al. 2004; Galle et al. 2005; Reber-Muller et al. 2006). The precise roles of BMP signaling, *msx*, and *piwi* in striated muscle transdifferentiation remain to be established by gene function perturbation approaches.

Third, an understanding of how cellular states are maintained in early-diverging animal groups is lacking. Some somatic cell types such as neurons appear to be stably differentiated across Bilateria and Cnidaria, although the possibility of transdifferentiation under specific conditions (e.g., during regeneration) remains to be investigated. The knowledge of the molecular mechanisms that maintain differentiated cellular states in early-evolving animal groups is key to gaining insights into how cellular plasticity was regulated—to prevent malignant cellular reprogramming that leads to the formation of cancer—in early animals.

4.8 CONCLUSIONS

In this chapter, we have reviewed developmental origins of postembryonic cell types that arise at life cycle transition and/or regeneration in early-branching animal lineages—Cnidaria, Ctenophora, and Porifera. Based on these data, we propose that *transdifferentiation is likely to be an ancestral mode to postembryonically generate new or lost cell types in animals.* It is possible that cellular reprogramming mechanisms even predated animal origin, as temporal alteration of differentiated cellular states occurs in unicellular relatives of animals. However, an alternative possibility of independent evolutionary origins of cellular reprogramming in choanoflagellates and Metazoa cannot be ruled out. Within Metazoa, the evolutionary history of pluripotent stem cells that are capable of generating new or lost cell types postembryonically appear more complex. They may have emerged independently in Demospongea/Hexactinellida (archeocytes), Hydrozoa (interstitial stem cells), and Bilateria. Alternatively, the last common ancestor of animals may have had pluripotent stem cells, which were later lost in some lineages—Homoscleromorpha/Calcarea and Cnidaria—followed by a reversal in Hydrozoa. We infer that regulation of cellular reprogramming was integral to the biology of early animals, and so a more comprehensive understanding of transdifferentiation is critical to an understanding of evolutionary history of stem cells and the evolution and diversification of animals.

REFERENCES

Alie, A., Leclere, L., Jager, M., Dayraud, C., Chang, P., Le Guyader, H., Quéinnec, E., and Manuel, M. 2011. Somatic stem cells express *Piwi* and *Vasa* genes in an adult ctenophore: Ancient association of "germ line genes" with stemness. *Dev. Biol.* 350: 183–197.

Amano, S., and Hori, I. 1993. Metamorphosis of calcareous sponges.2. Cell rearrangement and differentiation in metamorphosis. *Invertebr. Reprod. Dev.* 24: 13–26.

Amano, S., and Hori, I. 1996. Transdifferentiation of larval flagellated cells to choanocytes in the metamorphosis of the demosponge *Haliclona permollis*. *Biol. Bull.* 190: 161–172.
Amano, S., and Hori, I. 2001. Metamorphosis of coeloblastula performed by multipotential larval flagellated cells in the calcareous sponge *Leucosolenia laxa*. *Biol. Bull.* 200: 20–32.
Bergquist, P. R. 1978. *Sponges*. London: Hutchinson.
Bergquist, P. R., and Green, C. R. 1977. Ultrastructural-study of settlement and metamorphosis in sponge larvae. *Cah. Biol. Mari.* 18: 289–302.
Bode, H. R. 1996. The interstitial cell lineage of *Hydra*: A stem cell system that arose early in evolution. *J. Cell Sci.* 109: 1155–1164.
Boelsterli, U. 1977. Electron-microscopic study of early developmental stages, myogenesis, oogenesis and cnidogenesis in the anthomedusa, *Podocoryne carnea* M Sars. *J. Morphol.* 154: 259–289.
Borisenko, I. E., Adamska, M., Tokina, D. B., and Ereskovsky, A. V. 2015. Transdifferentiation is a driving force of regeneration in *Halisarca dujardini* (Demospongiae, Porifera). *Peer J.* 3: e12113.
Borojevic, R., and Lévi, C. 1965. Morphogènése expérimentale d'une Eponge a` partir de cellules de la larve nageante dissociée. *Z. Zellforsch.* 68: 57–69.
Boury-Esnault, N. 1976. Ultrastructure de la larve parenchymella d'*Hamigera hamigera* (Schmidt) (Demosponge, Poecilosclerida). *Cah. Biol. Mar.* 27: 9–20.
Boury-Esnault, N., Efremova, S., Bezac, C., and Vacelet, J. 1999. Reproduction of a hexactinellid sponge: First description of gastrulation by cellular determination in the Porifera. *Invertebr. Reprod. Dev.* 35: 187–201.
Boury-Esnault, N., Ereskovsky, A., Bezac, C., and Tokina, D. 2003. Larval development in the Homoscleromorpha (Porifera, Demospongiae). *Invertebr. Biol.* 122: 187–202.
Bradshaw, B., Thompson, K., and Frank, U. 2015. Distinct mechanisms underlie oral vs aboral regeneration in the cnidarian *Hydractinia echinata*. *Elife* 4: e05506.
Brien, P., and Meewis, H. 1938. Contribution a l'étude de l'embryogenése des Spongillidae. *Arch. Biol.* 49: 177–250.
Brien, P., and Meewis, H. 1973. Les Demosponges. In *Traité de Zool.* Vol 1(Ill), pp. 133–461. Paris: Mason Cie.
Burnett, A. L., Davis, L. E., and Ruffing, F. E. 1966. A histological and ultrastructural study of germinal differentiation of interstitial cells arising from gland cells in *Hydra viridis*. *J. Morphol.* 120: 1–8.
Carr, M., Leadbeater, B. S. C., Hassan, R., Nelson, M., and Baldauf, S. L. 2008. Molecular phylogeny of choanoflagellates, the sister group to Metazoa. *Proc. Natl. Acad. Sci. USA* 105: 16641–16646.
Collins, A. G., Schuchert, P., Marques, A. C., Jankowski, T., Medina, M., and Schierwater, B. 2006. Medusozoan phylogeny and character evolution clarified by new large and small subunit rDNA data and an assessment of the utility of phylogenetic mixture models. *Syst. Biol.* 55: 97–115.
Coonfield, B. R. 1936. Regeneration in *Mnemiopsis leidyi*, Agassiz. *Biol. Bull.* 71: 421–428.
Cummings, S. G., and Bode, H. R. 1984. Head regeneration and polarity reversal in *Hydra attenuata* can occur in the absence of DNA synthesis. *Wilhelm Roux's Arch. Dev. Biol.* 194: 79–86.
Davidson, E. H., Peterson, K. J., and Cameron, R. A. 1995. Origin of bilaterian body plans – evolution of developmental regulatory mechanisms. *Science* 270: 1319–1325.
Davis, L. E., Burnett, A. L., Haynes, J. F., and Mumaw, V. R. 1966. A histological and ultrastructural study of dedifferentiation and redifferentiation of digestive and gland cells in *Hydra viridis*. *Dev. Biol.* 14: 307–320.
Dayel, M. J., Alegado, R. A., Fairclough, S. R., Levin, T. C., Nichols, S. A., McDonald, K., and King, N. 2011. Cell differentiation and morphogenesis in the colony-forming choanoflagellate *Salpingoeca rosetta*. *Dev. Biol.* 357: 73–82.

de Caralt, S., Uriz, M. J., and Wijffels, R. H. 2007. Vertical transmission and successive location of symbiotic bacteria during embryo development and larva formation in *Corticium candelabrum* (Porifera: Demospongiae). *J. Mar. Biol. Assoc. U. K.* 87: 1693–1699.

Diaz, J.-P. 1977. Transformation histologiques et cytologiques post-traumatiques chez la demosponge *Suberites massa* Nardo. *Bull. Soc. Zool. France* 98: 145–156.

Dunn, C. W., Hejnol, A., Matus, D. Q., Pang, K., Browne, W. E., Smith, S. A., Seaver, E., Rouse, W. W. et al. 2008. Broad phylogenomic sampling improves resolution of the animal tree of life. *Nature* 452: 745–749.

Eguchi, G., and Shingai, R. 1971. Cellular analysis on localization of lens forming potency in the newt iris epithelium. *Dev. Growth Differ.* 13: 337–349.

Ereskovsky, A. V. 2010. *The Comparative Embryology of Sponges*. Dordrecht and New York: Springer.

Ereskovsky, A. V., Borisenko, I. E., Lapebie, P., Gazave, E., Tokina, D. B., and Borchiellini, C. 2015. Oscarella lobularis (Homoscleromorpha, Porifera) regeneration: Epithelial morphogenesis and metaplasia. *PLoS One* 1010(8): e0134566.

Ereskovsky, A. V., Konjukov, P., and Willenz, P. 2007. Experimental metamorphosis of *Halisarca dujardini* larvae (Demospongiae, Halisarcida): Evidence of flagellated cell totipotentiality. *J. Morphol.* 268: 529–536.

Ereskovsky, A. V., Tokina, D. B., Bezac, C., and Boury-Esnault, N. 2007. Metamorphosis of cinctoblastula larvae (Homoscleromorpha, Porifera). *J. Morphol.* 268: 518–528.

Erpenbeck, D., and Worheide, G. 2007. On the molecular phylogeny of sponges (Porifera). *Zootaxa* 1668: 107–126.

Evans, R. 1899. The structure and metamorphosis of the larva of *Spongilla lacustris. Q. J. Microsc. Sci.* 43: 363–477.

Fairclough, S. R., Chen, Z. H., Kramer, E., Zeng, Q., Young, S., Robertson, H. M., Begovic, E., Richter, D. J. et al. 2013. Premetazoan genome evolution and the regulation of cell differentiation in the choanoflagellate *Salpingoeca rosetta. Genome Biol.* 14: R15.

Fairclough, S. R., Dayel, M. J., and King, N. 2010. Multicellular development in a choanoflagellate. *Curr. Biol.* 20: R875–R876.

Farfaglio, G. 1963. Experiments on formation of combs in ctenophores. *Experientia* 19: 303–304.

Freeman, G. 1967. Studies on regeneration in the creeping ctenophore, *Vallicula multiformis. J. Morphol.* 123: 71–83.

Fritz, A. E., Ikmi, A., Seidel, C., Paulson, A., and Gibson, M. C. 2013. Mechanisms of tentacle morphogenesis in the sea anemone *Nematostella vectensis. Development* 140: 2212–2223.

Funayama, N. 2008. Stem Cell System of Sponge. In *Stem Cells: From Hydra to Man* (Bosch, T. C. H. ed.), pp. 17–35. Dordrecht: Springer.

Funayama, N., Nakatsukasa, M., Mohri, K., Masuda, Y., and Agata, K. 2010. *Piwi* expression in archeocytes and choanocytes in demosponges: Insights into the stem cell system in demosponges. *Evol. Dev.* 12: 275–287.

Gahan, J. M., Bradshaw, B., Flici, H., and Frank, U. 2016. The interstitial stem cells in Hydractinia and their role in regeneration. *Curr. Opin. Genet. Dev.* 40: 65–73.

Galle, S., Yanze, N., and Seipel, K. 2005. The homeobox gene *Msx* in development and transdifferentiation of jellyfish striated muscle. *Int. J. Dev. Biol.* 49: 961–967.

Gazave, E., Lapebie, P., Ereskovsky, A. V., and Borchiellini, C. 2012. No longer Demospongiae: Homoscleromorpha formal nomination as a fourth class of Porifera. *Hydrobiologia* 687: 3–10.

Gold, D. A., and Jacobs, D. K. 2013. Stem cell dynamics in Cnidaria: Are there unifying principles? *Dev. Genes Evol.* 223: 53–66.

Gold, D. A., Nakanishi, N., Hensley, N. M., Hartenstein, V., and Jacobs, D. K. 2016. Cell tracking supports secondary gastrulation in the moon jellyfish *Aurelia*. *Dev. Genes Evol.* 226: 383–387.

Harrison, F. W., and De Vos, L. 1991. Porifera. In *Microscopic Anatomy of Invertebrates* (Harrison, F. W. and Ruppert, E. W. eds). Vol. 1. New York: Wiley-Liss, Inc.

Haynes, J., and Burnett, A. L. 1963. Dedifferentiation and redifferentiation of cells in *Hydra viridis*. *Science* 142: 1481–1483.

He, J. R., Zheng, L. M., Zhang, W. J., and Lin, Y. S. 2015. Life cycle reversal in Aurelia sp.1 (Cnidaria, Scyphozoa). *PLoS One* 10(12): e0145314.

Helm, R. R., Tiozzo, S., Lilley, M. K., Lombard, S. F., and Dunn, C. W. 2015. Comparative muscle development of scyphozoan jellyfish with simple and complex life cycles. *Evodevo* 6: 11.

Henry, J. Q., and Martindale, M. Q. 2000. Regulation and regeneration in the ctenophore *Mnemiopsis leidyi*. *Dev. Biol.* 227: 720–733.

Hicklin, J., and Wolpert, L. 1973. Positional information and pattern regulation in *Hydra* – Effect of gamma-radiation. *J. Embryol. Exp. Morphol.* 30: 741–752.

Juliano, C. E., Voronina, E., Stack, C., Aldrich, M., Cameron, A. R., and Wessel, G. M. 2006. Germline determinants are not localized early in sea urchin development, but do accumulate in the small micromere lineage. *Dev. Biol.* 300: 406–415.

Kakinuma, Y. 1975. An experimental study of the life cycle and organ differentiation of *Aurelia aurita* Lamarck. *Bull. Mar. Biol. Stat. Asamushi* 15: 101–113.

King, N. 2004. The unicellular ancestry of animal development. *Dev. Cell* 7: 313–325.

King, N., Westbrook, M. J., Young, S. L., Kuo, A., Abedin, M., Chapman, J., Fairclough, S., Hellsten, U. et al. 2008. The genome of the choanoflagellate *Monosiga brevicollis* and the origin of metazoans. *Nature* 451: 783–788.

Korotkova, G. P. 1963. On types of restoration processes in sponges. *Acta Biol. Acad. Sci. Hung.* 13: 389–406.

Korotkova, G. P. 1970. Comparative morphological study of development of sponges from dissociated cells. *Cah. Biol. Mar.* 11: 325–334.

Leclere, L., Jager, M., Barreau, C., Chang, P., Le Guyader, H., Manuel, M., and Houliston, E. 2012. Maternally localized germ plasm mRNAs and germ cell/stem cell formation in the cnidarian *Clytia*. *Dev. Biol.* 364: 236–248.

Levin, T. C., and King, N. 2013. Evidence for sex and recombination in the choanoflagellate *Salpingoeca rosetta*. *Curr. Biol.* 23: 2176–2180.

Leys, S. P., and Degnan, B. M. 2002. Embryogenesis and metamorphosis in a haplosclerid demosponge: Gastrulation and transdifferentiation of larval ciliated cells to choanocytes. *Invertebr. Biol.* 121: 171–189.

Leys, S. P., Mackie, G. O., and Reiswig, H. M. 2007. The biology of glass sponges. *Adv. Mar. Biol.* 52: 1–145.

Love, G. D., Grosjean, E., Stalvies, C., Fike, D. A., Grotzinger, J. P., Bradley, A. S., Kelly, A. E., Bhatia, M. et al. 2009. Fossil steroids record the appearance of Demospongiae during the Cryogenian period. *Nature* 457: 718–721.

Maloof, A. C., Rose, C. V., Beach, R., Samuels, B. M., Calmet, C. C., Erwin, D. H., Poirier, G. R., Yao, N., and Simons, F. J. 2010. Possible animal-body fossils in pre-Marinoan limestones from South Australia. *Nature Geosci.* 3: 653–659.

Martindale, M. Q. 1986. The ontogeny and maintenance of adult symmetry properties in the ctenophore, *Mnemiopsis Mccradyi*. *Dev. Biol.* 118: 556–576.

Martindale, M. Q. 2016. The onset of regenerative properties in ctenophores. *Curr. Opin. Gen. Dev.* 40: 113–119.

Martindale, M. Q., and Henry, J. Q. 1996. Development and regeneration of comb plates in the ctenophore *Mnemiopsis leidyi*. *Biol. Bull.* 191: 290–292.

Martindale, M. Q., and Henry, J. Q. 1997a. Reassessing embryogenesis in the Ctenophora: The inductive role of el micromeres in organizing ctene row formation in the 'mosaic' embryo, *Mnemiopsis leidyi*. *Development* 124: 1999–2006.

Martindale, M. Q., and Henry, J. Q. 1997b. Experimental analysis of tentacle formation in the ctenophore *Mnemiopsis leidyi*. *Biol. Bull.* 193: 245–247.

Martindale, M. Q., and Henry, J. Q. 1999. Intracellular fate mapping in a basal metazoan, the ctenophore *Mnemiopsis leidyi*, reveals the origins of mesoderm and the existence of indeterminate cell lineages. *Dev. Biol.* 214: 243–257.

Medina, M., Collins, A. G., Silberman, J. D., and Sogin, M. L. 2001. Evaluating hypotheses of basal animal phylogeny using complete sequences of large and small subunit rRNA. *Proc. Natl. Acad. Sci. USA* 98: 9707–9712.

Meewis, H. 1939. Contribution a l'étude de l'embryogenése de Chalinulidae: *Haliclona limbata*. *Ann. Soc. R. Zool. Belg.* 70: 201–243.

Minchin, E. A. 1896. Note on the larva and the postlarval development of *Leucosolenia variabilis*, H. sp., with emarks on the development of other Asconidae. *Proc. R. Soc. London* 60: 42–52.

Misevic, G. N., and Burger, M. M. 1982. The molecular basis of species specific cell-cell recognition in marine sponges, and a study on organogenesis during metamorphosis. *Prog. Clin. Biol. Res.* 85: 193–209.

Misevic, G. N., Schlup, V., and Burger, M. M. 1990. Larval metamorphosis of *Microciona Prolifera*: Evidence against the reversal of layers. In *New Perspectives in Sponge Biology* (Rützler, K., Macintyre, K., and Smith, K. P. eds), pp. 182–187, Washington, DC: Smithsonian Institution Press.

Moroz, L. L., Kocot, K. M., Citarella, M. R., Dosung, S., Norekian, T. P., Povolotskaya, I. S., Grigorenko, A. P., et al. 2014. The ctenophore genome and the evolutionary origins of neural systems. *Nature* 510: 109–114.

Muller, W. A., Teo, R., and Frank, U. 2004. Totipotent migratory stem cells in a hydroid. *Dev. Biol.* 275: 215–224.

Nakanishi, N., Renfer, E., Technau, U., and Rentzsch, F. 2012. Nervous systems of the sea anemone *Nematostella vectensis* are generated by ectoderm and endoderm and shaped by distinct mechanisms. *Development* 139: 347–357.

Nakanishi, N., Sogabe, S., and Degnan, B. M. 2014. Evolutionary origin of gastrulation: Insights from sponge development. *BMC Biol.* 12: 347–357.

Okada, T. S. 1991. *Transdifferentiation: Flexibility in Cell Differentiation*. Oxford: Oxford University Press.

Passamaneck, Y. J. and Martindale, M. Q. 2012. Cell proliferation is necessary for the regeneration of oral structures in the anthozoan cnidarian Nematostella vectensis. BMC Dev. Biol. 12: art. 24.

Pehrson, J. R., and Cohen, L. H. 1986. The fate of the small micromeres in sea–rrchin development. *Dev. Biol.* 113: 522–526.

Peterson, K. J., Cameron, R. A., and Davidson, E. H. 1997. Set-aside cells in maximal indirect development: Evolutionary and developmental significance. *Bioessays* 19: 623–631.

Piraino, S., Boero, F., Aeschbach, B., and Schmid, V. 1996. Reversing the life cycle: Medusae transforming into polyps and cell transdifferentiation in *Turritopsis nutricula* (Cnidaria, Hydrozoa). *Biol. Bull.* 190: 302–312.

Piraino, S., De Vito, D., Schmich, J., Bouillon, J., and Boero, F. 2004. Reverse development in Cnidaria. *Can. J. Zool.* 82: 1748–1754.

Podar, M., Haddock, S. H. D., Sogin, M. L., and Richard Harbison, G. R. 2001. A molecular phylogenetic framework for the phylum Ctenophora using 18S rRNA genes. *Mol. Phylogenet. Evol.* 21: 218–230.

Putnam, N. H., Srivastava, M., Hellsten, U., Dirks, B., Chapman J., Salamov, A., Terry, A., Shapiro, H. et al. 2007. Sea anemone genome reveals ancestral eumetazoan gene repertoire and genomic organization. *Science* 317: 86–94.

Reber-Muller, S., Streitwolf-Engel R., Yanze, N., Schmid, V., Stierwald, M., Erb, M., and Seipel, K. 2006. BMP2/4 and BMP5–8 in jellyfish development and transdifferentiation. *Int. J. Dev. Biol.* 50: 377–384.

Rebscher, N., Volk, C., Teo, R., and Plickert, G. 2008. The germ plasm component Vasa allows tracing of the interstitial stem cells in the cnidarian *Hydractinia echinata. Dev. Dyn.* 237: 1736–1745.

Reddien, P. W., and Alvarado, A. S. 2004. Fundamentals of planarian regeneration. *Annu. Rev. Cell Dev. Biol.* 20: 725–757.

Reverberi, G., and Ortolani, G. 1963. On the origin of the ciliated plates and of the mesoderm in the ctenophores. *Acta Embryol. Morphol. Exp.* 6: 175–190.

Richmond, R. H. 1985. Reversible metamorphosis in coral planula larvae. *Mar. Ecol. Prog. Ser.* 22: 181–185.

Ruiz-Trillo, I., Roger, A. J., Burger, G., Gray, M. W., and Lang, B. F. 2008. A phylogenomic investigation into the origin of metazoa. *Mol. Biol. Evol.* 25: 664–672.

Ryan, J. F., Pang, K., Schnitzler, C. E., Nguyen, A. D., Moreland, R. T., Simmons, D. K., Koch, B. J., Francis, W. R. et al. 2013. The genome of the ctenophore Mnemiopsis leidyi and its implications for cell type evolution. *Science* 342: 1242592.

Sanchez Alvarado, A., and Yamanaka, S. 2014. Rethinking differentiation: Stem cells, regeneration, and plasticity. *Cell* 157: 110–119.

Schmid, V. 1978. Striated muscle: influence of an acellular layer on the maintainance of muscle differentiation in Anthomedusa. *Dev. Biol.* 64: 48–59.

Schmid, V., and Alder, H. 1984. Isolated, mononucleated, striated–muscle can undergo pluripotent transdifferentiation and form a complex regenerate. *Cell* 38: 801–809.

Schmid, V., Alder, H., Plickert, G., and Weber, C. 1988. Transdifferentiation from striated muscle of medusae in vitro. *Cell Differ. Dev.* 25(Suppl): 137–146.

Schmid, V., and Tardent, P. 1971. Reconstitutional Performances of *Leptomedusa campanularia jonstoni. Mar. Biol.* 8: 99–104.

Seipel, K., Yanze, N., and Schmid, V. 2004. Thegermline and somatic stem cell gene Cniwi in the jellyfish Podocoryne carnea. *Int. J. Dev. Biol.* 48: 1–7.

Seipp, S., Schmich, J., and Leitz, T. 2001 Apoptosis—a death-inducing mechanism tightly linked with morphogenesis in *Hydractina echinata* (Cnidaria, Hydrozoa). *Development* 128: 4891–4898.

Simion, P., Philippe, H., Baurain, D., Jager, M., Richter, D. J., Di Franco, A., Roure, B., Satoh, N. et al. 2017. A large and consistent phylogenomic dataset supports sponges as the sister group to all other animals. *Curr. Biol.* 27: 958–967.

Simpson, T. L. 1984. *The Cell Biology of Sponges*. New York: Springer-Verlag.

Sogabe, S., Nakanishi, N., and Degnan, B. M. 2016. The ontogeny of choanocyte chambers during metamorphosis in the demosponge *Amphimedon queenslandica. Evodevo* 7: 6.

Solana, J. 2013. Closing the circle ofgermline and stem cells: The primordial stem cell hypothesis. *Evodevo* 4: 4.

Steenkamp, E. T., Wright, J., and Baldauf, S. L. 2006. The protistan origins of animals and fungi. *Mol. Biol. Evol.* 23: 93–106.

Steinberg, S. N. 1963. Regeneration of whole polyps from ectodermal fragments of Scyphistoma larvae of *Aurelia aurita. Biol. Bull.* 124: 337–343.

Summers, R. G., and Haynes, J. F. 1969. Ontogeny of interstitial cells in *Pennaria tiarella. J. Morphol.* 129: 81–88.

Tuzet, O., and Paris, J. 1963. Recherches sur la régeénération de *Sycon raphanus* O. S. Vie Milieu 15: 285–291.

Usher, K. M., and Ereskovskyz, A. V. 2005. Larval development, ultrastructure and metamorphosis in *Chondrilla australiensis* Carter, 1873 (Demospongiae, Chondrosida, Chondrillidae). *Invertebr. Reprod. Dev.* 47: 51–62.
van Wolfswinkel, J. C. 2014. *Piwi* and potency: PIWI proteins in animal stem cells and regeneration. *Integr. Comp. Biol.* 54: 700–713.
Weber, C., Alder, H., and Schmid, V. 1987. *In vitro* transdifferentiation of striated muscle to smooth muscle cell of a medusa. *Cell Differ.* 20: 103–115.
Weis, V. M., and Buss, L. W. 1987. Ultrastructure of metamorphosis in *Hydractinia echinata. Postilla* 199: 1–20.
Woznica, A., Gerdt, J. P., Hulett, R. E., Clardy, J., and King, N. 2017. Mating in the closest living relatives of animals is induced by a bacterial chondroitinase. *Cell* 170: 1175–1183.
Wright, M. A., Mo, W., Nicolson, T., and Ribera, A. B. 2010. *In vivo* evidence for transdifferentiation of peripheral neurons. *Development* 137: 3047–3056.
Yuan, D., Nakanishi, N., Jacobs, D. K., and Hartenstein, V. 2008. Embryonic development and metamorphosis of the scyphozoan *Aurelia* (Cnidaria, Scyphozoa). *Dev. Genes Evol.* 218: 525–539.
Zapata, F., Goetz, F. E., Smith, S. A., Howison, M., Siebert, S., Church, S. H., Sanders, S. M., Ames, C. L. et al. 2015. Phylogenomic analyses support traditional relationships within Cnidaria. *PLoS One* 10(10): e0139068.

5 Macroalgae as Underexploited Model Systems for Stem Cell Research

David J. Garbary and Moira E. Galway
St. Francis Xavier University

CONTENTS

5.1 INTRODUCTION

Eukaryotic algae are a diverse polyphyletic assemblage assigned to the kingdoms Chromista, Plantae, and Protozoa (Guiry and Guiry 2018). Because they evolved multiple times independently of animals and land plants, they are natural experiments by which to explore the most diverse modes of cellular totipotency and stem cell ontogenies; algal multicellular body plans originated multiple times within diverse classes of Chromista and Plantae.

Three algal lineages stand out for their complex morphologies and high diversity: brown algae (class: Phaeophyceae, with over 2,000 species); red algae (phylum: Rhodophyta, with over 7,500 species); and green algae (subkingdom: Viridiplantae in part—the remainder being land plants). Algae in the subkingdom Viridiplantae are represented by over 5,000 species in the phylum Charophyta and almost 7,000 species in the phylum Chlorophyta. Following decades of research into development, ultrastructure, and molecular sequencing in which members of Charophyceae and Coleochaetophyceae were hypothesized as potential outgroups for land plants (Graham 1993), it has finally been determined that land plants evolved from organisms ancestral to the modern Conjugatophyceae (Charophyta; Puttick et al. 2018).

Of the multitudinous forms of red, brown, and green algae, only the brown algae are exclusively multicellular, although multicellularity evolved independently in multiple lineages of green algae and red algae. These multicellular algae and some of their unicellular, coenocytic relatives are the focus of this review. With their complex tissue differentiation and recent whole-genome sequencing, the filamentous brown alga *Ectocarpus* (Cock et al. 2010) and the blade-like green alga *Ulva* (sea lettuce) have been considered as model systems for plant organogenesis (Bogaert et al. 2013; De Clerck et al. 2018).

Understanding how gene expression regulates development and morphogenesis is becoming a reality for both *Ectocarpus* (e.g., Le Bail et al. 2011) and *Ulva* (De Clerck et al. 2018). Mitochondria and chloroplast genomes have been sequenced in many algal species along with some whole-genome sequencing, but complete annotations are lacking, which has hampered identification of molecular homologies associated with the regulation of development. The divergence of red, green, and brown algae no later than 1.6 billion years ago (Bengtson et al. 2017) represents two to three times the time span since the radiation of extant animal phyla in the Cambrian explosion and the origin of terrestrial plants. Fossils identified by Bengtson et al. (2017) include two new genera of multicellular red algae, one filamentous (*Rafatazmia*) and one pseudoparenchymatous (*Ramathallus*) with both assigned to crown florideophyte red algae. Thus, basal red algae—i.e., unicellular forms and simple filaments without pit plugs—must have existed even earlier, pushing back divergence from their green algal sister group. Therefore, *Rafatazmia* and *Ramathallus* provide the oldest example of multicellularity within eukaryotes. They predate the red alga *Bangiomorpha* from *c.* 1.2 billion years ago (Butterfield et al. 1990), known for its remarkable similarities to the modern genus *Bangia*. These fossil finds from India (Bengtson et al. 2017) and northern Canada (Butterfield et al. 1990) show that red algae must have achieved a global distribution prior to 1.6 billion years ago. This revised age for the evolution of complex red algae is particularly dramatic in that it is older than some estimates for the last eukaryotic common ancestor, assigned to 950–1,259 million years ago based on molecular clock dating methods (Douzery et al. 2004), or the more recent estimate of 1,000–1,600 million years ago (Eme et al. 2014).

The early origins of red and green algal lineages, including multicellular forms, indicate *an equally early origin for developmental specialization and the evolution of stem cells and developmental mechanisms*. Here, we focus on multicellular algal systems in which cell division results in two developmental pathways:

1. cell division associated with apical and nonapical meristems, and
2. "set-aside cells" in which both differentiated and nondifferentiated cells are quiescent and then resume development with the potential for totipotency or at least multipotency.

Laux (2003) defined plant stem cells as those in which "clonogenic precursors whose daughter cells can either remain stem cells or undergo differentiation," which can include apical cells in tip-growing algae as well as intercalary meristems located within or at the base of plant organs. In terrestrial plant biology, stem cells are largely considered in the context of meristems, i.e., "permanently active groups of pluripotent stem cells, embedded in specialized tissues" (Greb and Lohmann 2016). This is used elsewhere in this volume by Salvi et al. (2019, from Scheres 2008) with three general requirements for stem cells: (i) self-renewal, (ii) possession of undifferentiated characteristics, and (iii) ability to differentiate into an array of specialized cells. However, some algal examples considered here do not comply with the second criterion in that cells are fully differentiated. Unlike a typical differentiated animal cell, these cells can *dedifferentiate* and *redifferentiate* into other cell types. This phenomenon also occurs naturally or in response to injury in land plants; artificial cloning of algae and land plants from fully differentiated cells in the laboratory was achieved decades before the first successful animal clones.

Most complex multicellular algae usually have well-defined meristems that generate a diversity of differentiated cell types and tissues. These tissues typically include outer layers with protective epidermal cells and cells adapted for photosynthesis and interior cell layers that have more structural, reproductive, or transport functions. Examining these systems in the context of animal and land plant stem cells reveals what may be meaningful convergences and divergences in the fundamental nature of "plantness" and "animalness." This theme was highlighted by Bishop et al. (2012) in a survey of morphological and developmental differences between plants and animals that highlighted the constraints imposed by cell walls or other confining structures and compared the fractal nature of plant exteriors versus animal interiors. Rigid cell walls constrain algal and land plant cells, including their stem cells, obscuring their functional homology with animal stem cells. Nevertheless, many of the properties of animal stem cells are also found in terrestrial plants, e.g., those associated with root and shoot apical meristems (Laux 2003; Ivanov 2007; Dodueva et al. 2017; Warghat et al. 2018; elsewhere in this volume), as well as in the multicellular algae highlighted here.

In multicellular algae, ontogeny generally can follow one of two developmental patterns: diffuse growth in which cell divisions can occur more or less throughout tissues of the organism, or division of dedicated stem cells, either solitary or in meristems, mostly apical, but sometimes intercalary. Diffuse growth, whether it occurs in multicellular filaments (e.g., the water silk *Spirogyra*) or multicellular sheets (e.g., the sea lettuce *Ulva*), results in little cell diversity and no identifiable set-aside cells, although the cells do demonstrate virtual totipotency that is revealed through regeneration of a new thallus from thallus fragments or from artificially created protoplasts.

Terrestrial plant development is a highly integrated process regulated by a series of plant hormones, e.g., auxins, gibberellins, and cytokinins. These same compounds, or analogues, have been identified in all multicellular algal groups, where they presumably have equivalent regulatory functions (e.g., Tarakhovskaya et al., 2013 on photosynthesis in *Fucus*, a brown alga; Ohtaka et al., 2017 on auxin response in *Klebsormidium*, a green alga). While *Klebsormidium*, a filamentous member of the charophyte lineage whose ancestors gave rise to land plants, has diffuse filament growth and hence no set-aside cells, it does possess a relatively simple auxin-regulatory system that may resemble the ancestral form of this system from which the more complex canonical auxin system of terrestrial plants evolved. Even the unicellular green alga *Chlorella sorokiniana* has auxin-binding proteins and orthologs for intracellular auxin transport (Khasin et al. 2018). Bogaert et al. (2019) described auxin function in spore germination of the brown alga *Dictyota dichotoma* in which external auxin disrupted cellular polarity and induced both poles of the spore to germinate to form rhizoids. In only a few cases have the underlying molecular control mechanisms been investigated, and specific examples of gene regulation leading to morphogenesis via specific transcription factors (TFs) have been resolved. Whitewoods et al. (2018) highlighted the common function of the conserved CLAVATA peptide in regulating the plane of stem cell division in a moss and a flowering land plant, a process that is essential for the formation of a three-dimensional multicellular body in contrast to two-dimensional multicellular filamentous forms. The occurrence of this peptide-signaling pathway or its equivalent in red and brown algae with three-dimensional bodies would be another example of function driving convergent evolution.

The remainder of this chapter presents representative examples of the occurrence and function of *actively dividing stem cells, reactivated set-aside cells, algal cell totipotency, and capacity for regeneration among the multicellular algae.* The primary theme is the extensive totipotency and pluripotency of these algae from the perspective of development, regeneration, and tissue repair. We begin with subcellular totipotency, discuss the varieties of apical and intercalary meristems, and then examine various examples of natural and artificial regeneration. For the most part, these developmental patterns have only been described. Rarely are the underlying physiological and molecular regulatory processes even partly understood.

5.2 REGENERATION FROM CYTOPLASM

Many green algae are coenocytic, i.e., the entire thallus consists of a single, highly multinucleate cell, or else the thallus is multicellular with each cell possessing hundreds to thousands of nuclei. Unicellular forms include erect, highly branched structures one-cell wide (e.g., the marine green alga *Bryopsis*) as well as more complex structures in which the entire thallus consists of coalescent and interwoven filaments derived from a single highly branched cell (e.g., the seaweed *Codium*). When these giant cells are cut open in seawater, cytoplasm is extruded and small portions of cytoplasm containing multiple nuclei, chloroplasts, and mitochondria can form viable units that regenerate a new plant. This phenomenon was first described by Tatewaki and Nagata (1970) and later confirmed by LaClaire (1982 a,b), Pak et al. (1991) and Ye et al. (2005). Klotchkova et al. (2003) identified five green algal genera with this capacity: *Bryopsis*, *Chaetomorpha*, *Cladophoropsis*, *Ernodesmis*, and *Microdictyon*.

Kim et al. (2001) detailed a two-step regeneration process in *Bryopsis plumosa* in which a temporary gelatinous (polysaccharide) envelope formed and was then replaced by a lipid-based membrane. The initial envelope formed within 15 min with the membranes developing over the next 15 h. An interesting comparison here is the more rapid (at 22°C) cellularization of syncytial blastoderms during embryogenesis in *Drosophila melanogaster*, which takes 2.5–3 h (Wieschaus and Nüsslein-Volhard 1998). In *Bryopsis*, about 1,000 cytoplasmic fragments were produced from a single branch and the survival rate of the cytoplasmic fragments was about 40%. The molecules involved in regeneration are unknown although a 257-amino-acid lectin that weakly resembles certain animal lectins mediates the initial agglutination (Klotchkova and Kim 2006; Yoon et al. 2008; Niu et al. 2009).

All these observations are based strictly on laboratory studies, and it has yet to be determined to what extent these processes operate in nature. Also, it remains to be determined if this mode of regeneration is limited to multinucleate, unicellular, and multicellular green algae, or if multicellular coenocytic red algae, e.g., *Griffithsia*, or the yellow-green xanthophyte *Vaucheria* also has this regenerative potential. *Survival of these "cytoplasmic units" or "subcellular fragments" without a plasma membrane extends the normal definitions of viable cells and stem cells.*

5.3 APICAL CELLS AND MERISTEMS

Red algae are extremely variable in their morphology and apical systems. Vegetative development in large thalloid red algae was summarized by Gabrielson and Garbary (1986) and Coomans and Hommersand (1990) and will not be repeated here.

5.3.1 Apical Meristems with Apical and Sympodial Growth

The vast majority of red algae grow by means of strict apical cell division in which apical cells have indeterminate growth (Figure 5.1). While the resulting cells may produce lateral branches or filaments of various kinds, intercalary divisions in these axial files are typically absent. Thus, only the apical cells produce new cells in the primary axis. These apical cells enlarge via tip growth similar to the root hairs and pollen tubes of land plants. Their derivatives are initially very small and usually have either diffuse growth of their cell wall or a unique elongation pattern in which new cell wall material is deposited in one or two bands near the apex or base of the cell (Garbary et al. 1988; reviewed by Waaland 1990). The location of the elongation bands is under apical control; however, the underlying cell biology, physiology, and genetic regulatory mechanisms remain unexplored.

Extensive cell enlargement is usually accompanied by extensive nuclear endopolyploidy, resulting in cells that may have up to hundreds of times greater DNA content than apical cells (Goff and Coleman 1990; Garbary and MacDonald 1999). Cells derived from apical cells may subsequently divide to produce lateral filaments or branches. These lateral filaments may be determinate or indeterminate in growth and often differ morphologically from the primary axes. These patterns of cell division are so regular that computer models using Lindenmayer systems (L-systems, Figure 5.2) can be generated that grow virtual plants from single cells mimicking the actual cell division patterns of real species (Garbary and Corbit 1992). The challenge is to identify the molecular genetic interactions that correspond to the graphic rules of the L-systems. Of particular interest would be differences in the developmental regulation of apical cells in determinate axes relative to the corresponding cells in indeterminate axes.

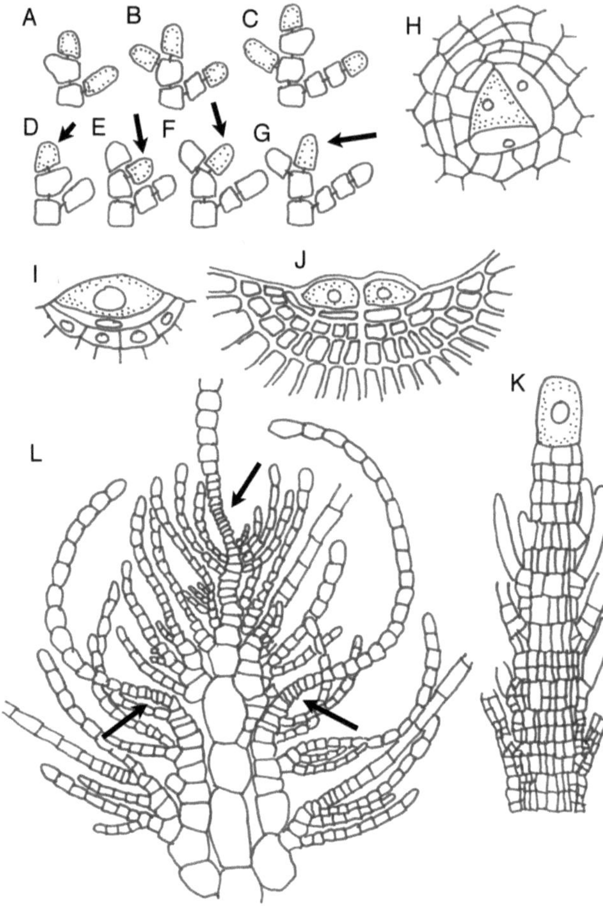

FIGURE 5.1 Examples of stem cell development associated with apical cells (stippled) of red and brown algae (A–K) and trichothallic growth (L). A–C, Monopodial growth in a filamentous red alga with initiation (A) and subsequent formation of lateral branches (B, C). D–G, Sympodial growth in a red alga with initial apical cell (D, short arrow) being pushed to the side (E–G) (B) by new apical cell (long arrows) that was initiated as a lateral on a sub-apical cell. H, Transverse section of apex in a fucoid brown alga showing three-sided apical cell. I–J, Apex of bladed brown alga *Dictyota dichotoma* showing prominent apical cell and derivatives (I) and early stage of dichotomous branch initiation (J). K, Prominent apical cells in parenchymatous brown alga with successively dividing cells. L, Trichothallic growth in member of Chordariales with meristems at the base of filaments (arrows). A–G redrawn from Parsons (1975), H redrawn from Jensen (1974), I–L redrawn from Fritsch (1945).

The red algal family Dasyaceae has an unusual apical system with well-defined sympodial or cellulosympodial growth (Norris et al. 1984) in which the derivative of an apical cell produces a lateral axis that displaces the original apical cell to one side and becomes the new apical cell (Figure 5.1). In this zigzag pattern of development, the original apical cell becomes a determinate lateral branch. Although vegetative

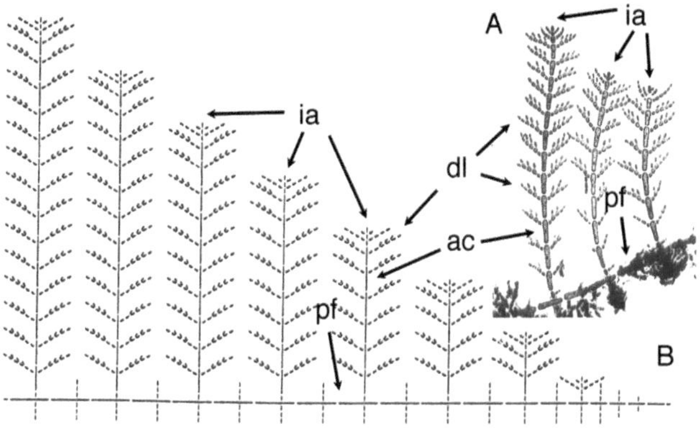

FIGURE 5.2 Determinate and indeterminate growth from apical cells in the filamentous red alga *Antithamnion percurrens.* A, Photograph with three upright, indeterminate axes (ia) growing from prostrate filament (pf) with opposite determinate laterals (dl) from each axial cell (ac). B, Lindenmayer system model of the same species at 24 iterations of the L-system that mimic ontogeny of the species. (Image redrawn from Garbary and Corbit (1992).

development of indeterminate axes has been well characterized, the associated underlying physiological mechanisms and cellular processes remain to be identified. This repetitive switching away from monopodial growth after each apical cell division provides an excellent model system in which to evaluate molecular processes that transform an apical cell in a lateral branch. At least one species in this family also has filaments in which intercalary cell division can occur (Parsons 1975).

It is not surprising that with the diversity of body plans, brown algae have a corresponding diversity of apical systems. Apical growth is considered ancestral in the class, as it is in land plants (bryophyte and vascular plants) and related green algae (Harrison 2017) and is typically generated by a single prominent apical cell at the apex of a filament or blade (e.g., *Dictyota*, Figure 5.1) or a band of apical cells at a blade apex (e.g., *Syringoderma*, *Padina*). In fucoids, this apical cell is maintained in an apical pit and cuts off derivatives from a mostly three-sided apical cell analogous to that in primitive mosses and liverworts (Renzaglia et al. 2018), which gives these algae the ability to generate three-dimensional forms like those of land plants. The convergent evolution of fucoid and land-plant apical systems results in similar regular patterns (phyllotaxy) of lateral-branch or lateral-organ placement around the main axis of the plant body below the central apical cell or meristem that conforms to the Fibonacci series (Peaucelle and Couder 2016).

Since similar three-dimensional growth occurs in many other brown, and red algae, as well as in land plants, identifying signaling mechanisms homologous or analogous to the *CLAVATA*-signaling pathway that regulates apical cell division planes in terrestrial plants (Whitewoods et al. 2018) is central to understanding algal morphogenesis. Recently, the well-known terrestrial plant growth hormone auxin has been implicated in polarization (establishment of an apical-basal axis) during embryogenesis in the brown alga *Dictyota dichotoma* (Bogaert et al. 2019). Homologues of many genes involved in auxin synthesis and transport in terrestrial plants are present in two diverse phaeophytes, namely, *Fucus* and *Ectocarpus* (Tarakhovskaya et al. 2013; Bogaert et al. 2019). Since Dictyotales are among the primitive assemblages in brown algae (Kawai and Henry 2017), this points to a *deep phylogenetic ancestry for auxin-regulatory pathways in plant systems.*

5.3.2 Intercalary Meristems of Red and Brown Algae

Brown algae have two kinds of intercalary meristems: (i) trichothallic meristems in which cell division occurs at the base of a multicellular hair (Figure 5.1L) to produce filamentous or syntagmatic thalli, and (ii) more elaborate meristems that give rise to parenchymatous systems in kelp (Kawai and Henry 2017 and references therein).

The intercalary meristems in kelp (i.e., in the order Laminariales) are analogous to certain types of terrestrial plant meristems. With elaborate differentiation yielding multiple cell types including outer epidermal cells, photosynthetic cells, and interior structural and transport cells, kelps resemble vascular plants in their complexity of cells and tissues.

In most kelp species, individual plants consist of a stipe or stem-like organ that supports a blade, a flattened leaf-like organ. At the junction of these organs is an intercalary multicellular meristematic region of stem cells that produces the cells required for elongation of both the stipe and the blade. These meristems can remain active for years. Perennial temperate to arctic species can exhibit seasonal growth cycles in which the blades detach above the meristem and a new one is regenerated de novo (e.g., *Laminaria hyperborea*). The kelp intercalary meristem is analogous to the vascular cambium (an interior ring of stem cells found in the stems of vascular plants), where cell division on one side of the ring produces cells that differentiate into water-transporting xylem and on the other side into photosynthate-transporting phloem, although functionally, they most resemble the intercalary meristem at the base of hornwort sporophytes or at the base of grass leaves.

Whereas red algae typically have apical meristems, individual members of the calcified red algal order Corallinales also have intercalary meristems (Garbary et al. 2012). These are prominent in perennial crusts where the upper surfaces of thalli are prone to herbivory and epiphyte settlement. The subsurface meristem produces new cells to replace those removed or shed from the exposed surface, in a process of continual regeneration reminiscent of the continuous shedding and replacement of epidermal cells via basal stem cell division in mammalian skin (Pueschel and Keats 1997 and citations therein). The closest analogy in land plants would be to the ring of vascular cambium that generates the lateral growth (thickening) of the stems in woody seed plants. As the crusts become thicker, the cells deeper in the thallus die, leaving calcified skeletons to maintain structural integrity of the thallus. These crusts can be very long-lived, and some living thalli of *Clathromorphum compactum* have been estimated to be 850 years old (Halfar et al. 2011).

5.3.3 Coenobia as Set-Aside Cells

Among the freshwater green algae are multicellular species in which each organism consists of a thallus of relatively unspecialized but totipotent cells known as a *coenobium* within which each cell is capable of producing a whole new organism

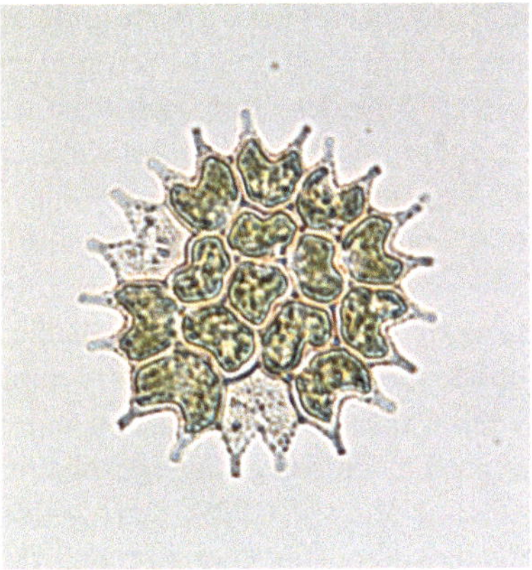

FIGURE 5.3 A single coenobial thallus of the green alga *Pseudopediastrum* in which, following mitosis and cytokinesis, each cell can produce a new coenobium. Note empty cells from which daughter coenobia have been released. (Photo courtesy of Joanna Lenarczyk.)

(Pickett-Heaps 1975 and references therein). Some coenobia are flat plates of cells in which the marginal cells are morphologically differentiated by forming projecting arms and spines (e.g., *Pseudopediastrum* and relatives, Figure 5.3). In other species, such as *Hydrodictyon*, the coenobia are large three-dimensional cylindrical nets, 20–30cm in diameter with hundreds to thousands of cells joined into pentagonal or hexagonal networks.

Each vegetative cell in a coenobium is the equivalent of a totipotent stem cell, which has the potential to form a new coenobium through an unusual form of asexual reproduction. The cell will initiate a variable number of mitoses to produce all the nuclei required for a new coenobium. This is followed by cytokinesis to subdivide the parental cell into much smaller cells, which then take on the fixed arrangement characteristic of the thallus of that species before being released through the parental cell walls. This release enables the cells of the young coenobium to grow back to the original parental cell size. In the nonmotile green alga *Pediastrum*, the newly formed cells remain attached throughout formation of the new thallus. However, in the asexual cycle of the water net *Hydrodictyon*, mitoses and cytokinesis generates separate flagellated cells, which, following a motile period inside the mother cell, lose their flagella and aggregate such that each cell is attached at anterior and posterior ends to two other cells. How the flagellated cells attach so as to recapitulate in miniature the network to which the parental cell belonged remains a developmental mystery but is likely to involve chemical signaling via ligand-receptor binding at the cell surfaces.

5.4 REGENERATION FROM VEGETATIVE FRAGMENTS

5.4.1 Totipotency of Vegetative Fragments

Little somatic cell differentiation occurs in many multicellular algae. In the bright green marine algae *Ulva* spp. (sea lettuce), there are only three cell types: blade cells (forming a simple, two-layered blade), stem cells, and rhizoid cells (Wichard 2015;

Wichard et al. 2015), but these species have a seemingly unlimited capacity for vegetative growth. Fragments of adult plants can be refragmented indefinitely, with each fragment able to regrow entire blades. Totipotency is exhibited by the previously nondividing cells at the center of blades, which transform into rapidly dividing cells following fragmentation. This regenerative capacity may be regulated by external factors such as microbiota, temperature, light, and/or nutrients.

The morphological development of *Ulva* is regulated by compounds secreted by surface-colonizing bacteria (Wichard et al. 2015; Ghaderiardakani et al. 2017 and references therein). Without the correct microbial symbionts, a callus-like tissue develops with abnormal cell walls, which can then resume differentiation when bacteria are reintroduced. Reproduction can be under exogenous (abiotic) or endogenous control (Wichard et al. 2015; Gao et al. 2015). During this process, vegetative cells transform into either flagellated zoospores or gametes, depending on the life cycle stage. Each vegetative cell protoplast transforms into many flagellated cells that are released, leaving empty cell walls that then decompose. The settlement, growth, division, and differentiation of these motile cells into a new plant complete with blade and rhizoids cannot occur without appropriate microbial chemical signals. In the red alga *Pyropia*, normal protoplast regeneration to form monostromatic blades also occurs but only when specific bacteria are present (Fukui et al. 2014). This convergence of developmental controls in red and green algae suggests that the presence of interspecies signaling to regulate morphogenesis may be widespread among algal systems.

The totipotency of marine macroalgal vegetative cells that allows for regeneration of plants from thallus fragments is a well-established tool for algal propagation. In the cultivation of kelp species with their heteromorphic life histories in which a microscopic, filamentous phase alternates with large kelp plants, long-term cultivation of clones is maintained with the filamentous phase. The filamentous phase with its intercalary nondividing cells and terminal dividing cells can be grown vegetatively. Masses of these microscopic plants can be fragmented, and the resulting slurry of separated cells left to regenerate.

Fragmentation is also a common approach when propagating macroalgae for aquaculture. Large highly branched fronds of the red algae *Eucheuma, Kappaphycus,* and *Gracilaria* are broken into small pieces, and each piece is attached to a rope (review by Kim et al. 2017). Fragmentation establishes new apical cells in these thallus fragments that then resume growth.

5.4.2 Algal Protoplasts as Analogues for Stem Cells

Plant biologists recognize that protoplasts—living plant cells devoid of a cell wall—provide a way to study plant and algal cell totipotency, so protoplasts have been a key tool in algal and plant biotechnology (Reddy et al. 2008; Baweja and Sahoo 2009; Baweja et al. 2009). From these cells without cell walls, artificial stem cells are generated that can be used to other cells, or used to induce somatic cell embryo formation (plant cloning), hybridize somatic cells, and genetically transform cells.

Algal protoplasts have not been used as extensively as terrestrial plant protoplasts but have still provided considerable insights into algal plant development and cell totipotency. Some early accounts of regeneration of multicellular algae from protoplasts are given by Gabriel (1970) and Marchant and Fowke (1977). Enzymatic treatments are typically used to break down cell walls and release protoplasts, which can be maintained in algal culture media with added osmotica, such as mannitol, to prevent protoplast rupture. Protoplasts of the filamentous green algae *Uronema*, *Mougeotia*, *Klebsormidium*, and *Stigeoclonium*, but not *Ulothrix*, regenerated filaments that were indistinguishable from the initial filaments. A surprising result with *Stigeoclonium* was that many of the protoplasts developed into bi- or tetra-flagellated cells after 12–16 h. This transformation of previously differentiated vegetative cells directly into presumed gametes or spores is unusual; however, Marchant and Fowke (1977) did not describe any subsequent development of the motile cells.

The totipotency of protoplasts obtained from a variety of red, green, and brown multicellular algae has now been demonstrated in culture (e.g., Millner et al. 1979; Kevekordes et al. 1993; Reddy et al. 2006, 2008 and references therein; Baweja and Sahoo 2009; Baweja et al. 2009; Huddy et al. 2015; Coelho et al. 2012b; Fukui et al. 2014). The conclusion is that *even differentiated cell types retain the potential to revert to stem cells with complete totipotency when removed from the constraints of adjacent cells and their own cell walls.* As pointed out above, normal development may, however, require signaling from microbial symbionts.

5.4.3 Set-Aside Cells in Complex Multicellular Algae

Fulcher and McCully (1969) described regeneration from transverse 1- to 2-mm-thick slices cut from the blades of *Fucus vesiculosus* (Figure 5.4). The exposed colorless medullary and cortical cells at the wound surface became brown and within 10 weeks formed a number of protuberances that gradually elongated, and, by 12–16 weeks, had flattened to resemble *Fucus* embryos without rhizoids or basal

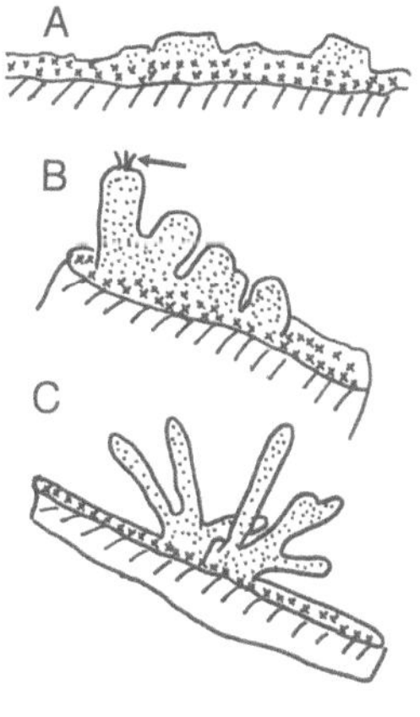

FIGURE 5.4 Regeneration of *Fucus* from cut surface of thallus segments. A, Initial, poorly differentiated mounds of cells on the cut surface. B, Young axes, one having hairs (arrow), emerging from apical slit. C, More mature axes with developing dichotomies having fully functional apical cells within the apex of each branch. Redrawn from Fulcher and McCully (1969).

root-like structures. At this point, they were up to 3 mm long and had developed normal apical hairs and an apical groove. The transformation of cells at the wound surface into embryonic branch initials combines cortical cell divisions followed by regeneration of interior filament cells into a new epidermis with transitory cells, followed by differentiation of subsurface cells to form the new branch. This capacity for regeneration was enhanced by adding the terrestrial plant hormones gibberellic acid (GA_3) and kinetin (Borowczak et al. 1977). Equivalent clusters of small branches have been observed in nature (DJG personal observation) at sites of apparent injury.

Regeneration from explants is a basic tool in laboratory experimental work and in field-based aquaculture (e.g., Dawes et al. 1993; Shen et al. 2010; Luhan and Mateo 2017). Small branches are cut off and either left free-floating in tumble culture or attached to ropes, whereupon the pieces simply resume growth. The ability of terrestrial plant hormones (in particular, auxins) to stimulate or modify regeneration of fragments in a laboratory setting, as shown, for example, for the red alga *Chondracanthus chamissoi* (Yokoya et al. 2014) suggests that underlying regulatory pathways have similar molecular underpinnings in terrestrial plants and algae.

Species in the genus *Spyridia* provide examples of set-aside cells in a complex red alga (West and Calumpong 1989). Two kinds of axes are formed: highly corticated indeterminate axes (up to about 10 cm) that grow from a prominent apical cell, and short determinate branches (up to 1.5 mm) in which the main axial file has distinct bands of cortical cells at nodes (Figure 5.5). Even when released from dominance by the primary axis by branch excision, apical cells of the laterals do not resume division. At reproductive maturity, these branches bear male gamete-forming gametangia in haploid plants or spore-forming tetrasporangia in diploid plants. Indeterminate axes may branch to produce similar axes; however, determinate axes do not rebranch. Once a determinate axis with cortical bands has reached its normal length, the apical cell ceases to divide. These short branches each have a well-defined apical cell, produce their distinctive branches, and then cease dividing. These determinate branches

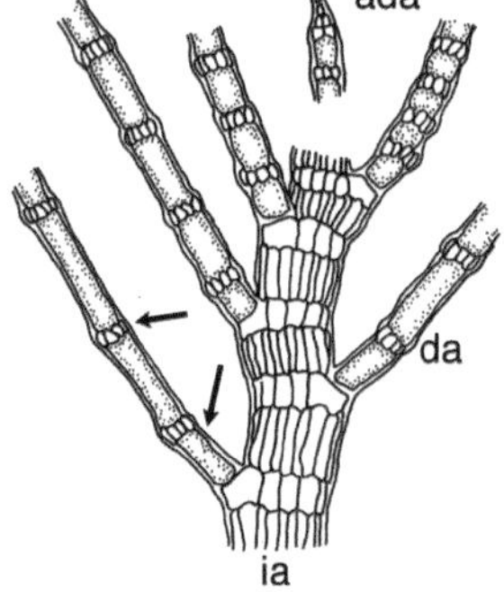

FIGURE 5.5 Morphology and regeneration in *Spyridia filamentosa* with differentiated primary, indeterminate axes (ia) developing from a single apical cell (not shown) and numerous differentiated determinate lateral branches (da) with apical cells that cease division (ada). Regeneration of an indeterminate axis can proceed from various cells in the da (arrows), but not from its apical cell. Redrawn from Coppejans (1983).

are often naturally deciduous. When excised from the parent plant, they regenerate new indeterminate axes. This regeneration occurs from cells of the main axial file and does not involve the original apical cell of the filament. Thus, *differentiated axial cells of the determinate axis can act as stem cells* by generating an apical cell that forms an indeterminate axis and regenerates the original plant morphology.

5.5 NATURAL REGENERATION FROM HOLDFASTS

Many perennial red and brown seaweeds produce basal crust-like or discoid holdfasts from which numerous (dozens to hundreds) of erect axes can emerge. The holdfasts are a permanent part of thallus architecture and may be able to grow from their margins to increase contact with the substratum on which the seaweed is growing. Many holdfasts initially produce one followed by more erect fronds as they expand. The erect fronds may grow for a number of years until reproduction, at which time the frond either senesces or becomes so large that increased drag and wave action remove it. In Irish moss (*Chondrus*), whole frond detachment allows the much smaller fronds to escape light- and nutrient shading and stimulates development (McLachlan et al. 1989). These smaller fronds have essentially been set aside until this event. Simultaneously, new fronds can emerge from the basal disc.

A similar process of development occurs in the common brown seaweed *Ascophyllum nodosum*, although it may extend over decades. Indeed, *A. nodosum* may be the longest lived noncalcified, noncrustose seaweed. With plants estimated to live up to a century, and individual fronds documented to survive 20 years or more, a frond may have almost indefinite vegetative growth (Baardseth 1970; Garbary et al. 2017 and references therein). The age of many fronds can be determined by their apical growth and annual production of air bladders as growth resumes in the spring. Counting air bladders down a primary axis provides a minimum age for the frond—as fronds enlarge, branch breakage from wave action and/or herbivory masks the true age of each plant (Eckersley and Garbary 2007). Another key feature of development is the annual crop of reproductive branches (i.e., receptacles) that can form along the entire frond except in the most recent annual growth increment at the frond apices. Receptacles are shed following 8–11 months of development and, after a short interval (weeks to months), new receptacles for next year's crop start to develop at the sites of receptacle detachment from vegetative tissue that was previously quiescent. At the same time, previously differentiated cortical cells from the most recent complete year of growth initiate numerous receptacles for the first time. Thus, *A. nodosum* has *three independent sets of cells that have been set aside* and that lead to reproductive development: (i) from the previous year's strictly vegetative growth, (ii) from the cells at the scar left by detachment of the previous year's receptacles, and (iii) from emergence of new vegetative axes from the basal holdfast.

5.6 GROWTH RINGS

While many kelp species (order: Laminariales) are annuals, some species are perennial and survive at least 2–5 years. While growth of blades and stipes results from intercalary—i.e., nonterminal—meristems, the stem-like portion of the stipes in some

FIGURE 5.6 Growth rings in a stipe of the kelp *Pterygophora* analogous to the growth in girth via vascular cambium in woody plants. (Photo courtesy of Jenn Burt.)

species in at least four genera (*Ecklonia*, *Laminaria*, *Saccharina*, and *Pterygophora*) have a growth pattern in which annual growth rings of tissue (Figure 5.6) develop to increase the diameter of the stem (e.g., Parke 1948; DeWreede 1984; Klinger and DeWreede 1988). Like the growth rings of woody vascular plants, algal growth rings represent transitions from a period with little to no cell division to one with more rapid cell division. Dividing cells are located below the outer epidermis subsurface, and previous outer layers are "cast off." This development starts at the base of the stipe above the holdfast and moves up the stipe (Parke 1948).

5.7 CONCLUSIONS

While the genomes of some multicellular algae have been sequenced, none has been fully annotated. These include species of *Gracilariopsis* (Lee et al. 2018), *Ectocarpus* (Cock et al. 2010), *Porphyra* (Brawley et al. 2017), *Pyropia* (Nakamura et al. 2013), *Chondrus* (Collén et al. 2013), and *Ulva* (De Clerck et al. 2018). These initiatives will further the development of the relevant species as model organisms (Coelho et al. 2012a) comparable to the establishment of the thale cress *Arabidopsis thaliana* as a model flowering land plant. Unlike animals, which typically develop through division of small undifferentiated stem cells, algal development (apart from undifferentiated apical stem cells) often involves fully differentiated cells that have low capacity for division and then revert to a state where nuclear mitoses and cell division resume. Based on the widespread presence of endogenous hormones resembling terrestrial plant hormones and/or algal cell responses to exogenous terrestrial plant hormones, analogous molecular genetic regulatory systems are anticipated. The detection of TALE-group TFs that regulate patterns of differentiation in apical meristems of plants and that occur across Metazoa and Archaeplastida (as KNOX- and BELL-class TFs; Gao et al. 2015; Joo et al. 2018) suggests that similar generalized control mechanisms will be found in multicellular algal systems. The limited number of TFs in these algal systems (de Mendoza et al. 2013) suggests that molecular cascades responsible for regulatory systems of stem cells in multicellular algae should be tractable and generate insights into developmental homologies across multicellular eukaryotic life.

REFERENCES

Baardseth, E. 1970. Synopsis of biological data on *Ascophyllum nodosum* (Linnaeus) Le Jolis. *FAO Fish. Synop.* 38: 1–70.

Baweja, P., and Sahoo, D. 2009. Regeneration studies in *Grateloupia filicina* (J.V. Lamouroux) C. Agardh – an important carrageenophyte and edible seaweed. *Algae* 24: 163–168.

Baweja, P., Sahoo, D., Garcia-Jimenez, P., and Robaina, P. R. 2009. Seaweed tissue culture as applied to biotechnology: problems, achievements and prospects. *Phycol. Res.* 57: 45–58.

Bengtson, S., Sallstedt, T., Belivanova, V., and Whitehouse, M. 2017. Three-dimensional preservation of cellular and subcellular structures suggests 1.6 billion-year-old crown-group red algae. *PLoS Biol.* 15(3): e2000735.

Bishop, C. D., Galway, M. E., and Garbary, D. J. 2012. Architecture and design among plants and animals: convergent and divergent developmental mechanisms. In *Origin(s) of Design in Nature* (Eds.) L. Swan, R. Gordon, and J. Seckbach, pp. 325–341. Dortrecht, Netherlands: Springer.

Bogaert, K. A., Arun, A., Coelho, S. M., and De Clerck, O. 2013. Brown algae as a model for plant organogenesis. In *Plant Organogenesis: Methods and Protocols* (Ed.) I. De Smet, pp. 97–125. New York: Springer.

Bogaert, K. A., Blommaert, L., Ljung, K., Beeckman, T., and De Clerck, O. 2019. Auxin function in the brown alga *Dictyota dichotoma*. *Plant Physiol.* 179: 280–299.

Borowczak, E., Kentzer, T., and Potulska-Klein, B. 1977. Effect of gibberellin and kinetin on the regeneration ability of *Fucus vesiculosus* L. *Biol. Plant.* 19: 405–412.

Brawley, S. H., Blouin, N. A., Ficko-Blean, E., Wheeler, G. L., Lohr, M., Goodson, H. V., Jenkins, J. W. et al. 2017. Insights into red algae and eukaryotic evolution from the genome of *Porphyra umbilicalis* (Bangiophyceae, Rhodophyta). *Proc. Natl. Acad. Sci. USA* 114: E6361–E6370.

Butterfield, N. J., Knoll, A. H., and Swett, K. 1990. A bangiophyte red alga from the Proterozoic of arctic Canada. *Science* 250: 104–107.

Cock, J.M., Sterck, L., Scornet, D., Allen, A., Amoutzias, G., Anthouard, V., Artiguenave, F. et al. 2010. The *Ectocarpus* genome and the independent evolution of multicellularity in brown algae. *Nature* 465: 617–621.

Coelho, S. M., Scornet, D., Rousvoal, S., Peters, N. T., Dartevelle, L., Peters, A. F., and Cock, J. M. 2012a. *Ectocarpus*: a model organism for the brown algae. *Cold Spring Harbor Protoc.* 2012: 193–198. doi: 10.1101/pdb.emo065821.

Coelho, S. M., Scornet, D., Rousvoal, S., Peters, N. T., Dartevelle, L., Peters, A. F., and Cock, J. M. 2012b. Isolation and regeneration of protoplasts from *Ectocarpus*. *Cold Spring Harbor Protoc.* 2012: 361–364. doi: 10.1101/pdb.prot067959.

Collén, J., Porcel, B., Carré, W., Ball, S. G., Chaparro, C., Tonon, T., Barbeyron, T. et al. 2013. Genome structure and metabolic features in the red seaweed *Chondrus crispus* shed light on evolution of the Archaeplastida. *Proc. Natl. Acad. Sci. USA* 110: 5247–5252.

Coomans, R. J., and Hommersand, M. H. 1990. Vegetative growth and organization. In *Biology of the Red Algae*. (Eds.) K. M. Cole and R. G. Sheath, pp. 275–304. Cambridge: Cambridge University Press.

Coppejans, E. 1983. *Iconographie d'algues méditerranénnes*. Bibliotheca Phycologica, Vol. 63. Vaduz: J. Cramer.

Dawes, C. J., Trono, G. C., and Lluisma, A. O. 1993. Clonal propagation of Eucheuma denticulatum and Kappaphycus alvarezii for Philippine seaweed farms. *Hydrobiologia* 260/261: 379–383.

De Clerck, O., Kao, S.-M., Bogaert, K. A., Blomme, J., Foflonker, F., Kwantes, M., Vancaester, E. et al. 2018. Insights into the evolution of multicellularity from the Sea Lettuce genome. *Curr. Biol.* 28: 2921–2933.

de Mendoza, A., Sebé-Pedrós, A., Šestak, M. S., Matejcic, M., Torruella, G., Domazet-Loso, T., and Ruiz-Trillo, I. 2013. Transcription factor evolution in eukaryotes and the assembly of the regulatory toolkit in multicellular lineages. *Proc. Natl. Acad. Sci. USA* 110: E4858–E4866.

DeWreede, R. E. 1984. Growth and age class distribution of *Pterygophora californica* (Phaeophyceae). *Mar. Ecol. Prog. Ser.* 19: 93–100.

Dodueva, I. E., Tvorogova, V. E., Azarakhsh, M., Lebedeva, M. A., and Lutova, L. A. 2017. Plant stem cells: unity and diversity. *Rus. J. Genet.: Appl. Res.* 7: 385–403.

Douzery, E. M., Snell, E. A., Bapteste, E., Delsuc, F., and Phillippe, H. 2004. The timing of eukaryotic evolution: does relaxed molecular clock reconcile proteins and fossils. *Proc. Natl. Acad. Sci. USA* 101: 15386–15391.

Eckersley, L. L., and Garbary, D. J. 2007. Developmental and environmental sources of variation on annual growth increments of *Ascophyllum nodosum* (Phaeophyceae). *Algae* 22: 107–116.

Eme, L., Sharpe, S. C., Brown, M. W., and Roger, A. 2014. On the age of eukaryotes: evaluating evidence from fossils and molecular clocks. *Cold Spring Harbor Perspec. Biol.* 6: a016139. doi: 10.1101/cshperspect.a016139.

Fritsch, F. E. 1945. *The Structure and Reproduction of the Algae*, Vol. 2. Cambridge: Cambridge University Press.

Fukui, Y., Abe, M., Kobayashi, M., Yano, Y., and Satomi, M. 2014. Isolation of *Hyphomonas* strains that induce normal morphogenesis in protoplasts of the marine red alga *Pyropia yezoensis*. *Microb. Ecol.* 68: 556–566.

Fulcher, R. G., and McCully, M. E. 1969. Histological studies of the genus *Fucus*. IV. Regeneration and adventive embryogeny. *Can. J. Bot.* 47: 1643–1649.

Gabriel, M. 1970. Formation, growth, and regeneration of protoplasts of the green alga, *Uronema gigas*. *Protoplasma* 70: 135–138.

Gabrielson, P. W., and Garbary, D. J. 1986. Systematics of red algae. *CRC Crit. Rev. Plant Sci.* 3: 325–366.

Gao, J., Yang, X., Zhao, W., Lang, T., and Samuelsson, T. 2015. Evolution, diversification, and expression of KNOX proteins in plants. *Front. Plant Sci.* 6: 882. doi: 10.3389/fpls.2015.00882.

Garbary, D. J., Belliveau, D., and Irwin, R. 1988. Apical control of band elongation in *Antithamnion defectum* (Ceramiaceae). *Can. J. Bot.* 66: 1308–1315.

Garbary, D. J., Brown, N. E., MacDonell, H. J., and Toxopeus, J. 2017. *Ascophyllum* and its symbionts – a complex symbiotic community on North Atlantic shores. In *Algal and Cyanobacteria Symbioses* (Eds) M. Grube, J. Seckbach, and L. Muggia, pp. 547–572. New Jersey: World Scientific.

Garbary, D. J., and Corbit, J. D. 1992. Lindenmayer-systems as models of red algal morphology and development. *Prog. Phycol. Res.* 8: 143–177.

Garbary, D. J., Galway, M. E., Lord, C. E., and Gunawardina, A. 2012. Programmed cell death in multicellular algae. In *Advances in Algal Cell Biology* (Ed.) K. Heimann and C. Katsaros, pp. 1–19. Berlin, Germany: De Gruyter.

Garbary, D. J. and MacDonald, A. R. 1999. Molecules, organelles and cells: fluorescence microscopy and red algal development. In *Molecular Approaches to the Study of the Oceans* (Ed.) K. Cooksey, pp. 409–422. London: Chapman & Hall.

Ghaderiardakani, F., Coates, J. C., and Wichard T. 2017. Bacteria-induced morphogenesis of *Ulva intestinalis* and *Ulva mutabilis* (Chlorophyta): a contribution to the lottery theory. *FEMS Microbiol. Ecol.* 93(8): 1–12. doi: 10.1093/femsec/fix094.

Goff, L. J., and Coleman, A. W. 1990. DNA: microspectrofluorometric studies. In: *Biology of the Red Algae* (Ed.) K. M. Cole and R. G. Sheath, pp. 43–71. Cambridge: Cambridge University Press.

Graham, L. E. 1993. *The Evolution of Land Plants.* New York: Wiley.

Greb, T., and Lohmann, J. U. 2016. Plant stem cells. *Curr. Biol.* 26: PR816–R821. doi: 10.1016/j.cub.2016.07.070.

Guiry, M. D., and Guiry, G. M. 2018. *AlgaeBase. World-Wide Electronic Publication.* Galway: National University of Ireland. www.algaebase.org; searched 23 November 2018.

Halfar, J., Hetzinger, S., Adey, W., Zack, T., Gamboa, G., Kunz, B., Williams, B., and Jacob, D. E. 2011. Coralline algal growth-increment widths archive North Atlantic climate variability. *Paleogeogr., Paleoclimatol., Paleoecol.* 302: 71–80.

Harrison, C. J. 2017. Development and genetics in the evolution of land plant body plans. *Philos. Trans. R. Soc. B* 372: 20150490. doi: 10.1098/rstb.2015.0490.

Huddy, S. M., Meyers, A. E., and Coyne, V. E. 2015. Regeneration of whole plants from protoplasts of *Gracilaria gracilis* (Gracilariales, Rhodophyta). *J. Appl. Phycol.* 27: 427–435.

Ivanov, V. B. 2007. Stem cells in the root and the problem of stem cells in plants. *Russ. J. Dev. Biol.* 38: 338–349.

Jensen, J. B. 1974. Morphological studies in Cystoseiraceae and Sargassaceae (Phaeophyceae) with special reference to apical organization. *Univ. Calif. Publ. Bot.* 68: 1–61.

Joo, S., Wang, M. H., Lui, G., Lee, J., Barnas, A., Kim, E., Sudek, S., Worden, A., and Lee, J.-H. 2018. Common ancestry of heterodimerizing TALE homeobox transcription factors across Metazoa and Archaeplastida. *BMC Biol.* 16: 136. doi: 10.1186/s12915-018-0605-5.

Kawai, H., and Henry, E. C. 2017. Phaeophyta. In *Handbook of the Protists.* (Ed.) J. Archibald, A. Simpson, and C. Slamovits, pp. 1–38. Cham, Switzerland: Springer.

Kevekordes, K., Beardall, J., and Clayton, M. N. 1993. A novel method for extracting protoplasts from large brown algae. *J. Exp. Bot.* 44: 1587–1593.

Khasin, M., Cahoon, R., Nickerson, K, and Riekhof, W. 2018. Molecular machinery of auxin synthesis, secretion, and perception in the unicellular chlorophyte alga *Chlorella sorokinana* UTEX 1230. *PLoS One* 13: e0205227.

Kim, G. H., Klochkova, T. A., and Kang, Y. M. 2001. Life without a cell membrane: regeneration of protoplasts from disintegrated cells of the marine green alga *Bryopsis plumosa. J. Cell Sci.* 114: 2009–2014.

Kim, J. K., Yarish, C., Hwang, E. K., Parke, M., and Kim, Y. 2017. Seaweed aquaculture technologies, challenges and its ecosystem services. *Algae* 32: 1–13.

Klinger, T., and DeWreede, R. E. 1988. Stipe rings, age, and size in populations of *Laminaria setchellii* Silva (Laminariales, Phaeophyta) in British Columbia, Canada. *Phycologia* 27: 234–240.

Klotchkova, T. A., Chah, O. K., West, J. A., and Kim, G. H. 2003. Cytochemical and ultrastructural studies on protoplast formation from disintegrated cells of a marine green alga *Chaetomorpha aerea* (Chlorophyta). *Eur. J. Phycol.* 38: 205–216.

Klotchkova, T. A., and Kim, G. H. 2006. Purification and characterization of a lectin, Bryohealin, involved in the protoplast formation of a marine green alga *Bryopsis plumosa. J. Phycol.* 42: 86–95.

LaClaire II, J. W. 1982a. Cytomorphological aspects of wound healing in selected Siphonocladales (Chlorophyceae). *J. Phycol.* 18: 379–384.

LaClaire II, J. W. 1982b. Wound-healing motility in the green alga Ernodesmis: calcium ions and metabolic energy are required. *Planta* 156: 466–474.

Laux, T. 2003. The stem cell concept in plants: a matter of debate. *Cell* 113: 281–283.
Le Bail, A., Billoud, B., Le Panse, S., Chenivesse, S., and Charrier, B. 2011. *ETOILE* regulates developmental patterning in the filamentous brown alga *Ectocarpus siliculosus*. *Plant Cell* 23: 1666–1678.
Lee, J., Yang, E. C., Graf, L., Yang, J. H., Qiu, H., Zelzion, U., Chan, C. X. et al. 2018. Analysis of the draft genome of the red seaweed *Gracilariopsis chorda* provides insights into genome size evolution in Rhodophyta. *Mol. Biol. Evol.* 35: 1869–1886.
Luhan, M.R.J., and Mateo, J.P. 2017. Clonal production of *Kappaphycus alvarezii* (Doty) Doty in vitro. *J. Appl. Phycol.* 29: 2339–2344.
Marchant, H. J., and Fowke, L. C. 1977. Preparation, culture, and regeneration of protoplasts from filamentous green algae. *Can. J. Bot.* 55: 3080–3086.
McLachlan, J. L., Quinn, J., and MacDougall, C. 1989. The structure of the plant of *Chondrus crispus* Stackhouse (Irish moss). *J. Appl. Phycol.* 1: 311–317.
Millner, P. A., Callow, M. E., and Evans, L. V. 1979. Preparation of protoplasts from the green alga *Enteromorpha intestinalis* (L.). *Planta* 147: 174–177.
Nakamura, Y., Sasaki, N., Kobayashi, M., Ojima, N., Yasuike, M., Shigenobu, Y., Satomi, M. et al. 2013. The first symbiont-free genome sequence of marine red alga, susabi-nori (*Pyropia yezoensis*). *PLoS One* 8(3): e57122.
Niu, J., Wang, G., Lü, F., Zhou, B., and Peng, G. 2009. Characterization of a new lectin involved in the protoplast regeneration of *Bryopsis hypnoides*. *Chin. J. Oceanol. Limnol.* 27: 502–512.
Norris, R. E., Wollaston, E. M., and Parsons, M. J. 1984. New terminology for sympodial growth in the Ceramiales (Rhodophyta). *Phycologia* 23: 233–237.
Ohtaka, K., Hori, K., Kanno, Y., Seo, M., and Ohta, H. 2017. Primitive auxin response without TIT1 and Aux/IAA in the charophyte alga *Klebsormidium nitens*. *Plant Physiol.* 174: 1621–1632.
Pak, J. Y., Solorzano, C., Arai, M., and Nitta, T. 1991. Two distinct steps for spontaneous generation of subprotoplasts from a disintegrated *Bryopsis* cell. *Plant Physiol.* 96: 819–825.
Parke, M. 1948. Studies on British Laminariaceae. I. Growth in *Laminaria saccharina* (L.) Lamour. *J. Mar. Biol. Assoc. U. K.* 27: 651–709.
Parsons, M. J. 1975. Morphology and taxonomy of the Dasyaceae and the Lophothalieae (Rhodomelaceae) of the Rhodophyta. *Aust. J. Bot.* 23: 549–573.
Peaucelle, A., and Couder, Y. 2016. Fibonacci spirals in a brown alga [Sargassum muticum (Yendo) Fensholt] and in a land plant [*Arabidopsis thaliana (L.) Heynh.*]: a case of morphogenetic convergence. *Acta Soc. Bot. Pol.* 85: 3526. doi: 10.5586/aspb.3526.
Pickett-Heaps, J. D. 1975. *Green Algae*. Sunderland, MA: Sinauer.
Pueschel, C., and Keats, D. K. 1997. Fine structure of deep-layer sloughing and epithallial regeneration in *Lithophyllum neoatalayense* (Corallinales, Rhodophyta). *Phycol. Res.* 45: 1–8.
Puttick, M. N., Morris, J. L., Williams, T. A., Cox, C. J., Edwards, D., Kenrick, P., Pressel, S. et al. 2018. The interrelationships of land plants and the nature of the ancestral embryophyte. *Curr. Biol.* 28: 733–745.
Reddy, C. R. K., Dipakkore, S., Rajakrishan, K. G., Jha, B., Cheney, D. P., and Fujita, Y. 2006. An improved enzyme preparation for rapid mass production of protoplasts as seed stock for aquaculture of macrophytic marine green algae. *Aquaculture* 260: 290–297.
Reddy, C. R. K., Gupta, M. J., Mantri, V. A., and Jha, B. 2008. Seaweed protoplasts: status, biotechnological perspectives and needs. *J. Appl. Phycol.* 20: 619–632.
Renzaglia, K. S., Villarreal, J. C., and Garbary, D. J. 2018. Morphology supports the setaphyte hypothesis: mosses plus liverworts form a natural group. *Bryophyte Diversity Evol.* 40: 11–17.

Salvi, E., Ioio, R. D., and Moubayidin, L. 2019. Meristems, stem cells and stem-cell niches in vascular land plants. In *Deferring Development: Setting Aside Cells for Future Use in Development and Evolution* (Eds.) C. D. Bishop and B. K. Hall, pp. 107–133. Boca Raton, Florida: CRC Press.

Scheres, B. 2008. Stem-cell niches: nursery rhymes across kingdoms. *Nature Rev. Mol. Cell Biol.* 8: 345–354.

Shen, S., Wu, X., Yan, B., and He, L. 2010. Tissue culture of three species of *Laurencia* complex. *Chin. J. Oceanol. Limnol.* 28: 514–520.

Tarakhovskaya, E., Kang, E. J., Kim, K. Y., and Garbary, D. J. 2013. Influence of phytohormones on morphology and chlorophyll a fluorescence parameters in embryos of *Fucus vesiculosus* (Phaeophyceae). *Russ. J. Plant Physiol.* 60: 166–173.

Tatewaki, M., and Nagata, K. 1970. Surviving protoplasts *in vitro* and their development in *Bryopsis*. *J. Phycol.* 6: 401–403.

Waaland, S. D. 1990. Development. In *Biology of the Red Algae* (Ed.) K. M. Cole and R. G. Sheath, pp. 259–273. Cambridge: Cambridge University Press.

Warghat, A. R., Thakur, K., and Sood, A. 2018. Plant stem cells: what we know and what is anticipated. *Mol. Biol. Rep.* 45: 2897–2905.

West, J. A., and Calumpong, H. P. 1989. On the reproductive biology of *Spyridia filamentosa* (Wulfen) Harvey (Rhodophyta) in culture. *Bot. Mar.* 32: 379–387.

Whitewoods, C. D., Cammarata, J., Venza, Z. N., Sang, S., Crook, A. D., Aoyama, T., Wang, X. Y. et al. 2018. *CLAVATA* was a genetic novelty for the morphological innovation of 3D growth in land plants. *Curr. Biol.* 28: 2365–2376.

Wichard, T. 2015. Exploring bacteria-induced growth and morphogenesis in the green macroalga order Ulvales (Chlorophyta). *Front. Plant Sci.* 6(86): 1–19. doi: 10.3389/fpls.2015.00086.

Wichard, T., Charrier, B., Mineur, F., Bothwell, J. H., De Clerck, O., and Coates, J. C. 2015. The green seaweed *Ulva*: a model system to study morphogenesis. *Front. Plant Sci.* 6(72): 1–8. doi: 10.3389/fpls.2015.00072.

Wieschaus, E. and Nüsslein-Volhard, C. 1998. Looking at embryos. In *Drosophila: A Practical Approach*, 2nd ed. (Ed.) D. B. Roberts, pp. 179–214. Oxford: Oxford University Press.

Ye, N. H., Wang, G. C., Wang, F. Z., and Zeng, C. K. 2005. Formation and growth of *Bryopsis hypnoides* Lamouroux regenerated from its protoplasts. *J. Integr. Plant Biol.* 47: 856–862.

Yokoya, N., Avila, M., Piel, M., Villanueva, F., and Alcapan, A. 2014. Effects of plant growth regulators on growth and morphogenesis in tissue culture of *Chondracanthus chamissoi* (Gigartinales, Rhodophyta). *J. Appl. Phycol.* 26: 819–823.

Yoon, K. S., Lee, K. P., Klochkova, T. A., and Kim, G. H. 2008. Molecular characterization of the lectin, bryohealin, involved in protoplast regeneration of the marine alga *Bryopsis plumosa* (Chlorophyta). *J. Phycol.* 44: 103–112.

6 Meristems, Stem Cells, and Stem Cell Niches in Vascular Land Plants

Elena Salvi and Raffaele Dello Ioio
University of Rome "Sapienza"

Laila Moubayidin
John Innes Centre

CONTENTS

ABBREVIATIONS

AC: apical cell
ACD: asymmetric cell division
ARR: ARABIDOPSIS RESPONSE REGULATOR genes
CEI: cortex/epidermis initial

CK: cytokinins
CZ: central zone
DZ: division zone
EDZ: elongation differention zone
GA: gibberellins
GMC: grand mother cell
MMC: meristemoid mother cell
OC: organizing center
PLT: PLETHORA genes
PZ: peripheral zone
QC: quiescent center
RAM: root apical meristem
RZ: rib zone
SAM: shoot apical meristem
SCN: stem cell niche
SCR: SCARECROW gene
SLGC: stomatal lineage ground cell
STM: SHOT MERISTEMLESS gene
TZ: transition zone
WOX: WUSCHEL HOMEOBOX RELATED genes

6.1 INTRODUCTION

All tissues and organs in vascular plants derive from a pool of undifferentiated, totipotent cells known as stem cells that divide repeatedly and asymmetrically to feed a growing organ with newly amplified cells. The evolution of stem cells and their organization within meristems of different types allowed the diversification of vascular land plant species, which have spread and diversified since the mid-Devonian period (about 400 Mya).

Stem cells are set aside very early during de novo organogenesis to sustain the development of leaves, roots, and flowers throughout plant life cycles. The evolution of stem cells was essential for plant survival and integrating external/exogenous stimuli with internal/endogenous mechanisms that allow coherent and plastic organ development and tissue replenishment. Stem cells are of pivotal importance for plant exploration of the surrounding space, both above and below the ground, for tissue repair and integration and to establish new generation during embryogenesis.

This chapter highlights the basic principles of plant stem cell biology and their deployment in the evolution in vascular land plants. We discuss the advances made by studying model plants, particularly thale cress *Arabidopsis thaliana*, focusing on specification of plant meristems during early stages of embryogenesis and maintenance of meristem integrity during undetermined organ growth. Also, we examine the evolutionary appearance of stem cells and their organization in extinct and extant vascular land-plant phyla, the different types of meristematic structures in lycophytes, ferns, gymnosperms and angiosperms,[1] and the importance of stem cells' activity for root and shoot evolution and for strategies of branching morphogenesis.

6.2 STEM CELLS AND STEM CELL NICHES IN VASCULAR PLANTS

Vascular plants develop, grow, and survive for long times because of the activity of their *stem cells*. Stem cells are unspecialized, primitive cells able to divide potentially an infinite number of times. Stem cells are located in specific environments within both animals and plants, known as *stem cell niches (SCNs)*. Within their niche, stem cells divide asymmetrically to generate two daughter cells with different fates: one cell retains undifferentiated features and generate new stem cells, while the other cell undergoes either a few additional rounds of mitotic cell division or differentiates.

The ongoing sequence of cell division and cell differentiation commitments undertaken by a stem cell and its daughter cells summarizes the three defining characteristics for a stem cell that are as follows:

1. self-renewal,
2. possession of undifferentiated characteristics, and
3. ability to differentiate into an array of specialized cells (Scheres, 2008).

In animals, the classification of stem cells is based on their ability to generate either a wide or a restricted pool of descendant cell types (Terskikh et al. 2006). The zygotic cell is the only mammalian *totipotent* cell that generates all embryonic and extraembryonic cell types. During embryonic development, *pluripotent* stem cells give rise to embryonic germ layers but can no longer produce an entire embryo. Finally, *multipotent* stem cells can only produce a range of different cells belonging to a single tissue. This concept is well illustrated by Waddington's metaphor that compared the irreversible restriction in stem cell potential to "a ball that falls down from the crest of a hill through many valleys and cannot go back to the top of the same hill" (Waddington, 1957).

Plants are different. Activation of cell fate is space dependent; every plant cell follows a developmental program that is driven by the position of the cell with respect to its surrounding, rather than by the lineage-based differentiation program seen in animals. For this reason, differentiated tissues in plants can regenerate a totipotent embryo or a callus.

The concept of a specialized microenvironment that sustains stem cell state was first postulated by Schofield in 1978 before strong experimental evidences were available. In his studies regarding hematopoietic stem cells in mice, Scofield

[1] Lycophytes (clubmosses) are vascular plants with single unbranched leaves known as microphylls or lycophylls. Allied to ferns, lycophytes do not produce seeds, fruit, flowers or woody tissues.
Ferns are vascular plants that reproduce via spores. They do not form seeds or flowers. Gymnosperms are seed-producing land plants with seeds ('naked' seeds) on the surfaces of scales or leaves. They do not form flowers or fruit.
Angiosperms are vascular land (flowering) plants with seeds enclosed in an ovary.

hypothesized that stem cells always occupy a specific environment that was named the *stem cell niche*, an environment that provides all conditions for stem cell functions and survival (Schofield, 1978). The concept of stem cells embedded in their specific environment (niche) was subsequently validated and widened to other metazoans and plants to elucidate the common principles that govern stem cell activities (Losick et al. 2011). In plants, although each niche can be different in a way that depends on its function and location, stem cells integrate different signals from the organism and the environment to guarantee plasticity and continuation of growth.

In the root tip of *A. thaliana* (Figure 6.1A), laser-induced ablation of specific cells demonstrated the existence of an organizer of the SCN, called the *quiescent center*

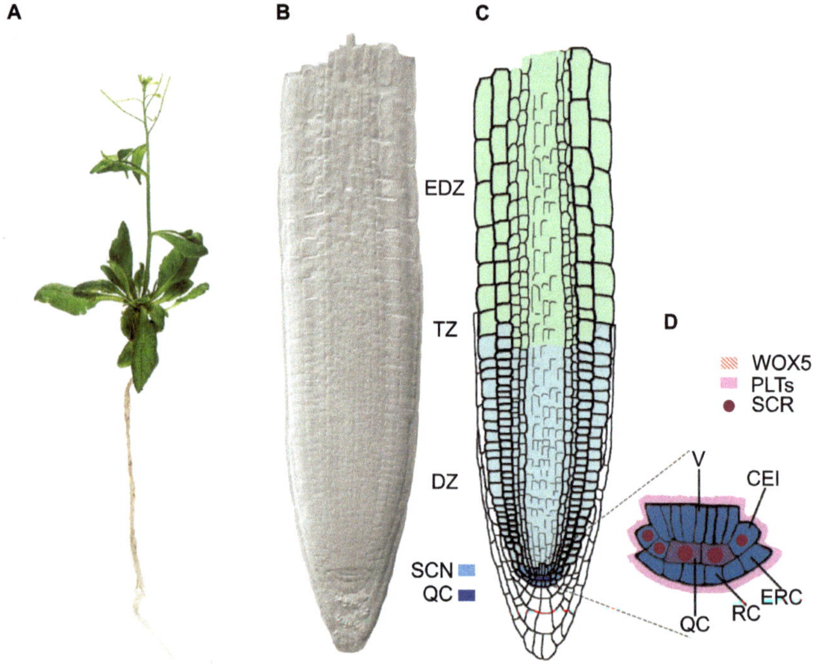

FIGURE 6.1 Organization of the root apical peristem (RAM) and the root stem cell niche (SCN). (A) Picture of an *Arabdopsis thaliana* plant. (B) Micrograph and (C) diagrammatic representation of an optical longitudinal section of an *Arabidopsis* RAM to show its zonation into an SCN and its quiescent center (QC), division zone (DZ, light blue), transition zone (TZ, light green), and elongation differentiation zone (EDZ, green). (D) Diagrammatic representation of the spatial organization of the SCN. The stem cells surrounding the QC initiate the root tissues and are the following: the vascular tissue initial (V), cortex and endodermis initial (CEI), epidermis/lateral root cap initial (ERC), and the columella/root cap stem cells (RC). Distribution of transcription factors in the SCN is also shown in (D): WUSCHEL-RELATED HOMEOBOX5 (WOX5 in red), SCARECROW (SCR, purple), and the mobile transcription factors PLETHORA (PLT, pink).

(QC)[2] that maintains the spatial organization of the SCN and the self-renewing activity of the surrounding stem cells (Figure 6.1B,C). The pool of stem cells in the root niche is specified on a per-cell basis, meaning that direct contact between the organizer of the root SCN, the QC, and the surrounding cells are the prime prerequisites for stem cell activity via short-range signals that travel from the QC to the neighboring cells (Van den Berg et al. 1995; Van den Berg et al. 1997).

At the shoot apex, a QC-similar set of cells, named the *organizer center* (OC), organizes layers of stem cells (Figure 6.2C). At the shoot tip, the stem cells pool size is maintained on a population basis that relies on the relative distance of stem cells from their organizer, the OC. This difference on stem cells between OC and QC activities is reflected by the shoot and root organization of their niches: in the root SCN, only a single layer of tissue-specific stem cells surrounds the QC; in the shoot SCN, three layers of stem cells are arranged in a small region above the OC (Heidstra and Sabatini, 2014) (Figures 6.1C and 6.2B, C). To guarantee coherent organ growth, the self-renewing activity of stem cells is often balanced with the distal differentiation input of the stem cell daughter cells. As two faces of the same medal, the ratio between cell division and cell differentiation inputs are tightly controlled via a balance of several molecular signals such as hormones, microRNAs, and peptides that influence stem cell fate, number, and position.

6.3 STEM CELLS AND MERISTEMS IN VASCULAR PLANTS

In higher plants, meristems are one of the driving forces of postembryonic organ growth and organism plasticity, being the keepers of plant plasticity (Steeves and Sussex, 1989).

Meristems supply the plant body with new tissues and organs through the newly amplified cells that are direct descendants of stem cells within the meristem's SCN. These meristematic daughter cells eventually expand and differentiate to acquire the functional features of the tissues to which they belong.

On the basis of meristem origin and position on the plant body, meristems can be grouped into three different categories (Mauseth, 2017):

1. The first group includes the two apical or primary meristems, *root apical meristem* (RAM) and *shoot apical meristem* (SAM). RAMs and SAMs, which are positioned at the tip of roots and the shoots respectively, greatly contribute to the entire postembryonic plant development; every part above and below the ground is clonally derived from a small pool of stem cells within the meristems. Both shoot and apical meristems are specified during embryogenesis and are maintained active during postembryonic plant growth. Coordination of cell division within the primary meristem with cell differentiation at the meristem's boundaries results in the growth of stems

[2] Abbreviations for the major regions of both root and shoot apical meristems (RAMs, SAMs) and other recurrent abbreviations in the text are listed in the ABBREVIATIONS list and in legends for Figures 6.1B and C.

and roots along the apical-basal plant axis, giving rise to all primary tissues and organs—leaves, flowers, and lateral roots.

2. The second group of meristems, *lateral (secondary) meristems*, originate within the vascular tissue and/or from the cambia (cambium, singular) and are generally active in areas of the plant where primary growth has ceased. Lateral meristems are secondary and not primary meristems because they are set after germination during the postembryonic growth of the plant and because they support the secondary growth of the plant body in the radial direction. Examples are the formation of the secondary vascular bundles (phloem and xylem) in the wood and cork cambia.
3. The third type of meristems, *intercalary meristems*, are positioned in adult tissues of the stems, at internodes where leaves are generally attached to the main body. Intercalary meristems are common in monocots (grasses), in which each stem possesses several intercalary meristems along its length, each of which produces new cells that feed into the stem growth, thus contributing to organ growth along the proximal-distal direction.

Outside these three categories of plant meristems, other cell types can display such features of stem cell as the ability to divide asymmetrically and generate daughter cells with a (limited) transit-amplifying potential, without being embedded within a meristematic niche.

Examples of these type of cells are *meristemoids*, which give rise to the stomatal pores on the aerial organ surface (Simmons and Bergmann, 2016), and the *founder cells* that support the development of branching lateral roots (Stoeckle et al. 2018). In both cases, a differentiated cell is reprogrammed and initiated as a stem cell. Epidermal cells form the precursor for meristemoids, while pericycle (or endodermis) cells form the founder cells of lateral roots.

6.4 PRIMARY APICAL MERISTEMS: STRUCTURE AND FUNCTIONAL REGULATION

Because of their primary importance for seedling establishment and plant development, the best characterized meristems in plants are the SAMs and RAMs. Most of our knowledge regarding the molecular mechanisms that control their activities comes from studies carried out in thale cress, *A. thaliana*, a small (20–25 cm tall) flowering plant in the mustard family. Its short life cycle (as little as 6 weeks), small genome—*Arabidopsis* was the first plant to have its genome sequenced—and ease of cultivation in greenhouses or laboratory have made *Arabidopsis* an ideal model plant species.

6.4.1 The Shoot Apical Meristem

The SAM of angiosperms is a highly organized structure that generates every cell of the aerial part of the adult plant body. The SAM guarantees the presence of a reservoir of cells as well as constantly initiating leaves and flowers at the meristem periphery from the seedling stage throughout the life span of the plant. Therefore, on one side, the SAM sustains undetermined postembryonic growth, while on the other, it determines the architecture and branching of the shoot by balancing cell division and cell differentiation in specific zones.

Three layers of pluripotent stem cells are present in the three outermost cell layers at the shoot apex (L1–L3 in Figure 6.2A, B). Stem cells within the first two layers (L1 and L2) divide anticlinally (Barton, 2010) so that cells in each layer are clonally separated from one another and easily distinguishable. Together, L1 and L2 form the so-called *tunica*. Differently, stem cells in the innermost layer (L3), also known as the *corpus*, divide in all directions (Figure 6.2B). The central zone (CZ in Figure 6.2C) is where the stem cells of each of these three layers divide, at a low rate, to form a pool of undifferentiated cells. Stem cell division also displays stem cell descendants into two neighboring zones: the peripheral zone (PZ; Figure 6.2C), where cells quickly divide and start their genetic commitments toward a differentiative program, and the rib zone (RZ; Figure 6.2C), the most hidden and secluded part of the SAM, where cell division of the meristematic cells controls stem elongation. As cells in the PZ transit amplify, the mitotic rate in the PZ is higher compared to that in the CZ (Laufs et al. 1998). Between the CZ and the RZ, the OC acts as the niche organizer sustaining stem cells activity (Figure 6.2C).

The function of the stem cells in the central zone relies on the noncell autonomous proliferative signal of the homeobox transcription factor WUSCHEL (WUS)

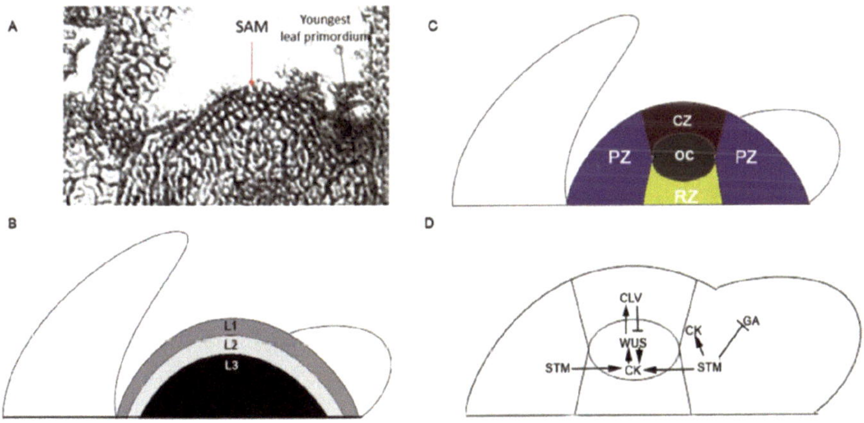

FIGURE 6.2 Organization of the shoot apical meristem (SAM) and shoot stem cell niche (SCN). (A) Histological longitudinal image of portion of a dichot shoot showing the SAM and the youngest leaf primordia. (B–D) Diagrams representing longitudinal section of a dicot shoot. (B) Stem cells are arranged in three consecutive layers: L1 (dark gray), L2 (light gray), and L3 (black). (C) Stylized representation of zonation of cells in the SAM showing the organizer center (OC) central zone, peripheral zone (PZ), and rib zone (RZ). (D) A model summarizing the gene network that maintains SAM function, shown within a diagram of an SAM in longitudinal section. Arrows indicate activation, ——| indicates repression. Genes shown are SHOOT MERISTEMLESS (STM), cytokinins (CK), WUSCHEL (WUS), CLAVATA (CLV), and the family of plant hormones known as gibberellins (GA). See the text for detailed discussion.

(Mayer et al. 1998; Laux et al. 1996). *WUS* mRNA is detectable in the organizer center (Schoof et al. 2000), but its protein moves through plasmodesmata to reach the stem cells–populated central zone (Yadav et al. 2011; Daum et al. 2014) (Figure 6.2D). Mutations in *WUS* causes loss of stem cells activity (Mayer et al. 1998; Laux et al. 1996); hence, *WUS* is necessary to maintain stemness in the three L1–L3 overlying layers (Mayer et al. 1998; Laux et al. 1996).

In the CZ, WUS protein directly promotes *CLAVATA 3* (*CLV3*) expression, which codifies for an extracellular peptide of the CLAVATA3/ESR-related (CLE) family (Perales et al. 2016; Ogawa et al. 2008). In the apoplast, CLV3 is bound by a receptor kinase complex of the leucine-rich repeat receptor kinase type (LRR receptor kinase) CLAVATA 1 (CLV1). At the plasma membrane in the central zone, CLV1 forms homodimers to bind its ligand CLV3. The CLV3-CLV1 complex triggers an intracellular signal that shuts down *WUS* expression in the neighboring zones, thus restricting *WUS* expression in the organizer center.

So, while WUS activates *CLV3* in the central zone, the CLV3/CLV1 complex represses *WUS* expression (Figure 6.2D). The control of WUS on *CLV3* expression is dose dependent. At low concentration of WUS in the central zone, WUS binds the regulatory region upstream and downstream of *CLV3* inducing its expression; at high intracellular concentration of WUS, such as in the organizer center, WUS homodimerizes repressing *CLV3* expression and sustaining its own expression in the organizer center (Perales et al. 2016). Because of this genetic network, *clv3* mutants display an enlarged meristem and an enlarged *WUS* expression domain (Fletcher et al. 1999; Brand et al. 2000, Schoof et al. 2000).

In addition to fine-tuning *CLV3* expression, WUS physically interacts with a transcription factor belonging to the GRAS family, HAIRY MERISTEM 1 (HAM1), to regulate the transcription of a set of target genes (Zhou et al. 2015). Interestingly, *CLV3* and *HAM1* are expressed as two opposite gradients within the central zone: *CLV3* expression pattern has a maximum in the L1 and decreases in L3, while *HAM1* is highly expressed in the L3 layer and lower in the L1. Mutants displaying loss of HAM activity indeed show ectopic expression of *CLV3* peptide, which expands throughout the SAM. This genetic network suggests that the HAM gradient acts as positional information that specifies the apical-basal distribution of CLV3 expression (Zhou et al. 2018).

The plant hormone *cytokinins* (CKs) also play a positive role in maintaining the stem cell pool of the SAM. CKs specify for *WUS* expression domain in the organizer center (Figure 6.2D). WUS in turn sustains CK-mediated cell division input in the central zone by repressing the CK negative regulators *type A ARABIDOPSIS RESPONSE REGULATORs* (*ARR-As*) (Leibfried et al. 2005). At the SAM lateral boundaries, another class of plant hormone called *gibberellins* (GAs) counteracts the proliferation inputs induced by CKs by promoting cell differentiation of the stem cell daughter cells. On the contrary, CKs activate degradation of GA by promoting the expression of specific GA catabolic enzymes to suppress the cell differentiation input (Jasinski et al. 2005).

At the crux of CKs and GAs cross-talk, the mobile transcription factor SHOOT MERISTEMLESS (STM), a *class I KNOTTED-like homeobox* (*KNOX*) gene, plays a fundamental role in SAM regulation by reducing GA contents on the one hand and promoting CK biosynthesis on the other (Jasinski et al. 2005; Yanai et al. 2005). *STM* expression, in turn, is positively regulated by CKs, thus specifying STM domain in all shoot meristematic cells where it prevents cell differentiation (Long et al. 2006; Balkunde et al. 2017; Xu et al. 2011) (Figure 6.2).

6.4.2 The Root Apical Meristem

The core of the RAM is its *SCN*, where the QC works as the organizer of the surrounding stem cells (Figure 6.1).

When a stem cell divides asymmetrically, the daughter cell remaining in direct contact with the QC retains its stem cell properties, while the distal daughter cell leaving the niche undergoes further mitotic cell divisions, shifting away from the niche toward the differentiation zone. The RAM is a powerful model for plant development studies, because cells are arranged in a spatiotemporal developmental sequence (Figure 6.1A, B). While undergoing mitosis along the longitudinal axis of the root, the transit-amplifying cells are part of the division zone (DZ). At a certain distance from the SCN, the meristematic cells pass through a zone called the transition zone (TZ) where cells are committed to differentiation, start acquiring differentiating landmarks, and lose their ability to divide. After this developmental checkpoint, in the elongation and differentiation zone (EDZ), cells start elongating, acquiring their terminal cell fate and shape on the basis of their relative position within the root tissues (Heidstra and Sabatini, 2014) (Figure 6.1A, B).

Root tissues are initiated in the SCN as concentric layers of cells around the QC: from the inside to the outside of the roots, tissues differentiate in vasculature, pericycle, endodermis, cortex, and epidermis. In addition, the RAM is protected by an external lateral root cap and a few layers of columella tissue at its distal end that provides the root with a gravity sensor system. Most of the tissues derive from one stem cell. Some of them such as the cortex/endodermis and the epidermis of the lateral root cap originate from a single progenitor stem cell (Dolan et al. 1993) (Figure 6.1C).

As in the SAM, hormonal cross-talks, transcriptional networks, microRNA, and peptides also regulate root development. The plant hormone *auxin* is the primary factor governing RAM organization and activity. Direct hormonal measurements and synthetic reporter analysis showed that auxin levels and its transcriptional response are distributed as an instructive gradient within the RAM. This gradient exhibits a maximum in the QC where it contributes to specification of the SCN, an intermediate level along the DZ where it promotes mitotic division of the meristematic cells, and ultimately a minimum at the TZ where it is interpreted as a cell differentiation input (Figure 6.3A and see (Petersson et al. 2009; Brunoud et al. 2012; Di Mambro et al. 2017).

Therefore, auxin has been proposed to work as a morphogen in plant development (Sabatini et al. 1999; Wolpert, 2011). Mathematical models coupled to experimental evidence demonstrated that the shape of this morphogen gradient is mainly due to the activity of the auxin efflux facilitators called PIN (PIN FORMED) proteins (Grieneisen et al. 2007). PINs are asymmetrically localized at the plasma membrane of meristematic cells where they orchestrate a polar auxin flux within the meristematic tissues, determining the auxin gradient (Figure 6.3A). The auxin maximum at the root meristem distal tip positions the QC cells, by selecting among those expressing the transcription factor SCARECROW (SCR). SCR is a key regulator of root development expressed in the QC and in the endodermis tissue, including its stem cells (Figure 6.1C). Only SCR-expressing cells experiencing a high level of auxin are specified as organizer cells (Sabatini et al. 1999, 2003) (Figures 6.1C and 6.3A).

In addition, another class of mobile transcription factors, called PLETHORA (PLT), whose family members have been shown to work redundantly, plays a pivotal role for RAM establishment and activity (Figures 6.1C and 6.3B). PLT proteins are expressed in SCN precursor cells since early embryonic stages. Combination of

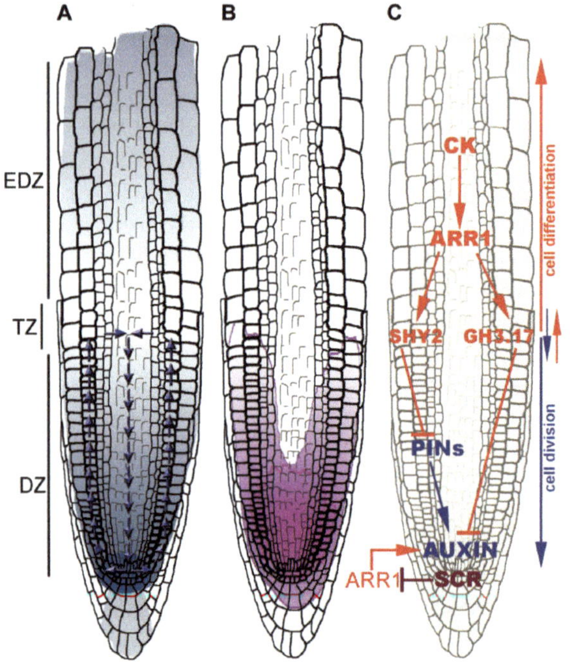

FIGURE 6.3 Molecular pathways controlling zonation and activity within the root apical meristem (RAM) shown as diagrams representing longitudinal sections of the RAM of *Arabidopsis*. The zones in (A), which apply to (B) and (C), are the division zone (D), transition zone (TZ), and elongation differentiation zone (EDZ). (A) The auxin gradient (blue) and direction of auxin flux (blue arrows) within the RAM are shown. (B) The gradient in PLETHORA genes is shown in violet. (C) A model illustrating the genetic mechanisms controlling RAM function and zonation and regulating cell division and cell differentiation. Arrows indicate activation, ——| indicates repression ARR1, type-B cytokinin responsive transcription factor; SHY2, SHORT HYPOCOTYL 2, an auxin repressor; GH3.17, *GRETCHEN HAGEN 3.17* gene, a gene that encodes an auxin catabolic enzyme; PINs, PIN FORMED family member genes, auxin efflux facilitators proteins; SCR, SCARECROW, a transcription factor; auxin, a plant hormone. See the text for discussion of the gene networks.

mutants in multiple *PLETHORA* genes shows loss of SCN activity and root meristem differentiation, leading to severe defects in root growth (Aida et al. 2004).

Similar to auxin, PLTs are expressed in a gradient along the root meristem axis, leading to the hypothesis that the auxin maximum readout was interpreted and carried out by PLT's signaling (Figure 6.3A, B). Several lines of evidences support instead reciprocal levels of PLTs and auxin genetic interactions; while auxin indirectly promotes *PLTs* expression in the SCN, PLTs promote auxin production and auxin polar transport to facilitate establishment of the auxin maximum in the SCN (Galinha et al. 2007; Blilou et al. 2005; Aida et al. 2004). The PLTs' gradient reaches its lowest levels at the TZ of the meristem as the result of two combined mechanisms: (i) diffusion of PLT proteins between cells via plasmodesmata and (ii) intracellular dilution caused by mitotic divisions along the RAM (Mähönen et al. 2014). Ultimately, PLTs' gradient orchestrates a transcriptional network that regulates meristematic cell state inducing cell proliferation genes and repressing cell differentiation (Santuari et al. 2016).

On the opposite end of the meristem, at the TZ, cell differentiation takes place, mainly driven by CKs (Dello Ioio et al. 2007). In contrast to their positive role in sustaining cell division in the shoot, CKs induce meristematic cell differentiation in the root via two principal mechanisms:

i. On the one hand, CKs activate the auxin repressor SHORT HYPOCOTYL 2 (SHY2) at the TZ, which in turn downregulates expression of the auxin transporters *PINs*, locally lowering auxin distribution (Dello Ioio et al. 2008) (Figure 6.3C).
ii. On the other hand, CK signaling induces the degradation of auxin via activating expression of the *GRETCHEN HAGEN 3* gene *GH3.17*, which encodes for an auxin catabolic enzyme, at the TZ (Di Mambro et al. 2017) (Figure 6.3C). This dual control on auxin levels mediated by CKs shapes the auxin gradient and guarantees an auxin minimum to set the boundary of the RAM (Sabatini et al. 1999; Grieneisen et al. 2007; Dello Ioio et al. 2008). The auxin minimum works as positional information that switches the cell status from meristematic to differentiating mainly triggered by a CK positive regulator, *type B ARABIDOPSIS RESPONSE REGULATOR 1* ARR1 (Dello Ioio et al. 2008; Di Mambro et al. 2017) (Figure 6.3C).

Auxin/CK cross-talk is also remarkably important for balancing cell division of the cells in the SCN with cell differentiation of the meristematic cells at the TZ and requires SCR activity. In the cells of the QC, SCR antagonizes the CK-mediated differentiation input by directly repressing the expression of *ARR1*. *scr* loss-of-function mutants display ectopic *ARR1* expression in the SCN that locally causes excessive auxin biosynthesis and, thus, cell differentiation (Moubayidin et al. 2013; Ding and Friml, 2010). Once auxin is synthetized in the SCN, it acts as key cell nonautonomous input for regulating *ARR1* levels at the TZ. Importantly, this feedback loop shows that the QC not only works as a local organizer for the stem cells but also as distal organizer of the meristematic daughter cells to fine-tune their rate of cell differentiation (Moubayidin et al. 2013) (Figure 6.3C).

As CKs play a key role in positioning the TZ, the regulation of their biosynthesis is fundamental for root patterning. CK synthesis is controlled in the meristem by the HD-ZIPIII transcription factor *PHABULOSA* (*PHB*) that directly activates a CK biosynthetic gene, *ISOPENTENYL TRANSFERASE 7* (*IPT7*) in the DZ. At the TZ, ARR1 represses *PHB* expression fine-tuning CK levels. Interestingly, ARR1 represses also the expression of *MIR165/6,* encoding for a family of *microRNA* that targets *PHB*, thus generating a regulatory circuitry that provides robustness against oscillations in CK contents in the RAM (Dello Ioio et al. 2012).

6.4.3 Embryonic Specification of Primary Meristems

In seed plants, the shoot and the root apical meristems are the only meristems specified during embryogenesis (Figure 6.4A). In vascular plants, stem cells' function has been extensively associated with the activity of WOX genes. Three clades of WOX genes, called ancient, intermediate, and modern, have been classified according to plant phylogenies, suggesting that the complexity of stem cells during plant evolution may be associated with diversification of the WOX gene family (Van Der Graaff et al. 2009).

In *Arabidopsis*, from early embryonic stages onward, apicobasal polarity and developmental programs are first instructed through the activity of the WOX2, WOX8, and WOX9 transcription factors (Figure 6.4B). Following the first asymmetric cell division of the zygote, the smaller upper cell gives rise to the proembryo, while the bigger basal cell will give rise to the suspensor (Figure 6.4A). After two rounds of cell divisions, the radially symmetric octant stage proembryo and the filamentous suspensor show different expression patterns of WOX genes (Figure 6.4). The upper and lower tier of the proembryo is where the shoot and the embryonic root will be positioned. The upper tier expresses *WOX2*, while the lower tier expresses *WOX9*. The suspensor cells exclusively divide anticlinally and express WOX8, with the exception of the first upper cell that directly contacts the embryo (named the *hypophysis*) that expresses WOX9 (Haecker et al. 2004; Brueninger et al. 2008) (Figure 6.4).

The hypophysis clonally generates the QC and the SCN. In the following stage, the embryo turns into a bilaterally symmetric heart-shaped body where the first morphological evidences of root and shoot SCNs and two cotyledon primordia are formed (Ten Hove et al. 2015). The hypophysis is incorporated into the embryo to become the lens-shaped cell that divides asymmetrically and expresses WOX5 (Sarkar et al. 2007) (Figure 6.4). From the asymmetric cell division of the lens-shaped cell, the apical daughter cell originates the QC, while the basal daughter cell originates the columella stem cells. At this stage, WUS is expressed in the precursor cell for the organizer center of the SAM (Ten Hove et al. 2015) (Figure 6.4B).

6.5 VASCULAR CAMBIUM DEVELOPMENT AND EVOLUTION

The evolution of radial growth has been proposed to have started during the Early Devonian (400Mya), coinciding with and supporting the transition from sea to land by adding efficient long-distance water conductance and mechanical support to the plant body.

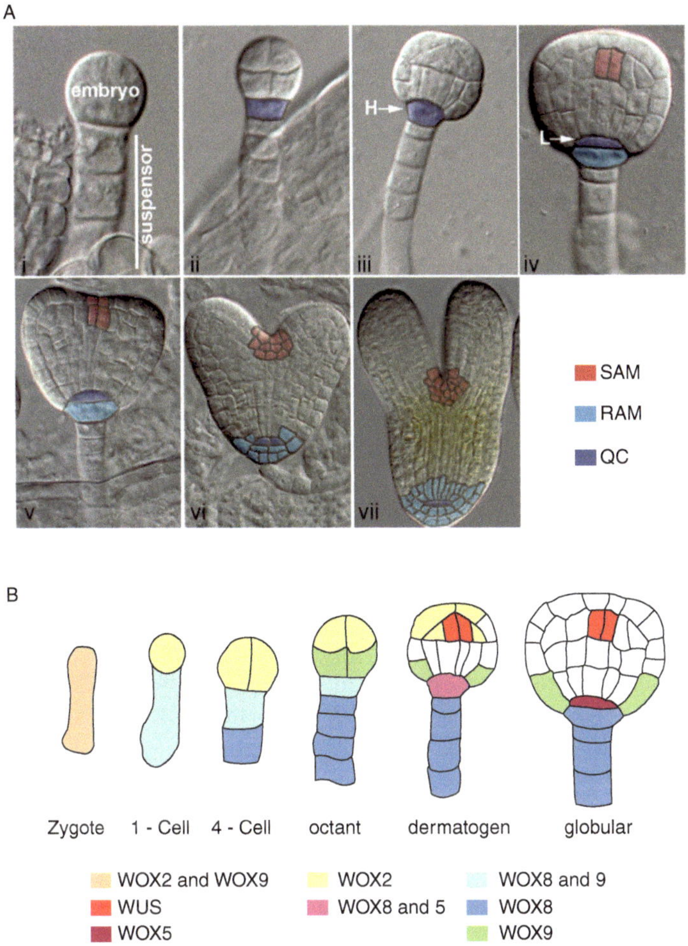

FIGURE 6.4 Embryonic specification of primary meristems and the role of *WOX* genes in their specification during embryogenesis in *Arabidopsis thaliana*. (A) False-colored optical micrographs of *Arabidopsis* embryos at four-cell stage (i), eight-cell or octant stage, (ii), dermatogen, (iii), globular, (iv), transition, (v), heart, (vi), and torpedo stages (vi). Red and blue highlight the position of the SAM and RAM during ontogenesis. Purple cells indicate the quiescent center (QC) and its precursor, the hypophyseal cell (H, white arrow), and the lens-shaped cell (L, white arrow). (B) Schematic diagrams of the zygote and six early embryonic stages to show expression domains of five *WOX* genes and of the homeobox transcription factor WUSCHEL (WUS) during embryogenesis.

To achieve radial secondary growth, concentric rings of stem cell–like tissues are organized at the organ peripheries. The cambial zone of the vascular cambium is the site of stem cell activity, where stem cells divide either anticlinally (perpendicularly to the organ surface to generate more initials along the apical-basal axis) or periclinally (parallel to the organ surface to generate the xylem and phloem vascular bundles in the radial direction).

Modern plants initiate xylem and phloem tissues in a bidirectional way, with the xylem-producing wood cells generally toward the inside of the plant body, and the phloem-producing bast fibers toward the outside (Figure 6.5). After periclinal asymmetric cell division of the cambium stem cells, few rounds of cell division rapidly increase the number of meristematic cells within the cambium. These fast-dividing cells, which are called *xylem and phloem mother cells* according to their position with respect to the mother stem cell, differentiate into tracheids, vessels, and parenchyma cells in the xylem or into sieve elements, companion cells, and fibers in the phloem (Spicer and Groover, 2010) (Figure 6.5).

The homeobox transcription factor WUSCHEL-RELATED HOMEOBOX4 (WOX4) promotes identity and proliferation of cambium stem cells by working cell autonomously in the cambium initials where it is expressed. Moreover, a tightly controlled peptide receptor module integrates activity of the cambium initials and initials' proximal derivatives, such as the phloem mother cells. The meristematic phloem cells express a small peptide, CLAVATA3/ESR-RELATED41 (CLE41) that travels back to the cambium initials where the peptide's receptor, PHLOEM INTERCALATED WITH XYLEM (PXY), is expressed in the same domain as is WOX4 (Hirakawa et al. 2008, 2010; Etchells et al. 2016).

Interestingly, the evolutionary origin of bidirectional growth of modern vascular cambium is not completely understood. Fossil records of extinct early vascular plants, such as clubmosses (lycophytes) or basal ferns, showed some cases of unifacial vascular cambium producing only xylem, while extant lycophytes have a bifacial cambium, i.e., a bidirectional cambium that externally produces phloem (inner bark)

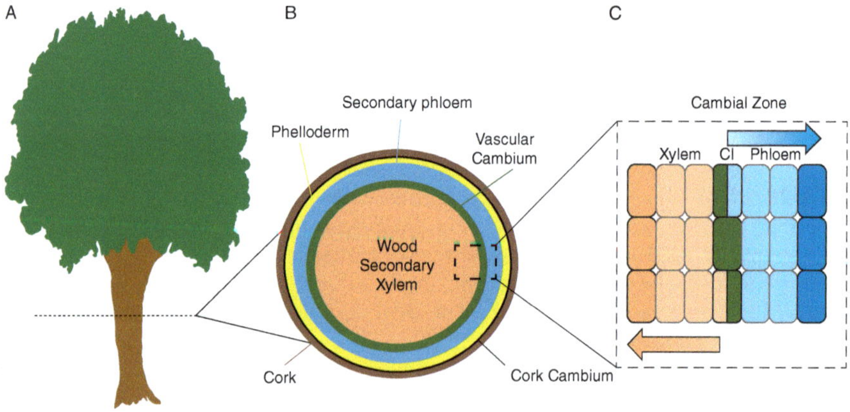

FIGURE 6.5 Cambium development from secondary meristem in angiosperms. (A) A tree with ongoing secondary growth. (B) Diagram of a hypothetical transversal section showing the zones of the vascular and cork cambia. (C) Higher magnification diagram representing the cell divisions in the cambial zone and the progressive differentiation in xylem tissues and phloem tissues, shown by the gradients and arrows in brown and blue, respectively. The cambium initial (CI, green) divides asymmetrically originating phloemic precursor cells (light blue) that will differentiate into phloem, and, on the other side, xylem precursor cells (light brown) that will differentiate into xylem vessels.

and internally xylem (wood). This raises the question of whether the bifacial mode of tissue production in vascular cambium emerged once during plant evolution or is a result of convergent evolution.

6.6 STOMATA DEVELOPMENT AND EVOLUTION

Stomata are structures that allow gas and water exchanges between the environment and the plant. They are present in the epidermis of all the aerial organs, including leaves, stems, and fruits. Details of stoma morphology varies among species, but the basic structure is unvaried among plants. Stoma consists of two epidermal guard cells flanking a pore and an airspace in the subtending tissue. Stomata are flanked by subsidiary cells that control the opening and closing of the guard cells.

Angiosperm stoma development is pretty well described. In dicots such as *Arabidopsis*, stomata derive from a protodermal cell that is committed to originate the stoma instead of a pavement cell. Thus, this protodermal cell acquires features of the *meristemoid mother cell* (MMC) (Figure 6.6). MMCs divide asymmetrically originating the meristemoid and a large cell called the *stomatal lineage ground cell* (SLGC) (Figure 6.6B). The SLGC can eventually divide asymmetrically producing additional meristemoids. Each meristemoid enters the amplifying division stage, undergoing up to three asymmetric rounds of division, increasing the number of epidermal pavement cells. Subsequently, the resulting meristemoid daughter cell differentiates as a *grand mother cell* (GMC) that divides symmetrically creating the two guard cells (Figure 6.6B). The GMC can also recruit neighboring subsidiary cells, providing functional assistance for guard cell movements (Zhao and Sack, 1999).

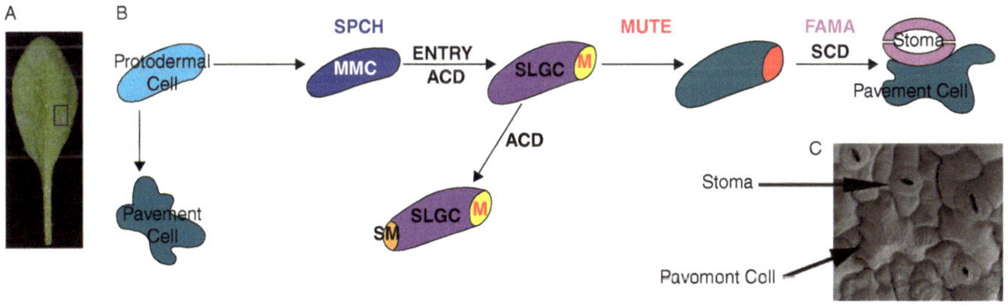

FIGURE 6.6 Model for dicot stomata development. (A) Picture of an *Arabidopsis* leaf where stomata are interspaced with epidermal cells and trichomes. (B) Diagram representing the sequence of events occurring during stomata initiation and development: a protodermal cell (light blue) that does not differentiate into an epidermal pavement cell (dark green) or a trichome (not shown) expresses *SPEACHLESS* (*SPCH*) is committed to form a meristemoid mother cell (dark blue). The meristemoid mother cell enters the stomata lineage by dividing asymmetrically (ACD) producing a stomatal lineage ground cell (SLGC, purple) and a meristemoid (M, yellow). The meristemoid divides asymmetrically generating a self-renewed meristemoid and a pavement cell. Expression of *MUTE* guides the meristemoid to transit into a guard mother cell (red) that will originate the stoma after the expression of *FAMA*. SLGC can divide asymmetrically producing a satellite meristemoid (orange). (C) Scanning electron micrograph showing fully differentiated stomata and pavement cells.

In monocots, the GMC derives directly from an asymmetric cell division occurring in a protodermal cell. The protodermal cells in contact with the GMC divide asymmetrically, giving rise to the subsidiary cells (Peterson et al. 2010).

The molecular mechanisms controlling stomatal development are well characterized in *Arabidopsis*. Three basic helix-loop-helix (bHLH) transcription factors named SPEECHLESS (SPCH), MUTE, and FAMA are required for stomata formation (Figure 6.6B, and see Ohashi-Ito and Begman, 2006; MacAlister et al. 2007; Pillitteri et al. 2007). These three TFs are expressed in different developmental windows and control cell fate transitions.

SPCH promotes the transition from protodermal cell to MMC. Corroborating this function, *spch* loss-of-function mutants lack stomatal lineage. Once that MMC divides, *SPCH* expression is detectable only in the meristemoid.

MUTE promotes the transition from meristemoid to the GMC. *MUTE* is expressed in the late stages of meristemoid development, and its expression coincides with the termination of the amplifying division stage. Transition from GMC to guard cells is promoted by FAMA, which is first expressed in the latest stages of GMC development (Figure 6.6B).

Whether stomata have a monophyletic origin or whether they originated from convergent evolution is still debated. Nowadays, genomic studies suggest that stomata had a monophyletic origin. Indeed, SPCH, MUTE, and FAMA are well conserved in all land plants and show similar functional activities (MacAlister et al. 2007; Liu et al. 2009; Liu and Xu, 2018).

6.7 LATERAL ROOT DEVELOPMENT AND EVOLUTION

Lateral roots form postembryonically and are initiated from the endodermis in ferns and from the vascular pericycle in seed plants. In both cases, cells from these inner tissues within the parent roots dedifferentiate and become newly specified as *lateral root founder cells*. Extant ferns usually develop *adventitious roots*, which form from the aerial plant body and which branch into lateral roots. In contrast, seed plants have a primary root that develops from a bipolar embryo, in addition to adventitious and lateral roots.

During fern lateral rooting, as in *Ceratopteris richardii*, a member of the intermediate clade of the WOX genes, *CrWOXA*, is specifically expressed in those endodermal cells that initiate lateral roots. The endodermal cells expressing *CrWOXA* change fate to become the lateral root mother cells. Subsequently, the lateral root mother cell divides asymmetrically, and concomitantly, expression of *CrWOXA* is turned off (Liu and Xu, 2018). Division of the lateral root mother cell results in the formation of a tetrahedral root apical cell that has four division planes (see Section 6.8A, Single-Celled Apical Meristem). Subsequent asymmetric cell divisions set aside the root cap initial cell (distal merophyte) and three other proximal initials of all other root tissues (proximal merophytes).

In seeds plants, including the dicotyledonous angiosperm *Arabidopsis*, lateral roots initiate from the pericycle, specifically from specialized xylem pole cells that are known as *lateral root founder cells*. From initiation in the inner root layers to its emergence outside the parental root body, the lateral root primordium grows through its neighboring tissues, the endodermis, the cortex, and the epidermis. In *Arabidopsis*, the stereotypical pattern for lateral root initiation and growth has been divided into eight stages. Particularly interesting is stage 1, where initiation of the lateral primordium occurs (Stoeckle et al. 2018). At this stage, the formative cell division of the xylem pole cells, which set aside the lateral root founder cells, takes place. This formative cell division is preceded by a coordinated migration of the xylem pole nuclei toward the common cell wall (in the apical-basal direction), also known as the nuclear polarization event (Vermeer et al. 2014). This latter is positively controlled by auxin, which works as a morphogen in promoting cell division and reestablishing an SCN in the lateral root meristem (Dubrovsky et al. 2008).

6.8 EVOLUTION OF APICAL MERISTEMS IN VASCULAR PLANTS

The earliest land plants arose about 470 million years ago (Figure 6.7). Fossil records provide evidence that the earliest land plants were morphologically simple, composed of small multicellular bodies. The transition from an aqueous to a gaseous terrestrial medium exposed these pioneer plants to new physical conditions that resulted in key physiological and structural changes. Starting around 430 million years ago, the innovations of stems with fluid transport mechanisms (vascular tissue), structural tissues (such as wood), epidermal structures for respiratory gas exchange (stomata), and various kinds of leaves, roots, and seeds, all began (Figure 6.7).

Vascular plants radiated in two branches: the lycophytes (clubmosses, spikemosses, and quillworts) and the euphyllophytes (ferns, gymnosperms, and angiosperms). These two branches diverged from one another 400 millions of years ago, expanding into an enormous number of different species and forms (Figure 6.7).

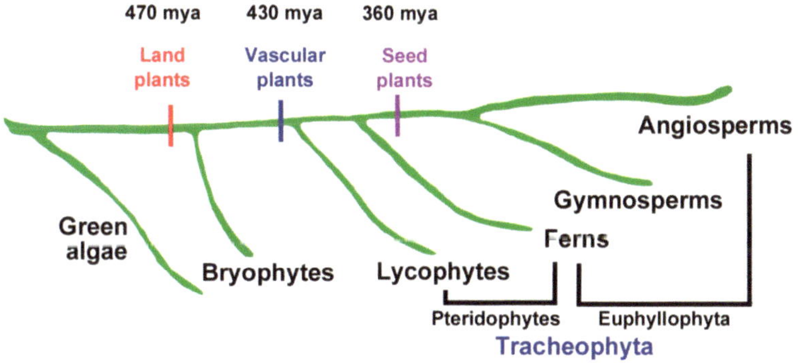

FIGURE 6.7 Evolution of vascular land plants. Green algae (charophytes) made the transition and colonized the space above the sea 470 million years ago (Mya), resulting in the evolution of bryophytes (mosses, hornworts, and liverworts) on land. Around 430 millions of years ago, evolution of the vascular system (xylem and phloem) allowed the diversification of larger species that could sustain transport of water and minerals at a distance. Vascular plants (Tracheophyta) includes pteridophytes and euphyllophytes. Pteridophytes, such as lycophytes (clubmosses, spikemosses, and quillworts) and some ferns, are reproduced by dispersing spores. On the other hand, euphyllophytes (most ferns, gymnosperms, and angiosperms (flowering plants)) are reproduced by the production and dispersal of seeds.

The huge variability in shoot and root morphology of vascular plants was accompanied by evolution in the organization and activity of their apical meristems. Indeed, all the shoots of the vascular plant sporophytes (the diploid multicellular stage) derive from a meristem located at the apex of each shoot. These apical meristems evolved into different sizes and structures that sustained the development of an enormous variability of plant body architectures that is determined in large part by events that occur within the meristem itself. Within the meristems, the plant stem cell organization evolved progressively, leading to diverse organization of apical meristems' structures.

Ferns (monilophytes) and some lycophytes (including members of the spike moss genus *Selaginella*), collectively referred to as pteridophytes (Figure 6.7), have apical meristems that are unusual compared to the apical meristems of other vascular plants: Pteridophytes' apical meristems usually have a single distinct apical, pie-shaped cell (Figure 6.7).

Single-cell apical meristems appeared first during land plant evolution in the nonvascular bryophytes and eventually in vascular nonseed pteridophytes supporting the development of primitive tridimensional shapes. Conversely, seed plant meristems of gymnosperms and angiosperms are multicellular with distinct zones. The number of initials in their root and shoot apices is variable, with several initials organized within a multicellular meristem arranged in radial layers around the meristem organizer. In angiosperms, a distinct sets of initials give rise to individual tissue types, as described previously (Figure 6.1C). In contrast, in gymnosperms, the initials can be temporary, supporting multiple type of tissues.

Branching, a way for shoots and roots to expand in the surrounding space, is one of the body structures most strictly dictated by the meristem architecture. While single-cell apical meristems support dychotonoumous (or exogenous) branching of the distal organ tip via meristem bifurcation and the establishment of two new daughter meristems, multicellular meristems in gymnosperms and angiosperms support lateral (or axillary) branching where a new organ primordium forms at the periphery of the meristematic zone. During dichotomous branching, the single-cell apical meristem structure reorganized entirely to accommodate two new distal apical meristems at the organ tip. In contrast, in lateral branching, the axillary buds and roots are positioned, specified, and organized without affecting the primary meristem's structure and/or activity, forming axillary secondary meristem buds that branch laterally.

6.8.1 Single-Celled Apical Meristems

A simple single-celled histogenetic apical meristem (the apical cell, AC) appeared very early in land plant evolution. This large cell resembles an inverted pyramid (whose base is directed toward the apex surface) and represents the structural and functional unit of the apical meristem (Figure 6.8). Single-celled meristem occurs in modern plants that are considered to be early divergent, including small, nonvascular bryophytes (liverworts, hornworts, mosses), simple rootless plants, as well as simple vascular plants that produce spores rather than seeds. In some pteridophytes (lycophytes, clubmosses) such as *Selaginella*, more than one apical

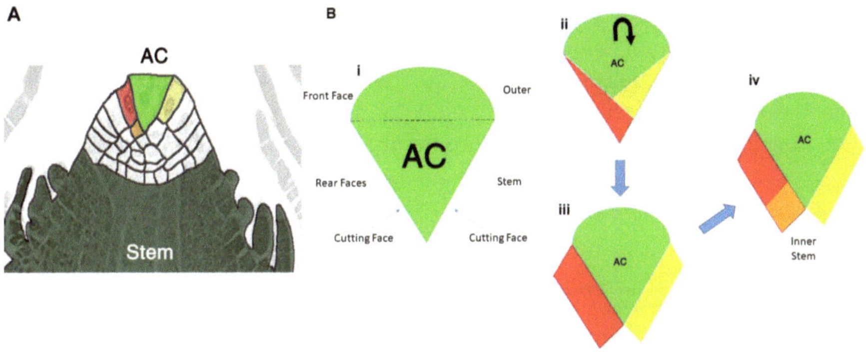

FIGURE 6.8 Apical cell (AC) meristems. (A) Representation of a fern shoot single-cell apical meristem. (B) Diagrams representing (i) the core unit of a single-cell meristem, the apical cell (AC, green), and in (ii–iv) its descendants. (i) Dashed line represents the border between the front and the rear faces of the AC. (ii) Red and yellow areas indicate stem cell divisions taking place parallel to the AC cutting faces. The curved black arrow indicates self-renewal of the AC through cell division. (iii) The first two cells derived from the division of the AC are indicated in red and yellow, respectively. (iv) After cell expansion, the derived cells (red and yellow) increase their volume and divide again, giving rise to the inner daughter cell (orange).

cell can be distinguished within the meristem; these are called *apical initials* and, similar to the apical cell, they occupy the entire shoot apex.

Clubmosses are the most ancient living group of vascular (nonseed) plants, and they were the first group of plants to evolve roots. Therefore, the study of their morphology and evolution has been a major goal for evolutionary scientists interested in understanding the origin of RAMs. As in other vascular plant sporophytes, shoots and roots in lycophytes are derived from a single-celled apical meristem located at their distal apex (Figure 6.8A). The single apical cell is generally tetrahedral in shape and divides in such a way that the new cells are formed on all sides (or cutting faces) except the surface (Figure 6.8B). The apical cell repeatedly divides, in the downwardly pointed faces of the apical cell, by an asymmetric division in a helical succession, giving rise to consecutive narrow and flat cells that differentiate into several tissues. The apical cell's daughter cells are therefore merophytes since they are all produced from the same initial cell. The daughter cells formed at the cutting faces of the apical cell subsequently divide and form a large packet of cells that differentiate and form different segments of the plant body, in a clonal fashion (Figure 6.8B). Interestingly, SAMs and RAMs belonging to the same species can display similar organization but different properties of the apical cell, as shown for the lycophyte *Selaginella uncinata*, in which the shoot apex has a single lens-shaped apical cell with two cutting faces, while the rhizophore (a root-producing axis) apex has a tetrahedral apical cell with three cutting faces (Imaichi and Kato, 1989).

One of the most distinguishing features of *Selaginella* is its pattern of shoot and root branching. *Selaginella* SAMs and RAMs regularly bifurcate to produce two new apices, each bifurcation forming a Y-shaped branch junction. During the bifurcation, the original apical cell of the branching meristem becomes unidentifiable before a new apical cell is initiated in each of the bifurcated apical daughter meristems. Interestingly, in *Selaginella kraussiana*, application of auxin promotes, while CKs inhibit dichotomous branching of the roots, highlighting the early recruitment and importance of this hormonal cross-talk for root meristem morphology and function (Sanders and Langdale, 2013). Similar to *Selaginella*, other pteridophytes such as ferns (monilophytes) display one (or few) apical initial(s) in their root meristems, and they branch dychotonoumously.

6.8.2 Evolution of Shoot Apical Meristems

Evolution of SAMs as a new land plant feature played a central role in the body plan rearrangement, providing new tissues and continuous growth to the aerial organs. All land plants possess a SAM at their growing apex; indeed, this meristem is considered the most important innovation of land plants when compared with the ancestral algae (Graham and Wilcox, 2000). SAMs can be composed of a single apical tetrahedral stem cell (pteridophyte type) or of several cells with a dome structure (seed plant type). Despite the modern classification of two main types of SAM (pteridophyte and seed plant types), differences in function and architecture are known. Newman's classification of SAMs (Newman, 1965) has been the most accepted classification for a long time. He classified the seed plant SAM into two types: (i) *simplex* with the initial cells in one zone (gymnosperms) and (ii) *duplex* with initial cells divided into two zones (angiosperms). Despite Newman's classification, other interspecific differences are present; for example, in the apical cells in Selaginellaceae, cutting faces number can expand from two to four. In the gymnosperm *Ginkgo biloba*, the SAM is divided into five different morphological and functional zones. Those differences in SAM structure suggest that SAMs independently arose several times during vascular plant evolution.

Laser capture microdissections combined with RNA sequencing of the SAM structures of *Selaginella moellendorffii* (Selaginellacaea), *Equisetum arvense* (fern), and *Zea mays* (angiosperm) have shown that several genes involved in meristematic activity display conserved expression (Frank et al. 2015). These data support the idea that the apical cell is functionally equivalent to the SAM. Accordingly, phytohormones such as CKs and GA show a conserved function in SAM activity. In *S. moellendorffii*, upregulation of genes involved in damping GA contents and response in the apical cell suggests a conserved function for GA in inducing cell differentiation during the evolution of shoot apexes (Frank et al. 2015). Similarly, exogenous applications of CK have been shown to be sufficient to promote apical cell fate and callus formation in the bryophyte *Physcomitrella patens* and the fern *C. richardii*, respectively (Frank and Scanlon, 2015; Coudert et al. 2015).

Expression of *WOX* genes is found in all the SAM structures in plants. WOX family complexity increased during evolution according to the increase of SAM complexity. Based on their appearance during plant evolution, *WOX* genes can be classified in three main clades: the ancient clade, the intermediate clade, and the WUS clade. Mosses and green algae display only the ancient clade, lycophytes possess both the ancient and the intermediate clade, whereas ferns, gymnosperms, and angiosperms possess all the three clades (Van der Graaff et al. 2009). In the moss *P. patens*, a *WOX* gene is preferentially expressed in apical cells and could be exploited to mark this cell. Nevertheless, its function is unknown, and CLV homologs are absent from *P. patens* genome, suggesting that the WUS/CLV mechanism is not acting to maintain the apical cell. Similarly, *KNOX* genes are found to be expressed in all the SAM structures of tracheophytes. In *S. moellendorffii* and *S. kraussiana*, *KNOX* genes expression is localized in the apical cell and absent in primordia. The apical cell of *C. richardii* expresses *KNOX* genes, and it was proven that their function is conserved with *A. thaliana*. Overexpression of *C. richardii KNOX* genes generates similar phenotypes to endogenous *KNOX* overexpression in *Arabidopsis* (Harrison et al. 2005; Sano et al. 2005).

6.8.3 Evolution of Root Apical Meristems

The structure of a modern RAM in extant vascular plants (i) is radially symmetric, (ii) forms an apical root cap, (iii) has an SCN with proliferation activity, and (iv) sustains root growth and formation of root hairs from the epidermis outward to the soil, as described above for the angiosperm eudicot *A. thaliana*.

Once plants colonized the land, rhizoid-bearing organs and roots evolved in several clades. Structures that carried out rooting functions were one of the finest adaptations that allowed anchorage, nutrient uptake, water absorption, and colonization of drier land spaces.

A long-standing question in plant evolution is whether the complex meristems of angiosperms (sporophytic roots in extant vascular plants) were derived from the simple meristems of early diverging land plants (such as members of the lycophyte phylum) as the result of millions of years of evolution. An alternative hypothesis is that true roots evolved independently in the major vascular plant branches via converging evolutionary strategies. Several types of organization of RAMs in flowering plants (angiosperms) have been described, based on the patterns of RAM organization in basal angiosperms, monocots, and dicots. The main differences among these several RAM types in angiosperms are based on the presence or the absence of boundaries between tissues, as well as the origin of the epidermal cell file.

In closed meristems, such as those of the eudicot *A. thaliana* or the monocot *Z. mays*, the inner and the outer tissues are separated by a sharp boundary that readily distinguishes the internal vasculature from the external cortex, epidermis, and lateral root cap. On the other side, in open meristems, cells in between the innermost vasculature tissue and the outermost root cap layer divide intermittently, generating neighboring tissues with common origins and unstable boundaries between adjacent tissues. Slightly more angiosperm families have exclusively closed RAMs rather than exclusively open RAMs, but many families have representatives with both open and closed meristems. In both closed and open meristems, the origin of the epidermis is associated either with the cortex (in some basal angiosperms and monocots) or with the lateral root cap (in eudicots such as *A. thaliana* and other basal angiosperms). In dicots, the inner cell layer of the lateral root cap lineage forms the epidermis, while roots of most monocots derive the epidermis as the outer layer of the cortex.

Altogether, it is reasonable to state that the RAM of the eudicot *Arabidopsis* does not represent the norm for angiosperm RAM structures and development.

In a shorter evolutionary scale, root anatomical diversity emerged from specific developmental domains absent in the *Arabidopsis* RAM. For example, the hairy bitter crest *Cardamine hirsuta*, a close relative of *Arabidopsis*, shows two layers of cortex tissues in contrast to the one layer in *Arabidopsis*. The second cortical layer in *Cardamine* derives from a periclinal cell division of the cortex/endodermis initial daughter cell and has both cortical and endodermal identities (Di Ruocco et al. 2018).

In contrast to the lycophytes, euphyllophytes do not branch by bifurcation but from internal tissues (from the endodermis in ferns and from the pericycle in seed plants) proximally to the apical meristem.

The root meristems of gymnosperms are distinctive from the meristems of other vascular plants because of the presence of a broad *promeristem*—the part of the meristem where undifferentiated cells actively undergo mitosis—with common initials for all or the majority of the mature tissues of the root. All mature tissues converge on a broad promeristem that takes the form of an upturned truncated pyramid. Distinct sets of initials for specific tissue types are not specified in gymnosperms. Instead, different types of organizations (generally three) of the initial cells originate different tissues. The three types of initials organization are as follows:

i. a common set of initials for all fundamental tissues or all fundamental tissues except the vasculature tissues,
ii. one set of common initials for the root cap, columella, and the procambium and another set that gives rise to the ground tissue, epidermis, and lateral root cap, and
iii. one set of common initials for ground tissue, epidermis, and lateral root cap and another set of initials for root cap, columella, and procambium.

The earliest rooting structures in vascular plants were broadly equivalent to aerial axes: radially symmetric leafless tubes, which were modified by the presence of tip-growing filamentous cells called *rhizoids*, at the interface between plant and soil. Plagiotropic (horizontal) axis developed unicellular rhizoids (similar to root hairs) on their underside, gradually accommodating polarized structures for rooting function and growth. Therefore, a radial-to-bilateral symmetry transition occurred prior to formation of a functional RAM. These bilaterally symmetrical rhizoid-bearing axes composed the rooting structure of the common ancestor of vascular plants, as documented by fossil preserved in the Rhynie chert in Scotland (Hetherington and Dolan, 2018). The oldest meristems of rooting axes, called rhizomes, belonged to the lycopsid *Asteroxylon mackiei* and date to the Early Devonian, 407-million-year old (Hetherington et al 2016; Hetherington and Dolan, 2017). Rhizomes of *A. mackiei* had radially symmetric axes that branched dichotomously, similar to the roots of extant lycophytes. However, unlike any other root meristems, the rooting axes of *A. mackiei* lacked a root cap and instead developed a continuous hairless layer of epidermis over the surface of the meristem.

The rooting axes and meristems of *A. mackiei* were unique among vascular plants, highlighting that roots evolved multiple times rather than having a single origin. In fact, roots are not homologous in vascular plants but had at least two separate evolutionary origins: once in lycophytes and independently in the euphyllophytes (ferns, gymnosperms, and angiosperms). The oldest example of active meristem of euphyllophytes belongs to *Radix carbonica*, an extinct gymnosperm that lived during the Carboniferous period (350–300 million years ago) (Hetherington et al. 2016). Fossil records showed that the promeristem of *R. carbonica* was quite large consisting of more than 100 cells arranged in more than ten tiers. All the mature cell files in this promeristem originate from only two sets of initials, and the lateral root cap developed discontinuously around the promeristem. Many anticlinal cell divisions are found in the columella-like promeristem, highlighting structural differences between old extinct and the modern extant root meristems in gymnosperms. Despite the aforementioned differences, the overall organization of stem cells and differentiating cells observed in *R. carbonica* suggests that the oldest root meristem of gymnosperms had a self-renewing populations and cellular dynamics similar to that observed in extant plants (Hetherington et al. 2016).

6.9 CONCLUSIONS

During plant exploration and colonization of the newly emerged land, the strategic innovation of self-renewal, totipotent, cells allowed a great adaptation and diversification of plant species. Stem cells guarantee highly plastic developmental strategies for sessile plants to address new challenges and cope with different environments and climate conditions, biotic competitors, and abiotic stresses, as well as mechanical growth constraints. The organization of plant meristems and the dynamics of their stem cells highlight how the molecular mechanisms for stem cell fate determination and activity are general and conserved, among extinct and extant plant species as well as when compared to the animal kingdom.

REFERENCES

Aida, M., Beis, D., Heidstra, R., Willemsen, V., Blilou, I., Galinha, C., Nussaume, L., Noh, Y. S. et al. 2004. The PLETHORA genes mediate patterning of the *Arabidopsis* root stem cell niche. *Cell* 119: 109–120.

Balkunde, R., Kitagawa, M., Xu, X. M., Wang, J., and Jackson, D. 2017. SHOOT MERISTEMLESS trafficking controls axillary meristem formation, meristem size and organ boundaries in *Arabidopsis*. *Plant J.* 90: 435–446.

Barton, M. K. 2010. Twenty years on: the inner workings of the shoot apical meristem, a developmental dynamo. *Dev. Biol.* 341: 95–113.

Blilou, I., Xu, J., Wildwater, M., Willemsen, V., Paponov, I., Friml, J., Heidstra, J., Aida, M., Palme, K., and Scheres, B. 2005. The PIN auxin efflux facilitator network controls growth and patterning in *Arabidopsis* roots. *Nature* 433: 39.

Brand, U., Fletcher, J. C., Hobe, M., Meyerowitz, E. M., and Simon, R. 2000. Dependence of stem cell fate in *Arabidopsis* on a feedback loop regulated by CLV3 activity. *Science* 289: 617–619.

Brueninger, H., Rikirsch, E., Hermann, M., Ueda, M., and Laux, T. 2008. Differential expression of WOX genes mediates apical-basal axis formation in the *Arabidopsis* embryo. *Dev. Cell* 14: 867–876.

Brunoud, G., Wells, D. M., Oliva, M., Larrieu, A., Mirabet, V., Burrow, A. H., Beeckman, T., Kepinski, S. et al. 2012. A novel sensor to map auxin response and distribution at high spatio-temporal resolution. *Nature* 482: 103–106.

Coudert, Y., Palubicki, W., Ljung, K., Novak, O., Leyser, O., and Harrison, C. J. 2015 Mar 25. Three ancient hormonal cues co-ordinate shoot branching in a moss. *Elife* 4: e06808. doi: 10.7554/eLife.06808.

Daum, G., Medzihradsky, A., Suzaki, T., and Lohmann, J. U. 2014. A mechanistic framework for non-cell autonomous stem cell induction in *Arabidopsis*. *Proc. Natl. Acad. Sci. U S A* 111: 14619–14624.

Dello Ioio, R., Galinha, C., Fletcher, A. G., Grigg, S. P., Molnar, A., Willemsen, V., Scheres, B., Sabatini, S. et al. 2012. A PHABULOSA/cytokinin feedback loop controls root growth in *Arabidopsis*. *Curr. Biol.* 22: 1699–1704.

Dello Ioio, R., Linhares, F. S., Scacchi, E., Casamitjana-Martinez, E., Heidstra, R., Costantino, P., and Sabatini, S. 2007. Cytokinins determine *Arabidopsis* root-meristem size by controlling cell differentiation. *Curr. Biol.* 17: 678–682.

Dello Ioio, R., Nakamura, K., Moubayidin, L., Perilli, S., Taniguchi, M., Morita, M. T., Aoyama, T., Costantino, P., and Sabatini, S. 2008. A genetic framework for the control of cell division and differentiation in the root meristem. *Science* 322: 1380–1384.

Di Mambro, R., De Ruvo, M., Pacifici, E., Salvi, E., Sozzani, R., Benfey, P. N., Busch, W., Novak, O. et al. 2017. Auxin minimum triggers the developmental switch from cell division to cell differentiation in the *Arabidopsis root*. *Proc. Natl. Acad. Sci. U S A* 114: E7641–E7649.

Ding, Z., and Friml, J. 2010. Auxin regulates distal stem cell differentiation in *Arabidopsis* roots. *Proc. Natl. Acad. Sci. U S A* 107: 12046–12051.

Di Ruocco, G., Bertolotti, G., Pacifici, E., Polverari, L., Tsiantis, M., Sabatini, S., Costantino, P., and Dello Ioio, R. 2018. Differential spatial distribution of miR165/6 determines variability in plant root anatomy. *Development* 145: dev153858.

Dolan, L., Janmaat, K., Willemsen, V., Linstead, P., Poethig, S., Roberts, K., and Scheres, B. 1993. Cellular organisation of the *Arabidopsis thaliana* root. *Development* 119: 71–84.

Dubrovsky, J. G., Sauer, M., Napsucialy-Mendivil, S., Ivanchenko, M. G., Friml, J., Shishkova, S., Celenza, J., and Benková, E. 2008. Auxin acts as a local morphogenetic trigger to specify lateral root founder cells. *Proc. Natl. Acad. Sci. U S A* 105: 8790–8794.

Etchells, J. P., Smit, M. E., Gaudinier, A., Williams, C. J., and Brady, S. M. 2016. A brief history of the TDIF–PXY signalling module: balancing meristem identity and differentiation during vascular development. *New Phytol.* 209: 474–484.

Fletcher, J. C., Brand, U., Running, M. P., Simon, R., and Meyerowitz, E. M. 1999. Signaling of cell fate decisions by CLAVATA3 in *Arabidopsis* shoot meristems. *Science* 283: 1911–1914.

Frank, M. H., Edwards, M. B., Schultz, E. R., McKain, M. R., Fei, Z., Sørensen, I., Rose, J. K., and Scanlon, M. J. 2015. Dissecting the molecular signatures of apical cell-type shoot meristems from two ancient land plant lineages. *New Phytol.* 207: 893–904.

Frank, M. H., and Scanlon, M. J. 2015. Transcriptomic evidence for the evolution of shoot meristem function in sporophyte-dominant land plants through concerted selection of ancestral gametophytic and sporophytic genetic programs. *Mol. Biol. Evol.* 32: 355–367.

Galinha, C., Hofhuis, H., Luijten, M., Willemsen, V., Blilou, I., Heidstra, R., and Scheres, B. 2007. PLETHORA proteins as dose-dependent master regulators of *Arabidopsis* root development. *Nature* 449: 1053.

Graham, L. E., and Wilcox, L. W. 2000. The origin of alternation of generations in landplants: a focus on matrotrophy and hexose transport. *Philos. Trans. R. Soc. Lond B* 355: 757–767.

Grieneisen, V. A., Xu, J., Marée, A. F., Hogeweg, P., and Scheres, B. 2007. Auxin transport is sufficient to generate a maximum and gradient guiding root growth. *Nature* 449: 1008–1013.

Haecker, A., Gross-Hardt, R., Geiges, B., Sarkar, A., Breuninger, H., Herrmann, M., and Laux, T. 2004. Expression dynamics of WOX genes mark cell fate decisions during early embryonic patterning in *Arabidopsis thaliana*. *Development* 131: 657–668.

Harrison, J., Möller, M., Langdale, J., Cronk, Q., and Hudson, A. 2005. The role of KNOX genes in the evolution of morphological novelty in *Streptocarpus*. *Plant Cell* 17: 430–443.

Heidstra, R., and Sabatini, S. 2014. Plant and animal stem cells: similar yet different. *Nat. Rev. Mol. Cell Biol.* 15: 301–312.

Hetherington, A. J., and Dolan, L. 2017. Bilaterally symmetric axes with rhizoids composed the rooting structure of the common ancestor of vascular plants. *Philos. Trans. R. Soc.,B* 373. doi: 10.1098/rstb.2017.0042.

Hetherington, A. J., and Dolan, L. 2018. Stepwise and independent origins of roots among land plants. *Nature* 561: 235.

Hetherington, A. J., Dubrovsky, J. G., and Dolan, L. 2016. Unique cellular organization in the oldest root meristem. *Curr. Biol.* 26: 1629–1633.

Hirakawa, Y., Kondo, Y., and Fukuda, H. 2010. TDIF peptide signaling regulates vascular stem cell proliferation via the WOX4 homeobox gene in *Arabidopsis*. *Plant Cell* 22: 2618–2629.

Hirakawa, Y., Shinohara, H., Kondo, Y., Inoue, A., Nakanomyo, I., Ogawa, M., Sawa, S., Ohashi-Ito, K. et al. 2008. Non-cell-autonomous control of vascular stem cell fate by a CLE peptide/receptor system. *Proc. Natl. Acad. Sci. U S A* 105: 15208–15213.

Imaichi, R., and Kato, M. 1989. Developmental anatomy of the shoot apical cell, rhizophore and root of *Selaginella uncinata*. *Bot. Mag. Tokyo* 102: 369–380.

Jasinski, S., Piazza, P., Craft, J., Hay, A., Woolley, L., Rieu, I., Phillips, A., Hedden, P., and Tsiantis, M. 2005. KNOX action in *Arabidopsis* is mediated by coordinate regulation of cytokinin and gibberellin activities. *Curr. Biol.* 15: 1560–1565.

Laufs, P., Grandjean, O., Jonak, C., Kiêu, K., and Traas, J. 1998. Cellular parameters of the shoot apical meristem in *Arabidopsis*. *Plant Cell* 10: 1375–1389.

Laux, T., Mayer, K. F., Berger, J., and Jürgens, G. 1996. The WUSCHEL gene is required for shoot and floral meristem integrity in *Arabidopsis*. *Development* 122: 87–96.

Leibfried, A., To, J. P., Busch, W., Stehling, S., Kehle, A., Demar, M., Kieber, J. J., and Lohmann, J. U. 2005. WUSCHEL controls meristem function by direct regulation of cytokinin–inducible response regulators. *Nature* 438: 1172–1175.

Liu, T., Ohashi-Ito, K., and Bergmann, D. C. 2009 Orthologs of *Arabidopsis thaliana* stomatal bHLH genes and regulation of stomatal development in grasses. *Development* 136: 2265–2276.

Liu, W., and Xu, L. 2018. Recruitment of IC-WOX genes in root evolution. *Trends Plant Sci.* 23: 490–496.

Long, J. A., Moan, E. I., Medford, J. I., and Barton, M. K. 2006. A member of the KNOTTED class of homeodomain proteins encoded by the SHOOTMERISTEMLESS gene of *Arabidopsis*. *Nature* 379: 66–69.

Losick, V. P., Morris, L. X., Fox, D. T., and Spradling, A. 2011. *Drosophila* stem cell niches: a decade of discovery suggests a unified view of stem cell regulation. *Dev. Cell* 21: 159–171.

MacAlister, C. A., Ohashi-Ito, K., and Bergmann, D. C. 2007. Transcription factor control of asymmetric cell divisions that establish the stomatal lineage. *Nature* 445: 537–540.

Mähönen, A. P., Ten Tusscher, K., Siligato, R., Smetana, O., Díaz-Triviño, S., Salojärvi, J., Wachsman, G., Prasad, K. et al. 2014. PLETHORA gradient formation mechanism separates auxin responses. *Nature* 515: 125–129.

Mayer, K. F., Schoof, H., Haecker, A., Lenhard, M., Jürgens, G., and Laux, T. 1998. Role of WUSCHEL in regulating stem cell fate in the *Arabidopsis* shoot meristem. *Cell* 95: 805–815.

Mauseth, J. D. 2017. *Botany: An Introduction to Plant Biology*. Burlington, Massachusetts, Jones and Bartlett Publishers.

Moubayidin, L., Di Mambro, R., Sozzani, R., Pacifici, E., Salvi, E., Terpstra, I., Bao, D., van Dijken, A. et al. 2013. Spatial coordination between stem cell activity and cell differentiation in the root meristem. *Dev. Cell* 26: 405–415.

Newman, I. V. 1965. Pattern in the meristems of vascular plants: III. Pursuing the patterns in the apical meristem where no cell is a permanent cell. *J. Linn. Soc. London, Bot.* 59: 185–214.

Ogawa, M., Shinohara, H., Sakagami, Y., and Matsubayashi, Y. 2008. *Arabidopsis* CLV3 peptide directly binds CLV1 ectodomain. *Science* 319: 294.

Ohashi-Ito, K., and Bergmann, D. C. 2006. *Arabidopsis* FAMA controls the final proliferation/differentiation switch during stomatal development. *Plant Cell 18*: 2493–2505.

Perales, M., Rodriguez, K., Snipes, S., Yadav, R. K., Diaz-Mendoza, M., and Reddy, G. V. 2016. Threshold-dependent transcriptional discrimination underlies stem cell homeostasis. *Proc. Natl. Acad. Sci. USA* 113: E6298–E6306.

Peterson, K. M., Rychel, A. L., and Torii, K. U. 2010. Out of the mouths of plants: the molecular basis of the evolution and diversity of stomatal development. *Plant Cell* 22:296–306.

Petersson, S. V., Johansson, A. I., Kowalczyk, M., Makoveychuk, A., Wang, J. Y., Moritz, T., Grebe, M., Benfey, P. N. et al. 2009. An auxin gradient and maximum in the *Arabidopsis* root apex shown by high-resolution cell-specific analysis of IAA distribution and synthesis. *Plant Cell* 21: 1659–1668.

Pillitteri, L. J., Sloan, D. B., Bogenschutz, N. L., and Torii, K. U. 2007. Termination of asymmetric cell division and differentiation of stomata. *Nature* 445: 501–505.

Sabatini, S., Beis, D., Wolkenfelt, H., Murfett, J., Guilfoyle, T., Malamy, J., Benfey, P., Leyser, O. et al. 1999. An auxin-dependent distal organizer of pattern and polarity in the *Arabidopsis* root. *Cell* 99: 463–472.

Sabatini, S., Heidstra, R., Wildwater, M., and Scheres, B. 2003. SCARECROW is involved in positioning the stem cell niche in the *Arabidopsis* root meristem. *Genes Dev.* 17: 354–358.

Sanders, H. L., and Langdale, J. A. 2013. Conserved transport mechanisms but distinct auxin responses govern shoot patterning in *Selaginella kraussiana*. *New Phytol.* 198: 419–428.

Sano, R., Juárez, C. M., Hass, B., Sakakibara, K., Ito, M., Banks, J. A., and Hasebe, M. 2005. KNOX homeobox genes potentially have similar function in both diploid unicellular and multicellular meristems, but not in haploid meristems. *Evol. Dev.* 7: 69–78.

Santuari, L., Sanchez-Perez, G. F., Luijten, M., Rutjens, B., Terpstra, I., Berke, L., Gorte, M., Prasad, K. et al. 2016. The PLETHORA gene regulatory network guides growth and cell differentiation in *Arabidopsis* roots. *Plant Cell* 28: 2937–2951.

Sarkar, A. K., Luijten, M., Miyashima, S., Lenhard, M., Hashimoto, T., Nakajima, K., Scheres, B., Heidstra, R., and Laux, T. 2007. Conserved factors regulate signalling in *Arabidopsis thaliana* shoot and root stem cell organizers. *Nature* 446: 811–814.

Scheres, B. 2008. Stem-cell niches: nursery rhymes across kingdoms. *Nat. Rev. Mol. Cell Biol.* 8: 345–354.

Schofield, R. 1978. The relationship between the spleen colony-forming cell and the haemopoietic stem cell. *Blood Cells* 4: 7–25.

Schoof, H., Lenhard, M., Haecker, A., Mayer, K. F., Jürgens, G., and Laux, T. 2000. The stem cell population of *Arabidopsis* shoot meristems in maintained by a regulatory loop between the CLAVATA and WUSCHEL genes. *Cell* 100: 635–644.

Simmons, A. R., and Bergmann, D. C. 2016. Transcriptional control of cell fate in the stomatal lineage. *Curr. Opin. Plant Biol.* 29: 1–8.

Spicer, R. and Groover, A. 2010. Evolution of development of vascular cambia and secondary growth. *New Phytol.* 186: 577–592.

Steeves, T. A., and Sussex, I. M. 1989. *Patterns in Plant Development*, 2nd edn, Cambridge, Cambridge University Press.

Stoeckle, D., Thellmann, M., and Vermeer, J. E. 2018. Breakout-lateral root emergence in *Arabidopsis thaliana*. *Curr. Opin. Plant Biol.* 41: 67–72.

Ten Hove, C. A., Lu, K. J., and Weijers, D. 2015. Building a plant: cell fate specification in the early *Arabidopsis* embryo. *Development* 142: 420–430.

Terskikh, A. V., Bryant, P. J., and Schwartz, P. H. 2006. Mammalian stem cells. *Pediatr. Res.* 59(4 Pt 2): 13R–20R.

van den Berg, C., Willemsen, V., Hage, W., Weisbeek, P., and Scheres, B. 1995. Cell fate in the *Arabidopsis* root meristem determined by directional signalling. *Nature* 378: 62–65.

van den Berg, C., Willemsen, V., Hendriks, G., Weisbeek, P., and Scheres, B. 1997. Short-range control of cell differentiation in the *Arabidopsis* root meristem. *Nature* 390: 287–289.

van der Graaff, E., Laux, T., and Rensing, S. A. 2009. The WUS homeobox-containing (WOX) protein family. *Genome Biol.* 10: 248.

Vermeer, J. E., von Wangenheim, D., Barberon, M., Lee, Y., Stelzer, E. H., Maizel, A., and Geldner, N. 2014. A spatial accommodation by neighboring cells is required for organ initiation in *Arabidopsis*. *Science* 343: 178–183.

Waddington, C. H. 1957. *The Strategy of the Genes*. London, Geo Allen and Unwin.

Wolpert, L. 2011. Positional information and patterning revisited. *J. Theor. Biol.* 269: 359–365.

Xu, X. M., Wang, J., Xuan, Z., Goldshmidt, A., Borrill, P. G., Hariharan, N., Kim, J. Y., and Jackson, D. 2011. Chaperonins facilitate KNOTTED1 cell-to–cell trafficking and stem cell function. *Science* 333: 1141–1144.

Yadav, R. K., Perales, M., Gruel, J., Girke, T., Jönsson, H., and Reddy, G. V. 2011. WUSCHEL protein movement mediates stem cell homeostasis in the *Arabidopsis* shoot apex. *Genes Dev.* 25: 2025–2030.

Yanai, O., Shani, E., Dolezal, K., Tarkowski, P., Sablowski, R., Sandberg, G., Samach, A., and Ori, N. 2005. *Arabidopsis* KNOXI proteins activate cytokinin biosynthesis. *Curr. Biol.* 15: 1566–1571.

Zhao, L., and Sack, F. D. 1999. Ultrastructure of stomatal development in *Arabidopsis* (Brassicaceae) leaves. *Am. J. Bot.* 86: 929–939.

Zhou, Y., Liu, X., Engstrom, E. M., Nimchuk, Z. L., Pruneda-Paz, J. L., Tarr, P. T., Yan, A., Kay, S. A., and Meyerowitz, E. M. 2015. Control of plant stem cell function by conserved interacting transcriptional regulators. *Nature* 517: 377–380.

Zhou, Y., Yan, A., Han, H., Li, T., Geng, Y., Liu, X., and Meyerowitz, E. M. 2018. HAIRY MERISTEM with WUSCHEL confines CLAVATA3 expression to the outer apical meristem layers. *Science* 361: 502–506

7 Planarian Neoblasts

Nondeferred, Multipurpose Stem Cells for Body Homeostasis, Growth, Degrowth, and Regeneration

Jaume Baguñà
Universitat de Barcelona

CONTENTS

7.1 STEM CELL PUZZLES

Stem cells (hereafter SCs) are small morphologically undifferentiated cells with large nuclear/cytoplasmic ratios and a scant and poorly differentiated cytoplasm. Their main properties are a potential to go through numerous cycles of cell division while maintaining their undifferentiated state (*self-renewal*) and the potential to differentiate into one or more differentiated cell types (*potency* or potential). Consequently, SCs are key elements that replenish dying differentiated cells during daily wear and tear (homeostasis) and replace/regenerate damaged tissues, organs, and entire body regions (regeneration).

All these features define the SCs we are most familiar with, which are SCs in adult organisms (adult stem cells: ASCs; also known as somatic SCs) such as the hematopoietic SCs, intestinal crypt stem cells (ISCs), epithelial SCs, and any of the SCs that renew several tissues and organs of most animals and plants. In kinetic terms, the most accepted model of cell renewal (Figure 7.1A) sees ASCs at the top of a hierarchy dividing asymmetrically into two cells, one of which maintains the ASC nature of the mother cell, the other of which becomes a transit amplifying (TA) or precursor cell. In turn, TA cells divide symmetrically to expand the number of cells of a particular lineage until they undergo terminal differentiation. Some behave as new SCs producing new TA cells with successively narrower potential (multi-, oligo-, bi-, and unipotency) before terminal differentiation.

7.1.1 Stem Cell Differentiation Can Be Reversed by Dedifferentiation, Transdetermination, and Transdifferentiation

Under normal homeostasis, renewal of tissues and organs proceeds unidirectionally: ASCs produce precursor (TA) cells, which in turn produce new differentiated cells that replace the old, dying, damaged cells (Figure 7.1). In morphologically simple organs such as the liver or pancreas, proliferation of resident differentiated cells does the job. However, under special circumstances (such as regeneration after asexual reproduction, amputation or injury, cell culture, stress, or poisoning), new tissues can be formed from differentiated cells that dedifferentiate (lose the characteristics of their origin), become undifferentiated, and after some proliferation (or not) redifferentiate into the original cell type. In some cases, dedifferentiated cells can change their determined state (transdetermination) and redifferentiate into a cell type different from the original (transdifferentiation). And in very rare cases, differentiated cells can differentiate directly into another cell type without apparent dedifferentiation; this is direct transdifferentiation or lineage switch, often known as metaplasia. These pathways all are shown and discussed in Figure 7.2.

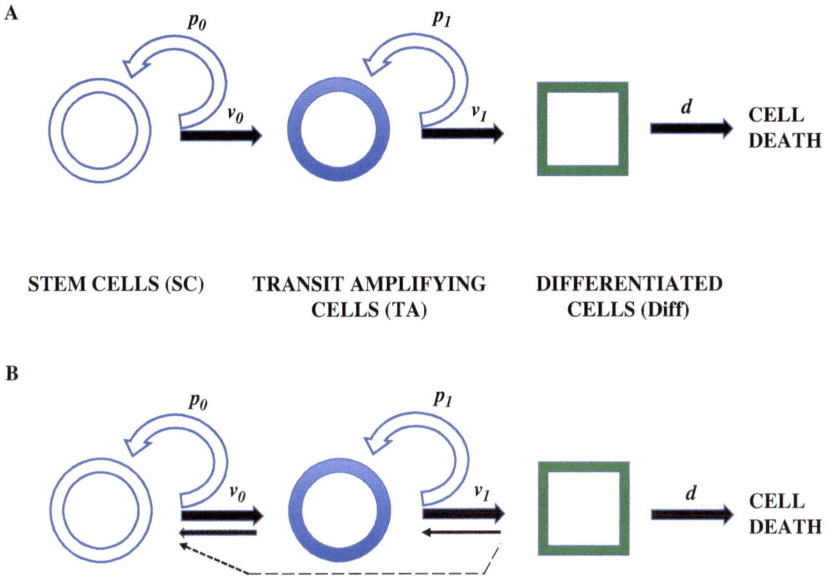

FIGURE 7.1 Diagrammatic representations of a simple unbranched two-stage cell lineage that starts with a stem cell (SC), progresses through a transit-amplifying (TA) cell compartment and ends as a postmitotic differentiated cell (Diff). Turnover of the Diff compartment is represented by a constant rate of cell death (d). Parameters v_0–v_1 and p_0–p_1 are the rate constants of cell cycle progression and the replication probabilities, respectively. For the sake of simplicity, no control mechanisms are depicted. (A) The "classical" directional, linear, and irreversible model (black arrows) that does not include reversals by dedifferentiation (modified after Lander 2009). (B) The most accepted model today incorporates reverse transitions from Diff to TA and TA to SC (thin solid black arrows) and even from Diff to SC (thin broken black arrow).

In several regenerative systems, dedifferentiation, redifferentiation, and transdifferentiation have been known for almost a century (reviewed in Okada 1991). Examples include limb amputation and regeneration in urodeles and axolotls, tail self-amputation and regeneration in lizards, and replacement of lens epithelial cells in salamanders, retinal pigment epithelial cells in chicken, and Schwann cells in birds and mammals (reviewed in Raff 2003). Of note, in nonbilaterian and simple bilaterian organisms, besides regeneration, dedifferentiation, and transdifferentiation, occur in normal daily homeostasis. Sponges and cnidarians (*Hydra* and *Hydractinia*) switch cell types (archeocytes to choanocytes and vice versa in sponges; Funayama 2013, and epithelial cell types in *Hydra*) during normal cell renewal. In flatworms (triclad Platyhelminthes), transdetermination from primordial germ cells and early germ cells to totipotent ASCs or neoblasts and from these to somatic cells were reported back in the 1980s (see Gremigni 1988 for a review). And in *Drosophila*, serially transplanting regenerating imaginal discs (set-aside larval cells) after several generations produced transdetermination of groups of imaginal discs cells into cells of other discs (Hadorn 1968).

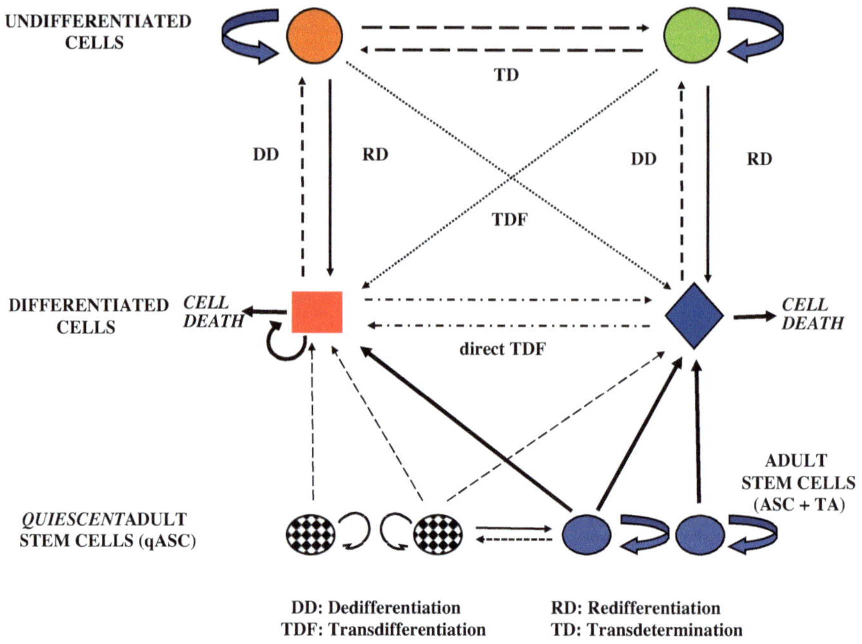

FIGURE 7.2 Diagrammatic representations depicting reported (proven and very likely) cell-cell transformations during tissue homeostasis and regeneration. In the middle tier, two differentiated cell types (red rectangle and blue diamond) represent most of the body mass kept in steady-state balancing cell losses due to cell death with the entrance of new cells differentiating from adult stem (ASC) and from transit-amplifying (TA) cells (lower tier, solid light blue ovals), which maintain themselves by cell proliferation (curved light blue arrows). The upper tier depicts reported transformations usually occurring during regeneration and in stressful conditions: dedifferentiation (DD; dashed black arrows) to undifferentiated cells (orange and green circles), followed by proliferation (light blue curved arrows), and redifferentiation (RD; solid black arrows) to the same cell type. After dedifferentiation and proliferation, cells might change their determination state (TD: transdetermination; long dashed black arrow) and redifferentiate to different cell types (TDF: transdifferentiation; thin point black arrows). Only rarely, differentiated cells transdifferentiate directly (without dedifferentiation and proliferation) into other cell types (direct TDF; dash and point black arrows), while some cells (liver, pancreas, and others) proliferate directly (curved black arrows in the middle-tier red rectangle). In the lower tier, the existence of rare, quiescent, adult stem cells (SCs) (checkered ovals) with low renewal potential (thin curved black arrows) is represented. When needed, these cells may produce adult SCs (thin black arrow) and, likely, could differentiate through TA cells into differentiated cells (thin long dash black arrows). The passage from ASC to quiescent cells (thin broken black arrow) is also contemplated.

In summary, the picture painted above strongly indicates that many precursor and some differentiated cells are not as irreversibly committed to a particular pathway of development as originally believed. The potential of precursor (committed or TA) cells to move up and down the hierarchy of differentiation is higher than previously thought (Figure 7.1B)

7.1.2 Terminology: Facts and Hypes

Surrogate, cache, reserve, deferred, set-aside, and quiescent cells are terms used in development and regeneration to refer to the presence and behavior of cells, often different than SCs, which perform some of the roles of SCs (Shostak 2008). *Surrogate cells* are differentiated cells present in static organs and populations that take the role of SCs (e.g., adult liver hepatocyte in the bile duct; adult endocrine, duct and acinar pancreatic cells) restoring damaged structures by limited proliferation. This is also called compensatory hyperplasia, with no relation to proper ASCs.

Reserve, cache, deferred, and set-aside cells are synonyms that refer to undifferentiated cells kept either in embryos or larvae, to be used later in ontogeny to produce adult structures. Examples of deferred and set-aside cells are the imaginal discs in the larvae of *Drosophila*, nemertean worms, and sea urchins. These cells, however, are not ASCs. *Drosophila* imaginal disc cells are undifferentiated but determined (committed) larval cells that produce specific structures of the adult (wing, legs, antennae). They divide symmetrically to produce many cells that later differentiate into a specific adult structure made by dozens of different cell types. They are more akin to precursor (TA amplifying cells) than to ASCs.

Instead, the terms reserve and cache cells have been repeatedly used to name "waiting" cells to be called and used as ASCs to produce all or most somatic cell types during regeneration—planarian neoblasts (Dubois 1949), sponge archeocytes (Funayama 2010), i-cells from hydrozoan cnidarians (Gahan et al. 2016)—or as the ultimate ASCs in most lineages of adult vertebrates for their daily homeostasis and for regeneration after trauma or radiation. Such cells are rare in number and divide infrequently, and when they do it, they divide asymmetrically, which led to them being considered as quiescent cells. Besides those mentioned above, examples of quiescent reserve cells are spermatogonia that becomes mitotically active only after damage to the seminiferous tubules; muscle satellite cells that become active following trauma or regeneration; hair follicle bulge cells and rare $Tert^+$ ISCs close to the crypt base activated after irradiation or SCs ablation; and cells making the quiescent center of the shoot apical, root apical, and cambium meristems in many plants (Greb and Lohmann 2016).

7.1.3 Where Do We Stand Now?

The SC field is in a state of flux. The traditional view of a one-way road from adult SCs to precursor cells to the entire spectrum of mature cell types has been replaced by a more versatile and diffuse one where cells gradually lose their ability to revert to SC fates (Lander 2009). And, as it is common in plants, differentiated animal cells may revert to undifferentiated ASCs (Figure 7.1B). While in normal homeostasis, the traditional view likely holds true, in organisms, organs, and tissues under stress, SCs could be recruited from TA cells, even from fully differentiated cells, and in very specific cases, differentiated cells proliferate directly to replace lost tissues (Figure 7.2). The mechanisms that underlie such variety, the organisms in which these mechanisms operate, and some sense of known phylogenetic trends are explored in the remainder of this chapter.

7.2 PLANARIANS: THE CELLULAR BASIS OF HOMEOSTASIS, REGENERATION, AND GROWTH-DEGROWTH

Planarians are bilaterally symmetric, triploblastic, unsegmented, acoelomate worms of the order Tricladida of the phylum Platyhelminthes (the flatworms) (Sluys et al. 2009). They have a clear anteroposterior polarity (head-tail) and usually are dorsoventrally flattened. They lack circulatory, respiratory, or skeletal structures. Adult planarians range in size from 4 to 5 mm in length and up to 1 m of some land planarians. They are hermaphrodites with cross-fertilization and sexual, asexual, and mixed (sexual/asexual, usually seasonal) modalities of reproduction. The space between the monolayered epidermis and the blind gut is filled by a solid mass of rather unstructured tissues and cells, called parenchyma or mesenchyme, made by different cell types, within which organs are embedded.

7.2.1 Cell Types and Kinetic Compartments

Planarians are made up by 15–20 cell types as seen under optical microscopy after cell maceration (Baguñà and Romero 1981) and up to 30–40 when single-cell sequencing is used (Plass et al. 2018). In kinetic and functional terms, planarians contain two cell compartments:

1. A *proliferative compartment* formed by a population of small, undifferentiated, heterogeneous albeit morphologically identifiable toti-, pluri-, and multipotent stem cells or neoblasts (approximately 20%–35% of the total cell number) that differentiate into all the differentiated cell types while maintaining their own density by cell proliferation (Baguñà 1981, 2012; reviewed in Baguñà et al. 1990).
2. *A functional compartment* made by 20–25 nonproliferating differentiated cell types (approximately 65%–80% of total cells) that turn over continuously during life and are replaced by differentiating neoblasts (Baguñà and Romero 1981; Romero and Baguñà 1991)

Transitions between these compartments are shown in Figure 7.3 and elaborated in the legend.

7.2.2 Regeneration

To most people, planarians are best known for the astounding feats of some species to regenerate a whole body from almost any tiny portion of it (down to 30–40,000 cells, or less than 0.5 mm in length). When a planarian is transversally cut along the A-P (anterior-posterior, or head to tail) axis, the epithelium closes the wound, and after 1 hour a thin film of epidermal cells from the stretched old epidermis covers it. Below the wound epithelium, groups of undifferentiated cells aggregate to form a regenerative blastema, which does not proliferate but increases in size by addition of new undifferentiated cells, such as neoblasts, produced by active cell proliferation in the old stump (Baguñà 1976; Saló and Baguñà 1984). Unidentified

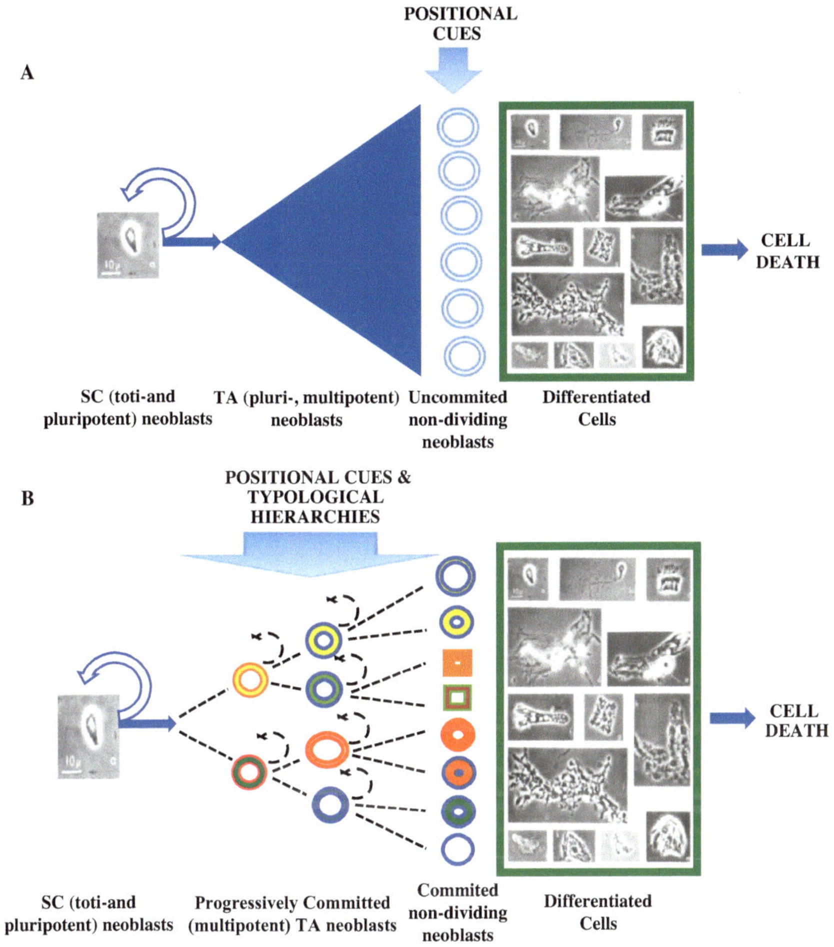

FIGURE 7.3 Alternative models of neoblast lineages from toti-, pluripotent stem cell (SC) neoblasts to terminal differentiated somatic cells, through an intermediate transit-amplifying (TA) compartment of proliferating neoblasts (light blue circle). Blue arrows indicate transit between each compartment; curved open arrows indicate proliferation. (A) Toti- and pluripotent SC neoblasts produce a TA compartment of equivalent proliferating uncommitted progenitor neoblasts and a final set of uncommitted nondividing precursor neoblasts (blue open double circles) from which, according to positional (topographical) cues (blue wide arrow), all differentiated cell types are produced (within the green square, cell type pictures after cell maceration; from Baguñà and Romero 1981). To the right, blue arrow and cell death means the continuous loss of differentiated cells. (B) An alternative model with a TA compartment made by proliferating neoblasts progressively committed to specific intermediate progenitor lineages (thin broken lines), likely proliferative (thin curved broken arrows), driven by autonomous (typological hierarchies) and nonautonomous (positional) mechanisms/cues. The final result is a large compartment of committed nondividing precursor neoblasts that give rise to specific differentiated cells (within the green square, cell type pictures after cell maceration). (Highly modified from Baguñà et al. 1990.)

wound-specific signals induce a first mitotic increase in neoblasts close to the wound peaking at 2–6 h (Saló and Baguñà 1984; Wenemoser and Reddien 2010). Such a very fast mitotic increase, also seen after feeding, suggests slowly cycling neoblasts in the G2 phase of the cell cycle (Baguñà 1974; Saló and Baguñà 1984; Rossi et al. 2008; Wenemoser and Reddien 2010).

After a relative minimum at 1 day, a second, higher, and temporally more sustained peak holds between 2 and 4 days. In parallel, starting within 24–40 h of regeneration, new structures are determined within the blastema and postblastema areas following a distal-proximal sequence shown using several grafting techniques (reviewed in Baguñà et al. 1990, 1994) and molecular markers (see Rojo-Laguna et al. 2019; Reddien 2018 for thorough reviews). The lost structural pattern is restored, and normal body proportions attained at 3–4 weeks of regeneration by cell proliferation, cell differentiation, and cell loss. This process of rescaling or body remodeling was earlier known as morphallaxis.

7.2.3 Homeostasis, Growth, and Degrowth

Besides regeneration, planarians are known for a much greater plasticity both in the growth of an individual and in its final size, and an enduring capacity to degrow in size/volume/number of cells when starved and to grow again when fed.

In steady state, cell proliferation and cell loss are always ongoing, are species specific, and depend on body length/size, nutritional state, and temperature (Baguñà and Romero 1981). During degrowth, planarians shrink from an adult size to or below their initial size at hatching. Degrowth results from more cells (differentiated cells and neoblasts) being lost than are born because of low proliferation due to lack of food. In turn, growth in size/volume is the result of more neoblasts being born by proliferation due to feeding than differentiated cells lost due to wear and tear (Baguñà 1974, 1976, 2012; Baguñà and Romero 1981; Baguñà et al. 1990). Rates of body growth and degrowth, cell proliferation, and cell death are species specific and depend on nutritional state, reproductive status, temperature, and body size (for further details, see Section 7.3.3)

Daily homeostasis and regeneration have a common cellular basis in neoblasts, which are central to both processes. If neoblasts are depleted by irradiation, cell proliferation stops, cell turnover and regeneration no longer hold, and the organism degenerates and fades away in 2–3 weeks.

7.3 PLANARIAN NEOBLASTS: MULTIPURPOSE STEM CELLS

7.3.1 Neoblast Features and Molecular Constituents

First described more than a century ago (Randolph 1897), the ins and outs of planarian neoblasts as daily SC and regenerative cell had a meandering history (reviewed in Baguñà 2012).

From 2011 (Wagner et al. 2011, refining previous work by Baguñà et al. 1989), neoblasts were accepted as the only dividing cells in planarians; they are the only cells incorporating DNA precursors, expressing cell division markers, and repopulating irradiated organisms when single cells are injected, restoring homeostasis and regenerative capacity (Wagner et al. 2011).

Besides components of cell division machinery, other genes expressed or required for neoblast function are a large number of evolutionary conserved proteins that make up the bulk of chromatoid bodies (CBs), which are large extranuclear ribonucleoprotein particles (Coward 1974; Auladell et al. 1993), and a large list of poorly known chromatin modifiers (Rink 2018). Of particular interest are scores of overexpressed RNA-binding proteins all involved in transcriptional repression and storage of mRNAs and posttranscriptional gene regulations (reviewed in Voronina et al. 2011; and Alié et al. 2015). Members of the *Piwi* family (*Vasa*, *Bruno*, *Ddx6*, *Mago-nashi*, *Pl-10*, and genes encoding the *Tudor* domain-containing proteins) are RNA-binding proteins with roles from transposon silencing to scaffold building and RNA docking (Salvetti et al. 2000, 2005; Reddien et al. 2005). Neoblast mRNAs are transiently blocked in extranuclear ribonucleoprotein particles, although basal translation rates hold during daily cell homeostasis. When neoblasts are "activated" by growth spurts, stress, or wounding, translation is unleashed and particular sets of these mRNAs are swiftly activated and translated, leading to a swift change of its proteome.

7.3.2 Heterogeneity of Neoblast Potency and Its Spatial Distribution

In cellular terms, the word neoblast embraces a heterogeneous collection of cells forming a hierarchical lineage from (i) toti-, pluripotent clonogenic cells (Salvetti et al. 2009; Wagner et al. 2011), (ii) to multipotent proliferating progenitor cells broadly committed to tissues and organs (Reddien 2018) down to (iii) a large population of nonproliferative fully committed precursor cells that differentiate into 30–40 differentiated cell types (Baguñà et al. 1990).

Knowledge of neoblast heterogeneity has benefited from the use of fluorescence-activated cell sorting (Hayashi et al. 2006) and neoblast-specific molecular markers. First, as it is for most SCs, *piwi-1* labels all neoblasts (Reddien et al. 2005). Second, *piwi-1*$^{+}$ neoblasts isolated from live dissociated cells by fluorescence-activated cell sorting had their transcriptome analyzed by single-cell RNA sequencing (sc-RNAseq). A wealth of markers were found for different sets of neoblasts (van Wolfswinkel et al. 2014); from totipotent clonogenic neoblasts, pluripotent sigma neoblasts, and committed progenitors to epidermis (zeta-neoblasts), gut (gamma-neoblasts), nerve (Ni-neoblasts), and more specific tissue/cell lineages (muscle, protonephridia, eye, prepharynx, and pharynx) (Plass et al. 2018; Zeng et al. 2018; Reddien 2018). Totipotent clonogenic neoblasts were found throughout the parenchyma but, like all neoblasts, absent from the epidermis, pharynx, and gut branches (Figure 7.4). Sigma-neoblasts are sparse but widely distributed. Beta-neoblasts were found in areas close to the epidermis and gamma-neoblasts namely at the base of gut branches. Costaining of clonogenic neoblasts with markers of sigma-, zeta-, and gamma-neoblasts showed coexpression with pluripotent sigma-neoblasts, but only a limited overlap with zeta-neoblasts and gamma-neoblasts, suggesting that clonogenic neoblasts have the potential to produce asymmetrically the last two classes of broad-lineage neoblasts (Zeng et al. 2018). The emergent picture of neoblast distribution is shown in Figure 7.4.

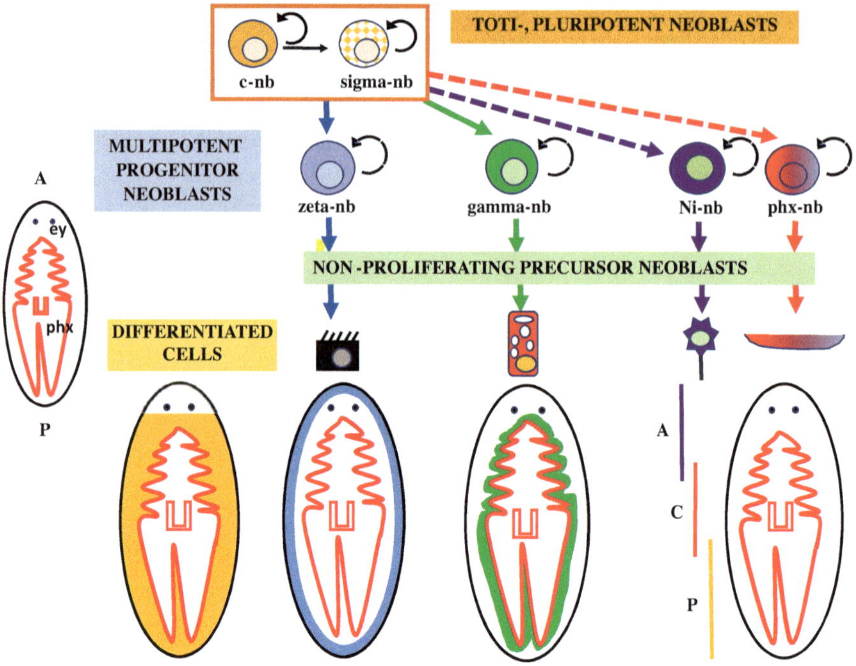

FIGURE 7.4 Diagrammatic representations of an updated, although simplified, view of neoblast heterogeneity, cell lineage, and spatial distribution. On the far left, schematic drawing of a planarian (dorsal view), featuring the outer epidermis (black), the inner gut (red), and the eyes (ey) and pharynx (phx). Upper half: toti- and pluripotent neoblasts self-renew (curved black arrows), produce the next class of multipotent progenitor neoblasts, likely proliferative (broken curved black arrows), which give rise to nonproliferative precursor neoblasts that lead to differentiated cell types (here represented, from left to right, by ciliated epidermal cells (black), gut cells (red), brain nerve cells (violet), and muscle cells of the pharynx and parapharynx (red and violet)). Lower half, from left to right, spatial distribution of clonogenic neoblasts (c-nb, in orange), zeta-neoblasts (blue) close to the epidermis (black), gamma-neoblasts (green) close to the gut (red), and Ni-neoblasts, regionally committed in the head region (anterior; A, violet vertical bar), and pharynx and parapharyngeal neoblasts (phx-nb) committed in the trunk region (central; C, red vertical bar). A: anterior; P: posterior. See text for details.

Another source of neoblast heterogeneity is linked to tissues and organs such as eyes, ovaries, and pharynx with restricted and regional distributions along the anterior-posterior (A-P) and/or dorsal-ventral (D-V) body axes. Whereas clonogenic neoblasts are broadly distributed along A-P and D-V axes, neoblast progenitors for eye cells and ovaries are only found anteriorly, while those for the pharynx and circumpharyngeal areas are localized in trunk (midbody) regions (Figure 7.4). Importantly, progenitor neoblasts are more broadly distributed than precursor neoblasts and differentiated cells. In the eye, progenitor neoblasts are specified between the centrally localized pharynx and the eyes. During regeneration, eye progenitors produce nondividing precursors that progressively migrate to target locations in the blastema and differentiate into photoreceptor cells (LoCascio et al. 2017). Similar behavior occurs for pharynx progenitors. In summary, neoblast progenitors of regionally located tissues and organs are found in a spatially broad but regionally restricted area from which they reach (by unknown mechanisms) narrower target destinations (reviewed in Reddien 2018).

7.3.3 The Topology of Neoblast Lineages and Basic Cell Kinetics

At this stage, several questions arise. Firstly, how are the different neoblast classes related as cell lineages? Secondly, how are they related kinetically and what is the molecular nature of control mechanisms? Finally, how are the different neoblast precursors determined into spatially broad, regionally restricted, and target-specific areas?

Two main models could answer how neoblasts lineages are related:

1. A classical linear hierarchical, and likely branching, model with clonogenic neoblasts and sigma-neoblasts as top-acting ASCs producing progressively committed, and likely proliferative, progenitor neoblasts, ending with fully committed nonproliferative precursor.
2. Top-acting ASCs as the only proliferative cells directly give rise, on demand, to specific nondividing progenitors and precursors for every organ, tissue, and the 40 different cell types.

Current evidence of the spatial heterogeneity of neoblast types favors the first model (see Reddien 2018, and Rink 2018, for thorough reviews). Unfortunately, lack of lineage tracing studies and lack of deep analyses of the simultaneous localization of these neoblasts classes within tissues and organs preclude providing a complete answer.

The second question has a bleaker answer. Despite all the tools available, no kinetic analyses of the proliferative and differentiation rates of different neoblast classes have been carried out, and sparse attention has been paid to potential activators and inhibitors. That is, how is the size of the neoblast pool during growth and degrowth (e.g., from 40,000 to 400,000 neoblasts between 3 and 11 mm body length in the freshwater planarian *Schmidtea mediterranea* and back during degrowth) controlled is poorly known.

Bulk kinetic data of total cells and neoblasts from Baguñà (1974, 1976, 2012), Baguñà and Romero (1981), Romero and Baguñà (1991)—reviewed in Baguñà et al. (1990) and in Section 7.2.3—produced some basic data:

i. neoblast proliferation is always on, even during degrowth by starvation (2,500 neoblasts born per day in starving 3 mm long individuals), increases three- to fourfold after feeding (up to 6,000 neoblasts per day), and drops to former levels at 4–5 days;
ii. cell loss is always on, even during optimal growth (1,700 cells lost per day), and increases up to 6,000–7,000 lost cells during starvation;
iii. accordingly, body growth after feeding results from newborn cells outnumbering lost cells, while the converse holds for degrowth by starvation;
iv. rates of cell proliferation and cell loss decrease with increasing body length, but, crucially, the slopes of the rates of cell loss are steeper than those of cell proliferation; and

v. hence, differences in steepness in these rates for each species, frequency of feeding, and temperature determine if the organism grows or degrows, the maximum size attainable being 30 mm in length and 25–30 million cells for *S. mediterranea* (J. Baguñà, unpublished data) and the minimum size compatible with life being 0.5 mm in length and 30,000 cells.

While asymmetric divisions—one SC neoblast, one committed progenitor—should predominate in steady-state homeostasis, growth spurts and regeneration demand increasing percentages of symmetric divisions (symmetric renewal or/and symmetric differentiation) and decreasing percentages of asymmetric divisions. The inverse should hold during degrowth.

The answer to the third question—how different neoblast precursors are set in different spatial patterns, one of the keystones of planarian homeostasis and regeneration—is linked to axial polarity, to how positional information is encoded along the body axes, and to how it is decoded. These processes are the key to understand the ever-changing phenotypes during growth, degrowth, and regeneration as discussed in the next section.

7.4 IS ADULT POSITIONAL INFORMATION AND PATTERN SIGNALING NEOBLAST-DEPENDENT OR INDEPENDENT?

The continuous renewal of all planarian cell types via loss of differentiated cells and replacement by proliferating and committed neoblasts poses two fundamental questions: (i) how the right differentiated cell types are produced in the right places and (ii) is the former question linked to the differential spatial distribution of neoblast types (Figure 7.4), or is additional information required to maintain axial polarity and pattern? These questions relate to the basic concepts of pattern and form. The main issue is how individual cells acquire the information or signal to become a specific type of cell. Because each cell locates in a specific axial position, this information or signal must have a nonuniform distribution along the axes, either quantitative or qualitative. This is the basis of *positional encoding* or *positional information*. The question then becomes "what is the nature of the signaling molecules fulfilling positional information in planarians?"

7.4.1 Position-Control Genes

Most pattern-formation genes are regionally expressed. In planarians, they were uncovered by homology to genes involved in pattern signaling in embryonic model systems (reviewed in Reddien 2018; Rink 2018). Not surprisingly, whole-mount—and cell—in situ hybridization found the usual suspects:

i. numerous *Wnt* ligands in decreasing transcription gradients in the tail and its inhibitors (secreted Frizzled-related proteins, *sFRP*s; *notum*) in the head, with *Wnt* downstream signaling components broadly distributed;
ii. the tandem BMP-AMP (bone morphogenetic protein-antidorsalizing morphogenetic protein) along the dorsoventral axis with *bmp* ligand in a

medial-to-lateral gradient in the dorsal side, *noggin-like* (*nlg*) dorsally, and its antagonists *admp*, and *noggin* (*nog*) among others, ventrally; and

iii. the tandem *Wnt5-slit*, with *Wnt* expressed laterally inhibiting the medially expressed *slit*, patterning the medial-lateral axis and bilateral symmetry.

All these molecules are coarsely or finely deployed from specific landmarks of the body (see Figure 7.3C, D in Reddien 2018). Functional analyses by RNA interference (RNAi) cause dramatic transformations of gene expression domains, organ positioning, and duplications and expansions of body regions and specific organs and affect specific inhibitions, attesting to their role as position-control genes (PCGs) (Witchley et al. 2013).

In summary, PCGs expression in planarians forms a molecular coordinate system of positional signals from which each cell acquires information of its three-dimensional position.

7.4.2 Position-Control Genes Are Expressed in the Two-Dimensional Body Muscle Grid

The quest for cells expressing PCGs is a classic in developmental biology and regeneration.

In adult planarians, PCGs were expected to be expressed in all neoblasts, in a subset of them, in all differentiated cells or only in a specific cell type. That differentiated cells and not neoblasts had axial positional information was first guessed from the repopulation and rescue of X-irradiated tail regions by neoblasts from unirradiated head parts (Saló and Baguñà 1985). However prescient, that was a mere guess. The key finding was that most PCG expression is restricted to the circular, longitudinal, and diagonal muscle cells of the body wall that form a two-dimensional grid between the epidermis and the parenchyma (where all neoblasts are found) (Witchley et al. 2013). RNAi of markers required for muscle maintenance and regeneration produced specific axial pattern defects (Scimone et al. 2017). Moreover, sc-RNAseq of muscle cells from different axial levels showed scores of additional genes displaying regional expression in muscle (Fincher et al. 2018). During regeneration, expression of PCG in muscle cells changes swiftly, with tail fragments expressing anterior PCG and posterior PCG shifting their expression more posteriorly. A converse process happens in head fragments. Further details of this model as applied to planarian regeneration are described in Reddien (2018).

7.4.3 The Relationship of Neoblast Lineage Hierarchy to PCG-Driven Lineage Choice in Daily Homeostasis

How do neoblasts fit within this model? That is, how the right differentiation of cells in the A-P, D-V, and M-L body axes relates to the general and restricted distribution of neoblast types (Figure 7.4) and to PCG expression? Because RNAi on PCG expression affects regeneration and tissue turnover, the cellular outcomes of neoblasts (cell differentiation) must somehow depend upon PCG expression in muscle cells. Recently, Wurtzel et al. (2017) reported a fine example. Epidermal (zeta)-neoblasts derived from clonogenic neoblasts and sigma-neoblasts within the parenchyma are the progenitors of different types of epidermal cells (ciliated and not bearing rhabdites ventrally; nonciliated and bearing rhabdites dorsally). sc-RNAseq showed dorsally located zeta-neoblasts to express the gene *prdm1-1* and those ventrally located to express the gene *kal1*, a difference passed onto the precursors of dorsal and ventral epidermal cells. Crucially, this difference is controlled by the D-V determinant *bmp4* from dorsal muscles that diffuse as BMP4 and turn zeta-neoblasts into dorsal precursors; absence of BMP4 determines ventral precursors.

Hence, combinations of PCG products diffusing from muscle cells impinge into the spatial identity and migratory behavior of nearby progenitor and precursor neoblasts influencing their final differentiated fate (Figure 7.5). The pluri-, multipotent lineage hierarchy of neoblasts with general, broad, and regionally spatial distribution seems aligned to a hierarchy of positional signals that goes from primary signals (A versus P, D versus V, M versus L, epidermis versus endoderm-mesoderm, endoderm versus mesoderm) to secondary signals (head/trunk/tail; circumpharynx/pharynx), and finally to tertiary ones (organ- and tissue-specific). That is, primary and secondary signals trim the branches of cell lineage not required in a given place. Later on, short-range tertiary signals from specific organs and tissues execute the choices for specific differentiated cell types. In turn, neoblast precursors differentiate into specific cell types, muscle cells among them, which produce new patterning signals. Thus, the end becomes the beginning and the pattern continuously replaces itself (for a thorough discussion, see Rink 2018).

7.4.4 The Effects of Growth and Degrowth

However provisional, this model fits pretty well normal homeostasis. But how does it fit growth after feeding and degrowth by starvation?

Feeding triggers a three- to fourfold increase in proliferation (Baguñà 1974, 1976) producing new neoblasts everywhere; firstly, multipotent and broadly distributed progenitor classes; secondly, tissue- and cell-specific precursors that lead to increasing numbers of differentiated cells. As tissues grow and the organism grows (Baguñà and Romero 1981), muscle cells also increase in number and PCG domains in muscle should scale to the new size. During prolonged starvation, neoblast proliferation rates decline, rates of cell loss increase, and total number of cells, body length, and area decrease. Muscle cells and PCG domains in muscle also decrease. In turn, total neoblasts decrease steadily but slowly in number although the percentage of neoblasts to total cells increases (Baguñà 1976). This is because rates of differentiation fall below rates seen during steady-state homeostasis and tissue size. In a progressively smaller body, tissue and organ proportions slowly change (body allometries, first studied in planarians by Abeloos 1930). That is, although the neoblast lineage pipeline always supplies progenitors, the rates of differentiation of neoblasts to specific cell types and, hence, their resulting percentages to total cells change with shrinking body lengths (Baguñà and Romero 1981).

The final result of prolonged degrowth (e.g., from 11 to 3 mm in length) is tiny organisms whose body length and volume, percentage of neoblasts, mitotic indices,

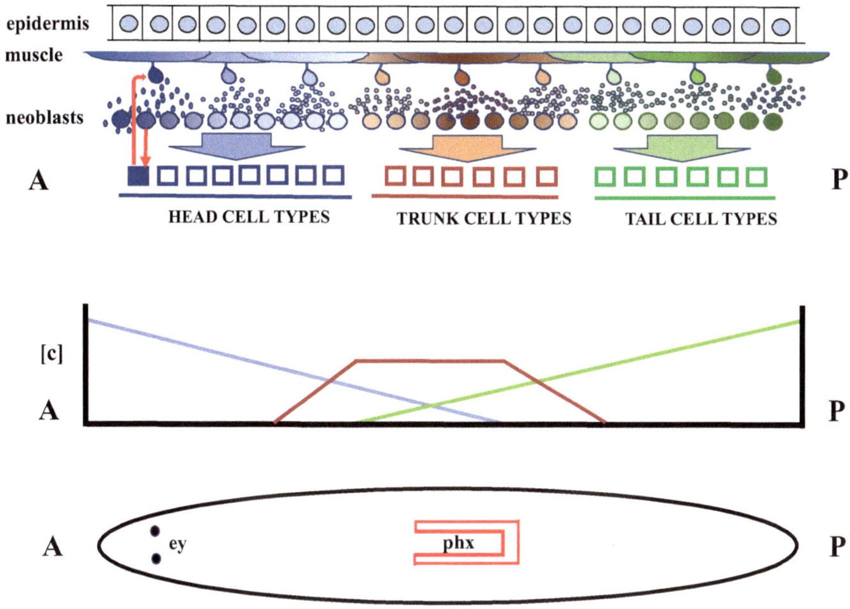

FIGURE 7.5 Axial polarity (here illustrated using the anteroposterior (A-P) axis) and pattern (here head, trunk and tail) driven by positional information stemming from differential expression of position-control gene (PCG) in the muscle system (Witchley et al. 2013) impinges onto adjacent neoblast to direct their lineage choices and differentiation into specific axial cell types. Upper tier: simplified diagram featuring the outer epidermis, the body wall muscle, neoblasts (filled colored circles), and differentiated cells (open colored squares). Variation in the concentration of PCG products in muscle cells is shown by the three colors, blue, brown, and green and as three graded intensities from A → P, P → A, and middle → A and P in head, tail, and trunk. Products are released into the intercellular space (colored dots) acting onto neoblasts. Different PCG combinations produce different neoblast types (and lineages) and, hence, different cell types that maintain the A-P body pattern. Continuous cell replacement (neoblasts-differentiated cells-neoblasts) is shown in the left by two red arrows going first from neoblasts to a particular differentiated cell (e.g., a muscle cell; filled dark blue square), which will later be incorporated (second arrow) within the muscle body wall. The middle tier represents the suggested and hypothetical molecules whose constitutive regional expression (e.g., decreasing posterior to anterior gradients of Wnt signaling (e.g., *β-catenin*, *Wnt-1*, green line); anterior to posterior decreasing gradient of Wnt inhibitors (*sFRPs*, *notum*); and decreasing central gradients (brown lines) of still uncharacterized products are proposed to control regionalization (and polarity) of the A-P axis. The lower tier represents a simplified view of a planarian from the dorsal side. ey: eyes; phx: pharynx.

percentages of differentiated cell types, axial positions and volumes of several organs and tissues, and rates of regeneration are very similar to those of newborn individuals. In turn, they are quite different from those of the initial 11-mm-long organism. Unfortunately, the allometric mechanisms that link changes in cell turnover and cell number to shifting organ axial position and tissue and cell type percentages as related to body length are today completely unknown.

7.5 REGENERATION IN PLANARIANS IS RARE: TRADE-OFFS BETWEEN REGENERATION, LIFE CYCLE, REPRODUCTION, CELL TURNOVER, AND AXIAL SIGNALS

In nature, planarians in regeneration are rare to spot. Only species (or populations) relying in *asexual reproduction* by fission are poised to regenerate, and they do it very well. Some of those populations or species with sexual reproduction regenerate very well, but a fair percentage do not regenerate at all or do so poorly (reviewed in Brønsted 1969).

When sectioned at any level along the A-P axis, asexual and a fair number of sexual species regenerate the head, the rate of head regeneration forming a decreasing gradient from anterior to posterior. Nonregenerating species do not regenerate from any level. In between, several species (e.g., the dendrocoelid *Dendrocoelum lacteum)* have a decreasing capacity to make heads that falls off just anterior to the pharynx; fragments cut posterior to the pharynx never regenerate a head. Interestingly, species that regenerate the head as well as those that do not regenerate the head or do it poorly, regenerate the tail all along the A-P axis. There is no obvious phylogenetic pattern to this ability.

7.5.1 The Proximate Causes of Regeneration Ability

Differences in regeneration ability were in the past interpreted as resulting from low number of neoblasts, low mitotic activity, or some "feature" of differentiated cells and tissues.

In all planarians studied, including some defective in regeneration after amputation, the percentage of neoblasts goes from 15%–20% to 35% of total cells, and mitotic activity is high and sustained (Lange 1967; Baguñà and Romero 1981; Romero et al. 1991). Neoblasts are thus not the culprits explaining inability to regenerate. In tail pieces of *D. lacteum* cut ahead of or behind the pharynx, neoblast proliferation increases and a blastema forms. A new head differentiates prepharyngeally, but when amputation is behind the pharynx, the blastema degenerates and the organism dies.

Irradiation eliminates neoblasts. Grafting irradiated anterior pieces into posterior regions rescued the ability to produce a head from a tail fragment. The converse experiment does not produce heads in anterior regions. Therefore, some property in differentiated cells but not in neoblasts from anterior regions has the potential for regeneration (Stéphan-Dubois and Gilgenkrantz 1961). More than 50 years later, the molecular basis for such differences was sought in the poorly regenerating species *D. lacteum* and *Procotyla fluviatilis* (Liu et al. 2013; Sikes and Newmark 2013). The starting point was the known role of Wnt signaling (e.g., beta-catenin, *ß-catenin*) in planarian regeneration: high levels of Wnt activity—present in decreasing gradients in the tail (Figure 7.5)—specify tail regeneration, while low levels seem necessary and sufficient for head formation. Regeneration-defective species might have higher levels of *ß-catenin* than good regenerators. Indeed, inhibiting the activity of several components of the Wnt pathway by RNAi restored head regeneration in tail fragments of regenerative-defective species. Therefore, *the stem-cell "niche" of PCGs is the proximate cause of such differences*, and levels of Wnt signaling are the likely molecular basis.

7.5.2 Trade-Offs and the Ultimate Causes of Regeneration

The ultimate (evolutionary) mechanism(s) responsible for differences in the ability of planaria to regenerate has(have) been the subject of much speculation. Most arguments revolve around *trade-offs* between regenerative capacity, life cycles, and reproduction (Calow 1978).

Asexual species or populations, with no expense in reproduction, regenerate very well. Sexual species instead invest energy and resources into reproduction (ovaries, testes, yolk glands, copulatory complex, sexual behavior) that otherwise could have been used to maintain somatic tissues (cell turnover). Because resources (energy, food) are limited, a trade-off exists between allocation of resources to reproduction and to maintain somatic tissues. If reproduction is favored, body maintenance may be at risk. Conversely, if body maintenance is kept at optimal rates, reproduction might be at a stake. This is the basis of two reproductive strategies: *semelparity* and *iteroparity*.

In semelparity, energy is mainly channeled to reproduction, and to egg and offspring production, even in starving conditions. This results in high number of offspring, often of large size, at the expense of less than optimal body maintenance (cell turnover and mitotic rates decline), increased tissue damage, and the final death of the organism.

Iteroparous species channel resources to both reproduction and body maintenance when food is abundant. When food declines, division of germ cells decreases, reproductive tissues regress, and reproduction halts, while somatic cell turnover rates are unchanged. The result is a low number of offspring, usually of small size, but with adults that survive and reproduce for several years. Why one or another strategy is taken is a question of evolutionary population genetics and depends on the chances of adult versus offspring and juvenile survival and on age-specific survivorship and mortality (Calow and Read 1986; Crow and Kimura 2009).

While a trade-off between reproduction and body maintenance as the ultimate explanation of planarian life span (semelparity versus iteroparity) is rather consistent, does it explain differences in regenerative ability? In general, iteroparous species (high cell turnover and midlevels of reproductive output) have high regeneration abilities, while most semelparous species (low-mid cell turnover and high reproductive output) have poor or no regenerative abilities. It has been suggested that poor regenerative ability may result from selection of higher Wnt signaling levels that halts head regeneration and enhances the location and differentiation of the reproductive system (e.g., increase the mass of testes and yolk glands), leading to increased reproduction (Vila-Farré and Rink 2018).

An alternative explanation stems from differences in cell turnover and semelparity/iteroparity within populations of the same species (Romero and Baguñà 1988, Romero et al. 1991). Populations of *D. lacteum* from Scotland and Sweden are semelparous, while populations in southern France are iteroparous. Southern populations have higher mitotic activity and cell turnover rates, lower cell turnover time (= time of residence of cells within the body), lower reproductive output (number and mass of offspring), and adults live longer (iteroparous) than do semelparous northeners. Most importantly, when cut at postpharyngeal levels, planaria from southern populations regenerate, albeit at low percentages (Romero and Baguñà 1991). Could it be that populations, with slow somatic cell renewal regeneration is not favored and, if environmental conditions are favorable, such populations adopt a semelparous strategy that favors juvenile versus adult survivorship, whereas fast somatic cell turnover, by keeping better-off adults, favors regeneration and iteroparity (Calow and Read 1986)?

The first hypothesis could be tested by comparing Wnt levels between null/poor regenerative species and good/very good ones. Were they similar, the hypothesis would not hold; if there is a negative correlation, the hypothesis holds. A further test is to determine whether asexual species (no expense in reproduction and excellent regeneration) and iteroparous sexual species (average or high expense in reproduction and excellent regeneration) have similar or different Wnt levels. If levels are similar, reproductive expense and regeneration ability are not directly related. If levels in asexuals are lower than in sexual iteroparous, the Wnt hypothesis holds; if they are higher it would not hold. In that case, explanations more related to trade-offs between adult versus newborn/juvenile survivorship/mortality rates with levels of somatic cell turnover must be looked for and their molecular and physiological basis explored.

7.6 WHY HAVE DEDIFFERENTIATION/ TRANSDIFFERENTIATION IF YOU HAVE STEM-CELLS?

In a short comment entitled "Who needs stem cells if you can dedifferentiate?" Malcom Maden raised again the question of how adult organisms obtain undifferentiated cells (Maden 2013). Too often (e.g., Slack 1980), stem cells and cell dedifferentiation have been presented as mutually exclusive processes. Hence, Maden's title could easily be turned upside down to "Who needs dedifferentiation if you have stem cells?" Then, how planarians dealt with it?

7.6.1 Dedifferentiation in Planarians

The question of whether *dedifferentiation* occurs in planarians has raged for more than a century (reviewed in Baguñà 2012).

In the first third of the twentieth century, dedifferentiation of differentiated cells was considered the leading mechanism of planarian regeneration, although the evidence was limited (references in Brønsted 1969). After a relapse, when neoblasts and the French School took the lead (reviewed in Baguñà 2019), dedifferentiation was resuscitated in the last 1960s. Vital dyes, enzyme staining, and transmission electron microscopy were used on intact and regenerating planarians to claim "transformation" of intestinal, goblet, and secretory cells into "neoblasts" (main references in Hay and Coward 1975). In their view, neoblasts could not be considered a permanent population of undifferentiated cells but, at the most, transient undifferentiated cells produced by cell dedifferentiation. To be fair, the methods used were too coarse to allow such a conclusion, and the evidence produced too weak to merit further consideration.

At the end of the 1970s and early 1980s, Vittorio Gremigni and coworkers in Pisa (Italy) revived dedifferentiation in planarians when they reported a set of simple and beautiful experiments using chromosomal markers (reviewed in Gremigni 1988). The experimental animal was a strain of the species *Schmidtea polychroa* that is a naturally occurring mosaic with triploid somatic cells, diploid male germ cells, and hexaploid female germ cells. After cutting through the gonadal (head) region, a blastema and a new head formed. Karyotype analyses of blastema and postblastema cells showed a majority of triploid cells but also diploid (from male germ cells) and/or hexaploid cells (from female germ cells) from which all sorts of somatic cells and tissues (e.g., pharyngeal muscle cells) formed (Gremigni and Miceli 1980). The inescapable conclusion was that dedifferentiation and transdifferentiation from germ cells to somatic cells, however limited, occurred during regeneration. These results were held as conclusive evidence for dedifferentiation in planarians and as an argument for the existence of similar mechanisms of blastema formation in most animal groups (Slack 1980).

A closer look, however, reinterpreted these results as dedetermination/transdetermination and not transdifferentiation (Baguñà 1981); because the loss of a haploid complement during spermiogenesis and its doubling during oogenesis is one of the first steps from neoblasts to germ cells and occurs in undifferentiated cells of the germinative epithelium, which are undistinguishable from somatic neoblasts. Moreover, differentiating and differentiated germ cells, such as spermatocytes, spermatids, and spermatozoa (and their counterparts in the female germ line), degenerate and lyse after transection and, therefore, cannot dedifferentiate to give blastema cells. Although in his later overview of planarian regeneration, Gremigni (1988) took an intermediate view assuming that both neoblasts and dedifferentiated cells take part in blastema formation, we believe his experiments do not validate transdifferentiation but rather dedetermination and transdetermination. To plead for transdifferentiation, two conditions must be met:

1. the differentiated state before and after the switch must be reliably described;
2. a direct ancestor-descendant relationship and a common developmental history between cells must be clearly demonstrated (Eguchi and Kodama 1993). None of the papers claiming dedifferentiation in planarians have fulfilled these conditions.

As of today, no evidence supports dedifferentiation in planarians. To be fair, however, transdetermination among progenitor neoblasts, and reversals from progenitors to pluripotent stem neoblasts (clonogenic neoblasts or sigma-neoblasts), as first shown by Gremigni and colleagues, is more than likely, although transdetermination and reversals need to be analyzed further using current molecular markers and new cell lineage techniques.

7.7 SOME CONCLUSIONS, ONE ENIGMA AND A HALF ENIGMA

1. Planarian neoblasts are a special and complex system of SCs with features unmatched in the Animal Kingdom. Despite the fact that other organisms

have cells with properties close to those of neoblasts, only the i-cells of the hydrozoans (Gahan et al. 2016) come close to the main property of neoblasts, which is to produce all somatic cells and all germ cells. This makes neoblasts a sort of totipotent embryonic cell (akin to mammalian embryonic SCs) kept in an adult body.

2. As is the case for true SCs, neoblasts are filled with RNA-binding proteins, microRNAs, and other molecules that repress scores of stable mRNAs ready for translation into proteins to start differentiation programs, making the molecular machinery of neoblasts complex, versatile, and ready to change.
3. Neoblasts cannot be considered reserve cells, set-aside cells, cache cells, or deferred cells because they are not embryonic, larval, or adult cells "waiting" to be called to do something. Neoblasts have an active and continuous main role as the source of cells for daily cell renewal (cell homeostasis). However, although no definitive evidence is at hand, quiescent neoblasts, which retain label for a long time and spend a prolonged time in G2 ready for mitosis, likely exist (see Figure 7.2).
4. Neoblasts have key roles in growth and degrowth. Even under stress from prolonged starvation, the neoblast pipeline is always on providing proliferating progenitors on a continuous basis independent of internal demand, although final neoblast numbers are likely modulated at the level of nondividing precursor cells.
5. Patterning molecules in adult planarians are expressed in constitutive and axial regional patterns. These molecules or PCGs are expressed in cells of the muscle system and convey both three-dimensional positional information and regional tissue identity to nearby progenitor and precursor neoblasts. This positional system is one necessary half of the renewal system in planarians. Neoblasts are the other half; only both working together allow the organism run and be self-sufficient.
6. Strange as it may seem and however enticing as it is, regeneration in planarians is a dispensable and, therefore, occasional process. Some planarian species and the populations of some species regenerate, others do not. In sheer contrast, all species need cell renewal for somatic body homeostasis.
7. Regenerative abilities in planarians are extremely variable. Some species regenerate easily, others fairly well, and a good deal poorly or not at all. The proximate causes of such differences may be variation in the axial levels of signaling molecules or variation in the turnover rates of somatic cells, both impinging on reproductive expense and output. The ultimate causes have to be sought in adult versus newborn/juvenile survivorship/mortality rates.
8. As of today, no clear evidence for dedifferentiation and/or transdifferentiation in planarians has been reported. It is more than likely, however, as attested by Gremigni's experiments, that reversals from progenitor neoblasts and even from precursor neoblasts to pluripotent stem neoblasts do occur. This needs to be further substantiated.

Finally, one enigma and a half. The half enigma concerns the embryonic origin of the neoblasts; the full enigma concerns their phylogenetic origin.

The half-enigma—when, where, and how neoblasts or neoblast-like cells originate during embryonic development—had a first answer in the 1960s from Le Moigne (1966). Using transmission electron microscopy, X-ray irradiation, and the ability to regenerate, they showed that some neoblasts were present at the end of the Stage 4B and increased in number from the fifth stage on after a burst of cell proliferation (planarian development is divided into eight stages; stages 4B and 5 are when the embryonic primordia of the future brain, epidermis, gut, and pharynx form). From the fifth stage onward, embryos are able to regenerate when cut along the A-P axis.

After a lapse of 50 years, the issue was tackled again using the plethora of new techniques and molecules now available. Briefly, toti-, pluripotent clonogenic neoblasts arise as a subpopulation of *piwi*-1⁺ blastomeres at stage 5. They are competent to proliferate and are incorporated when injected into the adult parenchyma where they rescue X-ray lethally irradiated organisms and give rise to all organ primordia (Davies et al. 2017). However, because the precise lineage between *piwi-1*⁺ blastomeres and the first clonogenic neoblasts has not been established at the cell level, we should qualify the ontogenetic origin of neoblasts as a half-enigma.

The full enigma is the evolutionary mechanisms that brought and maintain such a singular, embryo-like cell type as planarian neoblasts (akin to similar cells in nonbilaterian organisms) in the adults of a bilaterian phylum, the Platyhelminthes (flatworms), nested among other bilaterian phyla that do not have neoblasts.

ACKNOWLEDGMENTS

I am very grateful to the editors Brian K Hall and Cory Bishop for kindly inviting me to write this essay on neoblasts and for their unending patience waiting for the manuscript. All comments written down here are a reflection of my 40 years involved in the whereabouts of planarian regeneration and cell homeostasis. They are, therefore, personal views and thoughts for which I am completely responsible.

REFERENCES

Abeloos, M. 1930. Recherches expérimentales sur la croissance et la régénération chez les Planaires. *Bull. Biol. Fr. Belg.* 64: 1–140.

Alié, A., Hayashi, T., Sugimura, I., Manuel, M., Sugano, W., Mano, A., Satoh, N. et al. 2015. The ancestral gene repertoire of animal stem cells. *Proc. Natl. Acad. Sci. USA* 112: 7093–7100.

Auladell, C., García-Valero, J., and Baguñà, J. 1993. Ultrastructural localization of RNA in the chromatoid bodies of undifferentiated cells (neoblasts) in planarian by the RNAse-gold-complex technique. *J. Morphol.* 216: 319–326.

Baguñà, J. 1974. Dramatic mitotic response in planarians after feeding and a hypothesis for the control mechanism. *J. Exp. Zool.* 190: 117–122.

Baguñà, J. 1976. Mitosis in the intact and regenerating planarian *Dugesia mediterranea* n.sp. I. mitotic studies during growth, feeding and starvation. *J. Exp. Zool.* 195: 53–65.

Baguñà, J. 1981. Planarian neoblast. *Nature* 290: 14–15.

Baguñà, J. 2012. The planarian neoblast: the rambling history of its origins and some current black boxes. *Int. J. Dev. Biol.* 56: 19–37.

Baguñà, J. 2019. Planarian regeneration between 1960s and 1990s: from skilful baffled ancestors to bold integrative descendants. A personal account. *Sem. Cell Dev. Biol.* 87: 3–12.
Baguñà, J., and Romero, R. 1981. Quantitative analysis of cell types during growth, degrowth, and regeneration in the planarians *Dugesia(S) mediterranea* and *Dugesia(G) tigrina. Hydrobiologia* 84: 181–194.
Baguñà, J., Saló, E., and Auladell, M. C. 1989. Regeneration and pattern formation in planarians. III. Evidence that neoblasts are totipotent stem-cells and the source of blastema cells. *Development* 107: 77–86.
Baguñà, J., Romero, R., Saló, E., Collet, J., Auladell, M. C., Ribas, M., Riutort, M. et al. 1990. Growth, degrowth and regeneration as developmental phenomena in adult freshwater planarians. In *Experimental Embryology in Aquatic Plant and Animal Organisms.* NATO-ASI Series (Marthy, J. ed.), pp. 129–162. New York: Plenum Press.
Baguñà, J., Saló, E., Romero, R., García-Fernández, J., Bueno, D., Muñoz-Marmol, A. M., Bayascas-Ramírez, J. R. et al. 1994. Regeneration and pattern formation in planarians: cells, molecules and genes. *Zool. Sci.* 11: 781–795.
Brønsted, H. V. 1969. *Planarian Regeneration.* Oxford: Pergamon Press.
Calow, P. 1978. *Life Cycles. An Evolutionary Approach to the Physiology of Reproduction, Development and Ageing.* London: Chapman and Hall.
Calow, P., and Read, D. A. 1986. Ontogenetic patterns and phylogenetic trends in freshwater flatworms (Tricladida); constraint or selection? *Hydrobiologia* 132: 263–272.
Coward, S. J. 1974. Chromatoid bodies in somatic cells of the planarian: observations of their behavior during mitosis. *Anat. Rec.* 180: 533–546.
Crow, J. F., and Kimura, M. 2009. *An Introduction to Population Genetics Theory.* Caldwell, NJ: Blackburn Press.
Davies, E. L., Lei, K., Seidel, C. W., Kroesen, A. E., McKinney, S. A., Guo, L., Robb, S. M. et al. 2017. Embryonic origin of adult stem cells required for tissue homeostasis and regeneration. *Elife* 6: e21052.
Dubois, F. 1949. Contribution a l'etude de la migration des cellules de régénération chez les planaires dulcicoles. *Bull. Biol. Fr. Belg.* 83: 213–283.
Eguchi, G., and Kodama, R. 1993. Transdifferentiation. *Curr. Opin. Cell Biol* 5: 1023–1028.
Fincher, C. T., Würtzel, O., de Hoog, T., Kravarik, T. M., and Reddien, P. W. 2018. Cell type transcriptome atlas for the planarian *Schmidtea mediterranea. Science* 360: 874.
Funayama, N. 2010. The stem cell system in demosponges: insights into the origin of somatic stem cells. *Dev. Growth Differ.* 52: 1–14.
Funayama, N. 2013. The stem cell system in demosponges: suggested involvement of two types of cells: archeocytes (active stem cells) and choanocytes (food-entrapping flagellated cells). *Dev. Genes Evol.* 223: 23–38.
Gahan, J. M., Bradshaw, B., Flici, H., and Frank, U. 2016. The interstitial stem cells in *Hydractinia* and their role in regeneration. *Curr. Opin. Genet. Dev.* 40: 65–73.
Greb, T., and Lohmann, J. U. 2016. Plant stem cells. *Curr. Biol.* 26: R816–R821.
Gremigni, V. 1988. Planarian regeneration: an overview of some cellular mechanisms. *Zool. Sci.* 5: 1153–1163.
Gremigni, V., and Miceli, C. 1980. Cytophotometric evidence for cell transdifferentiation in planarian regeneration. *Wilhelm Roux's Arch. Dev. Biol.* 188: 107–113.
Hadorn, E. 1968. Transdetermination in cells. *Sci. Am.* 219: 110.
Hay, E. D., and Coward, S. J. 1975. Fine structure studies on the planarian, *Dugesia.* I. Nature of the "neoblast" and other cell types in the noninjured worm. *J. Ultrastruct. Res.* 50: 1–21.
Hayashi, T., Asami, M., Higuchi, S., Shibata, N., and Agata, K. 2006. Isolation of planarian X-ray-sensitive stem cells by fluorescent-activated cell sorting. *Dev. Growth Differ.* 48: 371–380.
Lander, A. D. 2009. The 'stem cell' concept: is it holding us back? *J. Biol.* 8: 70.

Lange, C. S. 1967. Quantitative study of the number and distribution of neoblasts in *Dugesia lugubris* (planaria) with reference to size and ploidy. *J. Embryol. Exp. Morphol.* 18: 199–213.

Le Moigne, A. 1966. Étude du développement embryonnaire et recherches sur les cellules de régénération chez l'embryon de la planaire *Polycelis nigra* (Turbellaries Triclades). *J. Embryol. Exp. Morphol.* 15: 39–60.

Liu, S.-Y., Selck, C., Friedrich, B., Lutz, R., Vila-Farré, M., Dahl, A., H. Brandl, H. et al. 2013. Reactivating head regrowth in a regeneration-deficient planarian species. *Nature* 500: 81–84.

LoCascio, S. A., Lapan, S. W., and Reddien, P. W. 2017. Eye absence does not regulate planarian stem cells during eye regeneration. *Dev. Cell* 40: 381–391.

Maden, M. 2013. Who needs stem cells if you can dedifferentiate? *Cell Stem Cell* 13: 640–641.

Okada, T. S. 1991. *Transdifferentiation. Flexibility in Cell Differentiation.* Oxford: Clarendon Press.

Plass, M., Solana, J., Wolf, F. A., Ayoub, S., Misios, A., Glazar, P., Obermayer, B. et al. 2018. Cell type atlas and lineage tree of a whole complex animal by single-cell transcriptomics. *Science* 360: 875.

Raff, M. 2003. Adult stem cell plasticity. *Annu. Rev. Cell Dev. Biol.* 19: 1–22.

Randolph, H. 1897. Observations and experiments on regeneration in planarians. *Arch. Entwicklungsmech. Org.* 5: 352–372.

Reddien, P. W. 2018. The cellular and molecular basis for planarian regeneration. *Cell* 175: 327–345.

Reddien, P. W., Oviedo, N. J., Jennings, J. R., Jenkin, J. C., and Alvarado, A. S. 2005. SMEDWI-2 is a PIWI-like protein that regulates planarian stem cells. *Science* 310: 1327–1330.

Rink, J. C. 2018. Stem cells, patterning and regeneration in planarians: self-organization at the organismal scale. In: *Planarian Regeneration: Methods and Protocols. Methods in Molecular Biology*, vol 1774 (Rink, J. C. ed.), pp. 57–171. Berlin, Germany: Springer.

Rojo-Laguna, J. I., García-Cabot, S., and Saló, E. 2019. Tissue transplantations in planarians: a useful tool for molecular analysis of pattern formation. *Semin. Cell Dev. Biol.* 87: 116–124.

Romero, R., and Baguñà, J. 1988. Quantitative cellular analysis of life-cycle strategies of iteroparous and semelparous triclads. In *Free-Living and Symbiotic Plathelminthes*, vol 36 (Ax, P., Ehlers, U., and Sopott-Elhers, B. eds.), pp. 283–289. *Fortschritte der Zoologie/ Progress in Zoology*. Stuttgart-New York: Gustav Fischer Verlag.

Romero, R., and Baguñà, J. 1991. Quantitative cellular analysis of growth and reproduction in freshwater planarians (Turbellaria; Tricladida). I. A cellular description of the intact organism. *Invertebr. Reprod. Dev.* 19: 157–165.

Romero, R., Baguñà, J., and Calow, P. 1991. Intraspecific variation in somatic cell turnover and regenerative rate in the freshwater planarian *Dendrocoelum lacteum*. *Invertebr. Reprod. Dev.* 20: 107–113.

Rossi, L., Salvetti, A., Batistoni, R., Deri, P., and Gremigni, V. 2008. Planarians, a tale of stem cells. *Cell Mol. Life Sci.* 65: 16–23.

Saló, E., and Baguñà, J. 1984. Regeneration and pattern formation in planarians. I. The pattern of mitosis in anterior and posterior regeneration in *Dugesia(G)tigrina*, and a new proposal for blastema formation. *J. Embryol. Exp. Morph.* 83: 63–80.

Saló, E., and Baguñà, J. 1985. Proximal and distal transformation during intercalary regeneration in the planarian *Dugesia (S) mediterranea*. Evidence using a chromosomal marker. *Wilhelm Roux's Arch. Dev. Biol.* 194: 364–368.

Salvetti, A., Rossi, L., Deri, P., and Batistoni, R. 2000. An MCM2-related gene is expressed in proliferating cells of intact and regenerating planarians. *Dev. Dyn.* 218: 603–614.

Salvetti, A., Rossi, L., Lena, A., Batistoni, R., Deri, P., Rainaldi, G., Locci, M. T. et al. 2005. *DjPum*, a homologue of *Drosophila* Pumilio is essential to planarian stem cell maintenance. *Development* 132: 1863–1874.

Salvetti, A., Rossi, L., Bonuccelli, L., Lena, A., Pugliesi, C., Rainaldi, G., Evangelista, M., and Gremigni, V. 2009. Adult stem cell plasticity: neoblast repopulation in non-lethally irradiated planarians. *Dev. Biol.* 328: 305–314.

Scimone, M. L., Cote, L. E., and Reddien, P. W. 2017. Orthogonal muscle fibers have different instructive roles in planarian regeneration. *Nature* 551: 623–629.

Shostak, S. 2008. Speculations on the evolution of stem cells. *Breast Dis.* 29: 3–13.

Sikes, J. M., and Newmark, P. A. 2013. Restoration of anterior regeneration in a planarian with limited regenerative ability. *Nature* 500: 77–80.

Slack, J. M. W. 1980. The source of cells for regeneration. *Nature* 286: 760.

Sluys, R., Kawakatsu, M., Riutort, M., and Baguñà, J. 2009. A new higher classification of planarian flatworms (Platyhelminthes, Tricladida). *J. Nat. Hist.* 43: 1763–1777.

Stéphan–Dubois, F., and Gilgenkrantz, F. 1961. Transplantiation et régénération chez la planaire *Dendrocoelum lacteum. J. Embryol. Exp. Morphol.* 9: 642–649.

Van Wolfswinkel, J. C., Wagner, D. E., and Reddien, P. W. 2014. Single–cell analysis reveals functionally distinct classes within the planarian stem cell compartment. *Cell Stem Cell* 15: 326–339.

Vila-Farré, M., and Rink, J. C. 2018. Ecology of freshwater planarians. In *Planarian Regeneration: Methods and Protocols*. Methods in Molecular Biology, vol 1774 (Rink, J. C. ed.), pp. 173–205. New York: Springer.

Voronina, E., Seydoux, G., Sassone-Corsi, P., and Nagamori, I. 2011. RNA granules in germ cells. *Cold Spring Harbor Perspect. Biol.* 3: a002774.

Wagner, D. E., Wang, I. E., and Reddien, P. W. 2011. Clonogenic neoblasts are pluripotent adult stem cells that underlie planarian regeneration. *Science* 332: 811–816.

Wenemoser, D., and Reddien, P. W. 2010. Planarian regeneration involves distinct stem cell responses to wounds and tissue absence. *Dev. Biol.* 344: 979–991.

Witchley, J. N., Mayer, M., Wagner, D. E., Owen, J. H., and Reddien, P. W. 2013. Muscle cells provide instructions for planarian regeneration. *Cell Reports* 4: 633–641.

Wurtzel, O., Oderberg, I. M., and Reddien, P. W. 2017. Planarian epidermal stem cells respond to positional cues to promote cell-type diversity. *Dev. Cell* 40: 491–504.

Zeng, A., Hua, L., Longhua, G., Peak, A., Box, A., and Sánchez Alvarado, A. 2018. Prospectively isolated tetraspanin$^+$ neoblasts are adult pluripotent stem cells underlying planarian regeneration. *Cell* 173: 1593–1608.

8 Skeletal Muscle Satellite Cells

Adult Stem Cells with Multipotential Capacity

Morten Ritso and Alexander Y.T. Lin
Ottawa Hospital Research Institute
University of Ottawa

J. Manuel Hernández-Hernández
Centre for Research and Advanced Studies
of the National Polytechnic Institute

Michael A. Rudnicki
Ottawa Hospital Research Institute
University of Ottawa

CONTENTS

8.1 STEM CELL POPULATIONS IN SKELETAL MUSCLE

The observational discovery of *satellite cells* in frog skeletal muscle (Mauro 1961) rightfully anticipated the future confirmation of the remarkable regenerative potential of these cells. The name satellite cell arises from the niche in which these cells reside: located between the myofiber plasma membrane and the extracellular basal lamina of skeletal muscle.

In addition to satellite cells, other skeletal muscle tissue-resident stem cells play diverse roles in the repair and rebuilding of functional muscle upon injury. Fibroadipogenic progenitors are mesenchymal stem cells responsible for extracellular matrix remodeling and facilitate myogenic progenitor differentiation during the regenerative process (Joe et al. 2010). Tissue-resident and infiltrating macrophages play a balancing role in recycling damaged muscle tissue, while also contributing key signaling cues to other cells during the regenerative process (Juban and Chazaud 2017). Endothelial progenitors are responsible for vascular remodeling during injury (Grenier et al. 2007; Latroche et al. 2015). Most recently, a skeletal muscle resident tenocyte progenitor population has been identified and characterized (Giordani et al. 2019). Together, these cell types are some of the key players in maintaining the homeostatic powerhouse that skeletal muscle is, as the biggest organ in many organisms. This chapter focuses on the developmental origins of satellite cells, the role they play in adult tissue regeneration and the multipotential characteristics these cells display.

8.2 DEVELOPMENTAL ORIGIN OF SATELLITE CELLS

Skeletal muscle stem cells are responsible for the repair and adult homeostasis of muscle throughout an organism's life span. These multipotential stem cells are wedged between the basal lamina and the differentiated myofiber, giving them the distinctive satellite shape (Figure 8.1), and thus are known as satellite cells (Mauro 1961). During homeostasis, satellite cells are quiescent and only reenter the cell cycle to self-renew or provide committed progenitors upon injury or stress (Chang and Rudnicki 2014). In this subsection, we explore the origins of satellite cells: how they arise from a common embryonic myogenic progenitor and how analogous satellite-like cells are observed and contribute to muscle repair in other organisms. In addition to frog muscle (Mauro 1961; Chen et al. 2006), satellite and satellite-like cells have also been reported and characterized in the fruit fly (Chaturvedi et al. 2017), zebrafish (Siegel et al. 2013; Berberoglu et al. 2017), other amphibians (Fei et al. 2017), birds (Halevy et al. 2004; Gros et al. 2005), mice (Seale et al. 2000), as well as in human skeletal muscle (Reimann et al. 2004) (see Figure 8.1).

8.2.1 Myogenesis in Amniotes

Mouse skeletal muscle derives from the paraxial mesoderm, which consist of two bilateral cell populations adjacent to the neural tube. Paraxial mesoderm—also known as unsegmented mesoderm—is a transient structure that further becomes

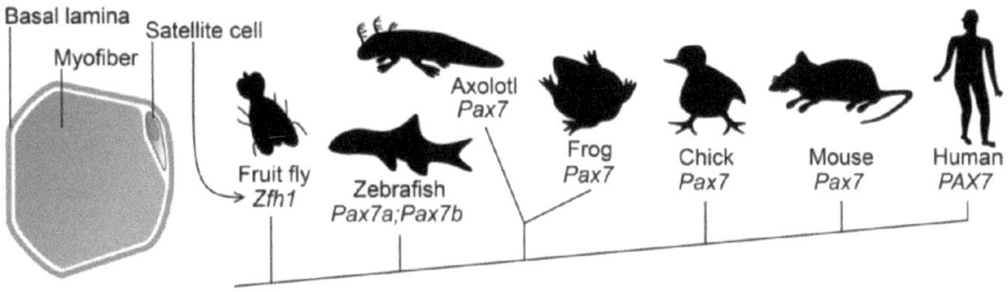

FIGURE 8.1 Skeletal muscle satellite cells are located between myofibers and the basal lamina. The phylogenetic cladogram representing evolutionary bifurcation of species with identified satellite and satellite-like cell populations, including the gene that characteristically defines the quiescent satellite cell population in each of them.

specified anterioposteriorly, generating the first segmented blocks of somites (Figure 8.2). The next phase of specification partitions somites dorsoventrally: ventromedial sclerotome gives rise to axial vertebrae, cartilage, and tendons, while the dorsolateral dermomyotome forms the dermis and skeletal muscle (Chal and Pourquié 2017; Yin et al. 2013). The spatial cell fates of the somites depend on secreted molecules and morphogen gradients from adjacent tissues, such as the neural tube and the dorsal ectoderm (Borycki et al. 1999; Hirsinger et al. 2000) (Figure 8.2). Using classical developmental biology experiments with chimeras, lineage tracing models, and genetic knockouts, *the dermomyotome, an epithelial structure, was shown to be the prospective origin and source of adult satellite cells.*

Within the dermomyotome, muscle progenitors (myoblasts) express the paired box transcription factors Pax3 and Pax7 (Galli et al. 2008; Kassar-Duchossoy et al. 2005; Relaix et al. 2005), which are evolutionarily conserved in muscle development with Pax7 playing a well-documented role in regulating the identity and function of adult satellite cells (Seale et al. 2000). At the lateral lips of the dermomyotome, myogenic precursors mature to postmitotic myocytes and become sandwiched between the dermomyotome dorsally and the sclerotome ventrally (Gros et al. 2004). These myocytes form the primary myotome, and the fusion of adjacent myocytes generates the primary myofibers, which provide the scaffold for future muscle growth (Murphy and Kardon 2011). As development continues, more myocytes migrate out and the dermomyotome loses its epithelial nature and disintegrates (Ben-Yair and Kalcheim 2005; Gros et al. 2005). The central region of the dermomyotome contains a population of $Pax3^+$ and $Pax7^+$ muscle progenitors that translocate into the myotome directly underneath (Kassar-Duchossoy et al. 2005; Relaix et al. 2005). These precursors either activate the expression of myogenic regulatory factors (MRFs), including basic helix-loop-helix (bHLH) transcription factors Myf5, MyoD, MRF4, and myogenin, or remain $Pax3^+/Pax7^+$ proliferative progenitors required for the later phases of myogenesis (Hernández-Hernández et al. 2017; Pownall et al. 2002; Tapscott 2005). The myotome is divided into the dorsomedial section, which forms the *epaxial myotome* and eventually builds the back muscles, and the ventrolateral *hypaxial myotome* that forms the limb and body wall muscles (Figure 8.2).

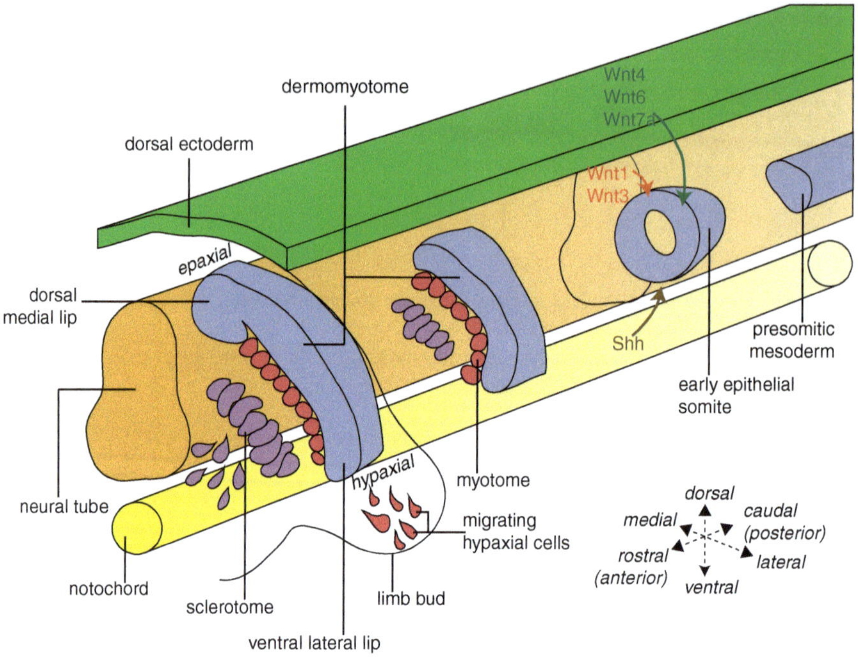

FIGURE 8.2 The embryonic origin of skeletal muscle as known from chicken and mouse embryos. A three-dimensional representation through the embryo during early (caudal end, right) and late (rostral end, left) stages of somitogenesis. Rostrocaudal, dorsoventral, and mediolateral axes are shown schematically. Dorsal ectoderm, green; mesoderm, blue; myotome/muscle cells, red; neural tube, orange; notochord, yellow; sclerotome, purple. The presomitic mesoderm becomes segmented into blocks of spherical structures, called somites. Morphogens pattern the somites to become the dermomyotome and the sclerotome under the control of Wnts emanating from the neural tube (Wnt1 and Wnt3) and the surface ectoderm (Wnt4, Wnt6, and Wnt7a) and Sonic Hedgehog (Shh) from the notochord. Within the dermomyotome, the hypaxial domain gives rise to trunk and limb muscles, including a population of hypaxial cells that migrate into the developing limb bud. The epaxial domain of the dermomyotome gives rise to the dorsal muscles of the back. (Figure adapted from Bentzinger et al. (2012) and Buckingham (2001).)

Altogether, the process of myogenesis can be divided into two phases. The first (E10.5–E12.5 in mouse embryos) involves embryonic myoblast differentiation into primary myotubes. The second phase (E14.5 to postnatal) involves fetal myoblast proliferation, fusion with the primary myotubes, or fusion with themselves to form secondary fibers (Biressi et al. 2007a). The discrete phases of myogenesis are not only temporally distinct: the myoblasts originating from each phase (embryonic, fetal, and eventually adult satellite stem cells) are intrinsically different and vary in their proliferative capacities, transcriptome, metabolic requirements, and ability to form myotubes (Biressi et al. 2007b; Pala et al. 2018; Tierney and Sacco 2016). In murine neonates, during fetal myogenesis, a subset of proliferating $Pax7^{+}$ progenitors are reserved and align juxtaposed with the myofibers in a sublaminar fashion, taking up the prospective satellite cell location (Kassar-Duchossoy et al. 2005; Relaix et al. 2005). Postnatal and prepubertal muscle growth depends on an increase in myonuclear content, also known as myonuclear accretion (Bachman et al. 2018; White et al. 2010). Fetal myogenic cells drive this phase of substantial growth, which ultimately results in a drastic decrease in their numbers. At around 3 weeks after birth, satellite cells enter quiescence as muscle growth reaches a homeostatic equilibrium.

Extensive work using mouse knockout models provided the genetic hierarchy for myogenesis. Pax3 and Pax7 are both expressed in dermomyotomal cells, with each paralog having a distinct function. Pax3$^{-/-}$ mutant mice lack limb muscles, a lack that is attributed to the loss of the migratory progenitors from the dermomyotome (Relaix et al. 2004). By contrast, Pax7$^{-/-}$ mice appear normal at birth but do not live past 2–3 weeks due to the progressive loss of satellite cells (Kuang et al. 2006; Oustanina et al. 2004; Seale et al. 2000). *Hence, Pax7 is dispensable for the specification of satellite cells during embryogenesis but is essential to maintain proliferation, survival, and identity of satellite cells in newborn mice.* Genetic ablation of Pax3 resulted in a loss of Pax7-expressing cells; yet Pax3 is downregulated in satellite cells before birth and is only expressed in select adult satellite cells of certain tissues, such as the diaphragm (Horst et al. 2006; Hutcheson et al. 2009; Kassar-Duchossoy et al. 2005; Montarras et al. 2005).

Both Pax3 and Pax7 genetically lie upstream of MRFs, as both can regulate and induce the expression of Myf5 and MyoD (Bajard et al. 2006; Hu et al. 2008; McKinnell et al. 2008). Individual knockouts of Myf5 and MyoD show only mild skeletal muscle developmental abnormalities, despite reported functional redundancy between the two factors. Interestingly though, MyoD and Myf5 nulls display delayed hypaxial and epaxial skeletal muscle development, respectively (Braun et al. 1992; Rudnicki et al. 1992, 1993). Myf5;MyoD double knockout mice, however, show a complete lack of skeletal muscle. Within the pool of Pax7^{+} fetal progenitors, a reserve population does not express any of the MRFs and is thought to become the prospective satellite cells (Kassar-Duchossoy et al. 2005).

8.2.2 Myogenesis in Other Model Organisms

Although the majority of studies have focused on amniote development, many other animals exhibit a discrete population of satellite or satellite-like cells that is crucial for muscle repair and regeneration (Sánchez Alvarado and Tsonis 2006). The Mexican axolotl (*Ambystoma mexicanum*) has the remarkable ability to regenerate whole limbs. Fate-mapping experiments in the context of limb amputation demonstrated that the axolotl utilizes Pax7^{+} (*AmPax7*) satellite cells derived from the regenerative blastema to rebuild missing muscle (Fei et al. 2017; Sandoval-Guzmán et al. 2014). Interestingly, the axolotl does not have the paralog Pax3 in its genome and solely relies on Pax7 (Nowoshilow et al. 2018). Mutations in *AmPax7* revealed embryonic phenotypes comparable to murine *Pax3* and *Pax7* knockouts, suggesting that indeed, the single *Pax3/7* homolog, *AmPax7*, has acquired the functional roles of Pax3 seen in other animals (Nowoshilow et al. 2018). This raises the question of

Pax3 and *Pax7* compensation in other vertebrate models and possible disease contexts. Future work is required to determine the necessity of *AmPax7* in adult tissues, specifically satellite cell function and repair.

In zebrafish (*Danio rerio*), there are two homologs of *Pax7*, *DrPax7a* and *DrPax7b*, a consequence of the whole-genome duplication in basal teleost fish (Seger et al. 2011). *DrPax7*$^+$ cells are somitic in origin and participate in the regeneration processes through proliferation and fusion, similar to mouse myogenesis (Chal and Pourquié 2017; Seger et al. 2011; Siegel et al. 2013). However, each Pax7 paralog seems to contribute differently to muscle regeneration: Pax7a$^+$ cells generate the primary nascent fibers, while Pax7b$^+$ cells predominantly contributed to fiber growth by fusing with existing fibers (Pipalia et al. 2016). Importantly, Pax7a/Pax7b double mutants show defective muscle repair (Berberoglu et al. 2017), similar to murine Pax7 mutants (Kuang et al. 2006; Oustanina et al. 2004; Relaix et al. 2005; Seale et al. 2000).

Eloquent time-lapse in vivo imaging of Pax7a$^+$ cells also revealed their ability to divide asymmetrically and contribute to a committed progenitor daughter or to self-renew and maintain a less committed state (Gurevich et al. 2016). In the same study, the authors used the Tg(*ubi*:zebrabow) fish crossed with various driver lines to assess the clonal relationships of satellite-like cells during repair and of embryonic myogenic progenitors during development and growth. Whereas clonal dominance was observed during regeneration, development was polyclonal, suggesting that a single developmental clone did not generate all satellite cells. Similar techniques have been used in murine studies by generating a *Pax7-CreER;R26R*$^{Brainbow2.1}$ mouse (Livet et al. 2007; Tierney et al. 2018). Similar to the zebrafish, repetitive regeneration cycles resulted in decreased clonal complexity and an increase in clonal dominance (Tierney et al. 2018).

Outside of vertebrates, tissue-resident uncommitted muscle progenitors or satellite-like cells show fewer similarities to canonical murine satellite cells. The fruit fly (*Drosophila melanogaster*) has small nucleated cells that sit juxtaposed to the basal lamina of differentiated multinucleated muscle cells, suggesting anatomical similarities to murine satellite cells. Upon tissue damage, these cells reenter the cell cycle and produce progeny to repair the missing tissue in a linear and clonal manner (Chaturvedi et al. 2017). Contrary to vertebrate satellite cells, these satellite-like cells do not express the homolog of *Pax3/Pax7*, *gooseberry* (Boukhatmi and Bray 2018). In nonbilaterian animals, the diploblastic cnidarian possesses the capacity to regenerate its "muscle"; cnidarian muscle is both ectodermally and endodermally derived (Leclère and Röttinger 2016). One cnidarian species, the starlet sea anemone *Nematostella vectensis*, contains a *Pax3/7* homolog (*NvPaxD*), but more investigations will be needed to determine whether *NvPaxD* marks an unidentified adult satellite-like population (Matus et al. 2007).

In summary, throughout the animal kingdom, many organisms display the ability to repair muscle tissue in response to stress and injury through a dedicated satellite or satellite-like cell population (Baghdadi and Tajbakhsh 2018). This deep evolutionary conservation of the cells, regenerative processes, and gene-regulatory networks suggest that insights can be gained from studying nonmammalian models. There is still a need to understand whether the satellite-like cells from zebrafish, axolotl, and *Drosophila* display a similar ontogeny to amniote satellite cells and whether a similar genetic toolkit is utilized to reserve them for adult tissue repair (Chaturvedi et al. 2017; Fei et al. 2017; Seger et al. 2011; Siegel et al. 2013).

To this end, the promise of transgenesis by CRISPR/Cas9 has opened the doors to answering these questions in all these organisms (Burgio 2018). For instance, building on the questions of clonality during development and regeneration, the combination of CRISPR barcoding cells, single-cell sequencing, and trajectory analyses will help deconvolute where and when satellite-like cells arise and identify possible regulators (Kalhor et al. 2018; Kester and van Oudenaarden 2018; McKenna et al. 2016; Raj et al. 2018). Moreover, the increase in knock-in efficiencies with CRISPR/Cas9—compared to zinc finger and transcription activator-like effector nucleases—expands the lineage tracing repertoire, which greatly benefits nonmammalian models due to their ex utero development and allows for powerful live imaging (Park et al. 2016). The promise to be able to temporally track unique populations by genetic marks (barcoding) or by fluorophores will greatly expand our knowledge of how adult satellite cells arise from an embryonic progenitor in normal, diseased, or genetically mutated/modified background.

8.3 THE REGENERATIVE POTENTIAL OF SATELLITE CELLS

8.3.1 Satellite Cells in Resting Muscle and Activation of the Stem Cell Pool

Adult regenerative myogenesis is largely governed by the satellite cell compartment. In resting muscles, satellite cells are in a quiescent state characterized by the absence of cell cycling, a low rate of metabolism, low RNA content, a low cytoplasmic volume, and a high degree of heterochromatin condensation (Schultz 1976; Cheung and Rando 2013). In response to environmental stress, such as exercise or injury, quiescent satellite cells are activated, whereby they reenter the cell cycle and proliferate extensively forming myoblasts that differentiate and fuse to repair myofibers.

Conditional and constitutive Pax7 knockout animal models have revealed that Pax7 acts as a nodal factor. On the one hand, Pax7 is developmentally essential for establishing and maintaining a myogenic satellite cell pool, while at the same time playing a key role in adult muscle as the enforcer of myogenic identity in activated proliferating precursors. Pax7 drives the expansion of the progenitor pool and initiates expression of downstream factors responsible for coordinating myofiber regeneration, preventing differentiation until Pax7 is downregulated (Seale et al. 2000; Lepper et al. 2009). A key difference between Pax7-null and inducible gene deletion models is the complete lack of a Pax7$^+$ adult satellite cell founding population in the null model compared with temporal selection of satellite cell pool depletion in the conditional knockouts, in which Pax7-deficient satellite cells cannot maintain the undifferentiated state and precociously fuse into myofibers, ultimately compromising the capacity for muscle regeneration (Kuang et al. 2006; Olguin and Olwin 2004; Relaix et al. 2004; Seale et al. 2000). Hence, *continued Pax7 expression is critical for maintaining quiescent satellite cells throughout life and for driving the proliferative expansion of activated satellite cells during regeneration* (Seale et al. 2000; von Maltzahn et al. 2013).

Comparative gene expression studies from different groups (Seale et al. 2004; Fukada et al. 2007; Liu et al. 2013) have revealed a set of more than 500 genes that are upregulated in quiescent satellite cells when compared with proliferating myoblasts. These include many negative regulators of cell cycle progression and myogenic inhibitors such as cyclin-dependent kinase inhibitors 1B (Cdkn1b) and 1C (Cdkn1c), the tumor suppressor protein retinoblastoma (Rb), and the negative regulator of fibroblast growth factor (FGF) signaling sprouty 1 (Spry1) (Fukada et al. 2007). Presumably, many of these are potentially targeted for downregulation through the process of activation. In fact, conditional knockout of Cdkn1b results in aberrant satellite cell activation and proliferation.

Among the highly expressed genes in quiescent satellite cells is the forkhead transcription factor *Foxo3*, which induces the expression of *Notch1* in satellite cells. Indeed, canonical Notch signaling is essential to maintain satellite stem cell quiescence. Perturbations of the downstream effectors of Notch signaling, such as the recombining binding protein suppressor of hairless (RBP-J), which promotes the expression of the Hes and Hey family of proteins, or double knockout of Hey and Hes, cause spontaneous activation of quiescent satellite cells and depletion of the satellite cell pool. (Bjornson et al. 2012; Fukada et al. 2011; Mourikis et al. 2012). Therefore, satellite cell quiescence is defined by an intricate genetic network and signaling mechanisms.

8.3.2 Expansion of Satellite Cells and Commitment to Myogenesis

Conversely, following muscle damage, activated satellite cells downregulate quiescence-associated pathways, such as Notch signaling, while upregulating a specific set of genes, including those involved in DNA and RNA synthesis and cell cycle progression. Additionally, a number of cell surface markers are thought to be involved in activation, including c-Met, FGFR1, FGFR4, syndecan-3, and syndecan-4 (Cornelison et al. 2001, 2004; Flanagan-Steet et al. 2000; Tatsumi et al. 1998).

Mechanical stress on muscle fibers triggers satellite cell activation through the release of nitric oxide (NO) and hepatocyte growth factor (HGF) (Anderson and Pilipowicz 2002). As a consequence of fiber damage, NO production by nitric oxide synthase (NOS) triggers the release of HGF from the extracellular matrix, which then rapidly associates with its receptor c-Met (Tatsumi et al. 2002, 2006). NOS activity also is required for the maintenance of quiescence in satellite cells; therefore, NO may play a dual role regulating satellite cell function (Wozniak and Anderson 2007). Although the direct role of c-Met activation in satellite cells remains elusive, HGF and FGF molecules can stimulate mitogen-activated protein kinase (MAPK) signaling pathways, which are known to regulate proliferation and differentiation in many cell types. It has been observed that signaling by p38α/β MAPK is required for satellite cell activation and differentiation of myogenic cells (Jones et al. 2005; Puri et al. 2000; Wu et al. 2000; Zetser et al. 1999), while p38γ inhibits differentiation (Gillespie et al. 2009; Chang et al. 2018).

For skeletal muscle to maintain its regenerative potential, there must be continual replenishing of satellite cells. Following activation, cell fate decisions specify satellite cell self-renewal or differentiation, an outcome achieved through asymmetric cell divisions. Studies from our group determined that satellite cells are a heterogeneous population of stem cells based on the expression of the two transcription factors, Pax7 and Myf5 (Kuang et al. 2007). Genetic analysis using a Cre-LoxP system (*Myf5-Cre/R26R-YFP*) revealed that about 10% of satellite cells never express YFP and, therefore, do no express Myf5 either. Prospective isolation and transplantation into injured muscle reveals that whereas $Pax7^{+}/YFP(Myf5)^{+}$ cells undergo terminal differentiation, $Pax7^{+}/YFP(Myf5)^{-}$ cells extensively contribute to the satellite cell reservoir. Satellite stem cells express Tie-2, and paracrine angiotensin-1 signaling from fibroblasts and vascular cells stimulates ERK activation to increase the number of quiescent satellite cells (Abou-Khalil et al. 2009). $Myf5^{-}$ satellite cells also express high levels of Fzd7, and signaling through the Wnt7a/Fzd7 planar cell polarity pathway drives the symmetric expansion of satellite stem cells to accelerate and augment muscle regeneration (Le Grand et al. 2009). In recent studies, our laboratory also found that binding of Wnt7a to Fzd7 in differentiated myofibers directly activates the AKT/mTOR pathway, thereby inducing hypertrophy (von Maltzahn et al. 2011). The balance between symmetric and asymmetric satellite cell divisions in muscle homeostasis is compromised in a muscular dystrophy mouse model, where lack of the structural protein dystrophin in satellite cells causes stem cell hyperplasia and impaired skeletal muscle regeneration (Dumont et al. 2015). This imbalance can in turn be ameliorated by activation of the epidermal growth factor and aurora kinase pathway, resulting in a higher number of asymmetric divisions and consequently improved muscle regeneration in the dystrophic mice (Wang et al. 2019).

Pax7 expression has the ability to maintain the undifferentiated state of satellite cells but also turn on myogenic commitment factors, such as Myf5, without permitting differentiation (Kuang et al. 2006; McKinnell et al. 2008; Soleimani et al. 2012a). Chromatin immunoprecipitation experiments performed in primary myoblasts showed that Myf5 expression is driven by Pax7 through two DNA regulatory elements located at −111 and −55 kb, relative to its transcriptional start site. Stable expression of Myf5 in committed satellite cells is achieved by methylation of Pax7 at the N-terminal domain by the arginine methyltransferase Carm1 (Kawabe et al. 2012). This methylation is required for Mll1/2 binding and for recruitment of the histone methyltransferase complex Wdr5-Ash2l-Mll1/2 (Kmt2) to the Myf5 locus, resulting in increasing levels of permissive chromatin modifications such as trimethylation of lysine 4 at histone 3 (H3K4me1). During asymmetric divisions, the interaction between Carm1 and Pax7 in the nucleus activates transcriptional expression of Myf5, thereby generating a committed daughter cell and a cascading series of molecular events directing final commitment and myogenic differentiation (Kawabe et al. 2012). Moreover, phosphorylation of Carm1 by p38γ prevents its interaction with Pax7, thereby inhibiting Kmt2 recruitment to the *Myf5* locus and inhibiting satellite cell commitment (Chang et al. 2018).

8.3.3 Myogenic Regulatory Factors and Differentiation

The process of adult myogenesis obeys a highly controlled molecular program driven by the sequential activation of transcription factors MRFs introduced above (Buckingham et al. 2003). MRFs form a select family of transcription factors whose function illustrated a paradigm *where a series of molecular switches determine the fate of an entire cell lineage.* MRFs are a group of four bHLH transcription factors that consist of MyoD, Myf5, myogenin, and MRF4 (expressed by the *Myod1*, *Myf5*, *Myog*, and *Myf6* genes, respectively). These factors act at multiple points in the muscle lineage to cooperatively establish skeletal muscle through regulation of proliferation, irreversible cell cycle arrest of precursor cells, followed by activation of sarcomeric and structural muscle-specific genes to facilitate differentiation and sarcomere assembly (Hernández-Hernández et al. 2017). Although MRF proteins contain conserved DNA-binding domains that recognize E-boxes, a motif containing the sequence CANNTG, posttranslational modifications and formation of protein complexes specify the activation of discrete target genes for each MRF (Rudnicki and Jaenisch 1995).

In vitro experiments show that Myf5 expression is only detected in myoblasts, whereas MyoD expression is maintained throughout the differentiation process (Ferri et al. 2009). Accordingly, the majority of quiescent adult satellite cells transcribe low levels of *Myf5* and *Myod1*, but both transcripts are translationally inhibited by either sequestration into mRNP granules or Staufen1, an RNA-binding protein, respectively (Crist et al. 2012; de Morrée et al. 2017). The difference in temporal expression of Myf5 and MyoD suggests that Myf5 has functions toward myoblast proliferation, whereas MyoD prepares myoblasts for entry into differentiation. MyoD is a stronger transcriptional activator of *Myog* than is Myf5, regardless of similarities in their DNA-binding domain (Ishibashi et al. 2005). The difference between MyoD and Myf5 relies on transcriptional activity rather than a preference for binding sites throughout the genome; genome-wide binding profiles of Myf5 and MyoD are identical. MyoD activates transcription of downstream genes responsible for differentiation more efficiently than does Myf5. These findings suggest that the transcriptional specificity of Myf5 and MyoD is determined by binding partners and accessibility of genomic binding sites.

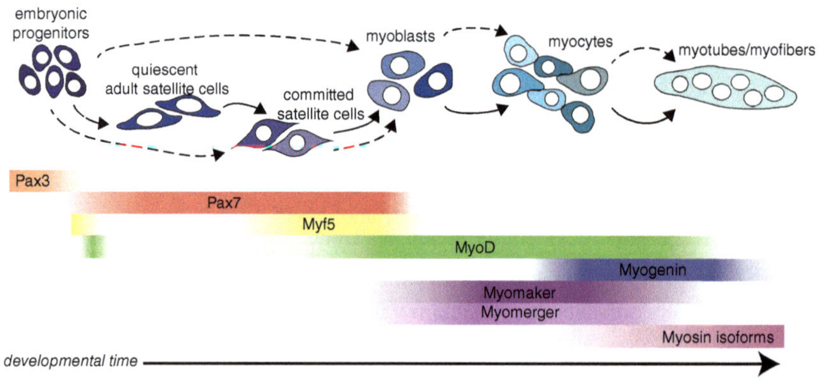

FIGURE 8.3 Comparison of myogenic gene expression over time, shown as developmental time from early (left) to later (right). Expression of myogenic genes involved in activation, differentiation, and fusion of myogenic cells is shown, along with the concomitant transition of satellite cells to myoblasts, myocytes, and ultimately fusing into myotubes or myofibers. Dotted lines show myogenic processes in embryonic development; solid lines show adult muscle regeneration following satellite cell activation.

A different study shows that Myf5 and MyoD share only ~30% of their gene targets (Soleimani et al. 2012b). Moreover, MyoD binds to a discrete set of myogenic target genes, which is blocked by the Snail1/2 repressor complex in proliferating myoblasts, to promote differentiation (Soleimani et al. 2012b). In agreement with this idea, myoblasts isolated from MyoD$^{-/-}$ mice have four-fold higher levels of Myf5, but show reduced differentiation potential (Sabourin et al. 1999). Furthermore, MyoD-deficient satellite cells resist differentiation and retain their proliferative status (Yablonka-Reuveni et al. 1999). This indicates that efficient progression through the myogenic program requires the coordinated action of Myf5 and MyoD to balancing commitment, proliferation, and differentiation. Therefore, Myf5 and MyoD hold functionally compensatory roles in determining the myogenic lineage upstream of myogenin and MRF4 (Figure 8.3) required for the expression of muscle-specific genes in terminal differentiation (Rudnicki et al. 2008).

All these studies show the properties of MRFs during myogenic differentiation and identify each of them in the hierarchical sequence of molecular events in the initiation, activation, and maintenance of muscle differentiation. The final step of myogenesis is the fusion of myocytes with one another, forming myotubes to create new myofibers, or fusion with existing myofibers to either repair damaged muscle or increase functional demand in response to exercise. The fusion process is facilitated by the recently discovered myomaker (Millay et al. 2013) and comprehensively characterized myomerger proteins (Quinn et al. 2017; Zhang et al. 2017; Bi et al. 2017), expressed by the *Tmem8c* and *Gm7325* genes, respectively. Throughout myogenic differentiation, some satellite cells maintain their regenerative capacity and, upon completion of the regenerative cycle, return to quiescence.

8.4 SATELLITE CELLS AS MULTIPOTENTIAL CELLS

8.4.1 Muscle Fiber Diversity and Myogenic Lineages

The function and concomitant metabolic requirements of skeletal muscle vary greatly. The diaphragm plays a vital role in mammalian, avian, and reptilian respiration, undergoing regular repeated cycles of contraction and relaxation, while other muscles are used on an ad hoc basis. These functional differences are possible due to *skeletal muscle fiber mosaicism*.

Although, in general, muscle fibers can be classified into type I or slow twitch fibers and type II or fast twitch fibers, there are further classifications depending on respiratory profile and myosin heavy chain composition, extensively reviewed by Schiaffino and Reggiani (2011). In mice, type I and type IIA fibers use oxidative energy production and therefore tend to be well vascularized with higher mitochondrial density; type I fibers express the *Myh7*-encoded slow myosin heavy chain, whereas type IIA fibers express *Myh2*. Type IIA fibers are also able to produce energy glycolytically, depending on oxygen availability and energy demand. Contrastingly, type IIB fibers, characterized by *Myh4* expression, and type IIX fibers that express *Myh1* are both fast twitch glycolytic fibers and therefore tend to be less vascularized and have lower mitochondrial densities (Schiaffino 2018).

Various muscles display proportional variability between these fiber types. For instance, the soleus muscle mostly comprises slow twitch fibers, the extensor digitorum longus is predominantly fast twitch, and the tibialis anterior muscle is a more balanced mosaic of the two types (Bloemberg and Quadrilatero 2012). Interestingly, the diaphragm is also a mosaic muscle; just under half the fibers are type I (Metzger et al. 1985). Moreover, satellite cell numbers vary between muscles with different fiber type composition; slow fiber-rich muscles have more resident satellite cells (Kelly 1978). The capability of satellite cells to regenerate the same fiber type as their muscle of origin has been extensively studied in a range of model animals (Kalhovde et al. 2005; reviewed by Biressi and Rando 2010).

Some of this predetermination has recently been attributed to the role of *Tbx1* expression (Motohashi et al. 2018); the variation in satellite cell abundance also could be due to the vascularization profile of muscle; interaction between adjacent satellite and endothelial cells is an essential satellite cell niche maintenance factor (Verma et al. 2018). Although some contradicting evidence exists for human satellite cells—clonally isolated cells from either slow or fast fibers create both types of myofibers (Bonavaud et al. 2001)—it may still be hypothesized that satellite cell intrinsic mechanisms dictate multiple myogenic lineages, which may yet be revealed at the epigenetic or mechanistic level in human satellite cells as well. *Whether the ability of satellite cells to generate several types of myofibers should be classified as multipotent potential is a question of interpretation*, considering that the yielding tissue is skeletal muscle, regardless of the myosin isoforms expressed in the individual fibers. However, the metabolic differences between myofibers and the abundance of functional variety resulting from fiber mosaicism, combined with satellite cell predisposition toward generating certain types of fibers, can be interpreted as evidence for multipotential capacity.

8.4.2 Adipogenesis and Osteogenesis from Satellite Cells

In addition to satellite cells inheriting their propensity toward creating a certain fiber type depending on the type of myofiber they are associated with, the muscle of origin can also affect the *adipogenic capacity* of these cells. Satellite cells isolated from muscles with a higher slow myofiber component differentiate more readily into the adipogenic lineage (Yada et al. 2006).

Indeed, (i) lineage tracing studies revealed a common Pax7-expressing progenitor for skeletal muscle and brown fat (Lepper and Fan 2010), (ii) the first observations of adipogenic differentiation from myogenic precursors were reported in studies using fatty acid treatment on myoblasts (Teboul et al. 1995), and (iii) in investigating how oxygen availability affects satellite cell proliferation on cultured myofibers (Csete et al. 2001). The proposed multipotential characteristics of satellite cells discussed in this section are summarized in Figure 8.4.

The key regulator for the switch between adipogenic and myogenic differentiation is PR domain zinc finger protein 16 (PRDM16) (Seale et al. 2008). Knocking down PRDM16 in brown fat resulted in spontaneous myogenesis, while in vitro transduction of PRDM16 into myogenic precursors leads to downregulation of MRFs with simultaneous upregulation of proteins involved in brown fat differentiation (normally

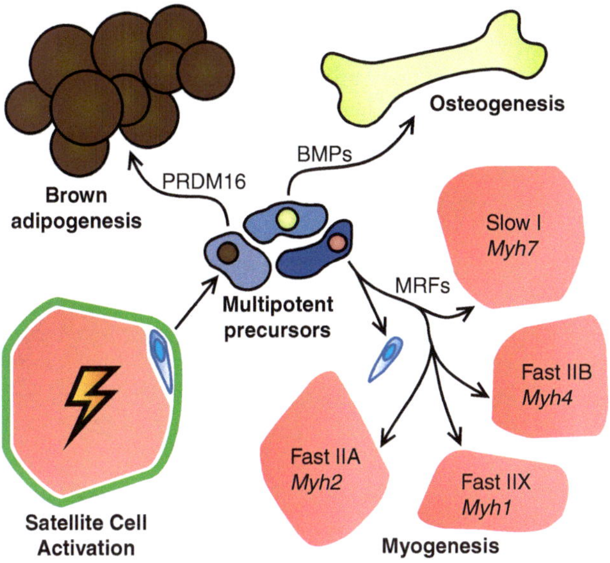

FIGURE 8.4 Multipotency of skeletal muscle satellite cells. Muscle injury, stress or regulatory signaling can lead to satellite cell activation and proliferation. Myogenic progenitors (dark blue, pink nucleus) derived from satellite cells can preferentially build different types of muscle fibers with characteristic myosin heavy chain gene (*Myh*) expression profiles. The transition into mature muscle fibers (four types depicted) is orchestrated by myogenic regulatory factors (MRFs). Activated satellite cells can return to quiescence, retaining their multipotency or they can differentiate into nonmyogenic lineages. Differentiation into adipocytes (brown) is mediated by PR domain zinc finger protein 16 (PRDM16), while differentiation into osteoblasts is promoted by bone morphogenetic proteins (BMPs).

repressed in myogenic precursors) and consequent brown adipogenesis. This switch mechanism is controlled by miR-133, which targets the 3'UTR of *Prdm16* mRNA; miR-133 downregulation can result in brown adipogenesis from satellite cells (Yin et al. 2013). Furthermore, conditional deletion of Pax7 with concomitant constitutive activation of Notch signaling results in copious brown adipogenic differentiation from satellite cells (Pasut et al. 2016).

Until a decade ago, it was widely believed that human brown adipose tissue was only present and metabolically active in infancy. However, three seminal publications published back-to-back in the same journal caused a paradigm shift in the field by identifying brown adipose tissue in adults and illustrating the inverse correlation of its abundance to body mass index (Cypess et al. 2009; van Marken Lichtenbelt et al. 2009; Virtanen et al. 2009). It was subsequently shown that some adult human brown adipocytes are derived from white adipose tissue, and thus are named, in a developmentally more appropriate way, as beige fat (Wu et al. 2012). The intertwined developmental relationship between skeletal muscle and metabolically active adipose tissue is also evident in glycolytic beige fat in humans. Fascinatingly, using a β-adrenergic receptor knockout model, it has now been illustrated that in response to thermal stress—exposure to low temperature—white adipose cells become beige fat through a myogenic intermediate stage (Chen et al. 2019). Taken together, these mechanistic discoveries in brown and beige adipose tissue development could lead to therapeutic avenues toward combating obesity and diabetes.

The *osteogenic capacity* of myogenic progenitors was first reported in bone morphogenetic protein (BMP) 2-treated myoblasts by Katagiri et al. (1994) and later in transplantation studies by Lee et al. (2000). Their clonally derived mouse skeletal muscle cells spontaneously differentiated into myotubes in vitro, fused into muscle fibers in vivo, as well as becoming osteogenic in response to BMP2 treatment both in vitro and by forming ectopic bone in vivo. This type of multipotent transdifferentiation from satellite cells into osteogenic cells was subsequently confirmed by BMP7 and BMP4 treatment of primary cells in culture along with spontaneous differentiation of cells that migrated away from isolated single myofibers in culture (Asakura et al. 2001). This osteogenic potential was shown to be enhanced by MyoD expression through validation of reduced osteogenic potential of $MyoD^{-/-}$ mouse muscle-derived primary myogenic cells (Komaki et al. 2004). Furthermore, satellite cells isolated from human fetal muscle exhibit osteogenic potential with BMP2 treatment (Ozeki et al. 2006).

Although mounting evidence has led to a consensus on the adipo- and osteogenic potential of skeletal muscle resident cells (reviewed by Owston et al. 2016), it remains to be shown whether satellite cells can directly contribute toward physiological repair in an in vivo bone injury setting or be coaxed into generating brown adipocytes to potentially burn excess energy and attenuate risk factors in the rising diabetic and prediabetic patient cohorts. While some contradicting datasets have been added to the ongoing debate regarding satellite cell multipotency, in some cases they are refutable, either because they used myoblasts, which are committed to the myogenic lineage, or because of inappropriate lineage tracing models to identify satellite cells and their fate (Starkey et al. 2011). In fact, there is some evidence to support the conclusion that satellite cells can even spontaneously contribute toward adipo- and osteogenesis and that fate determination happens before the first division postactivation of the satellite cells (Shefer et al. 2004). All of these multipotential capacities of satellite cells are depicted in Figure 8.4.

It could well be that neither spontaneous, growth factor, nor nutrient-driven plasticity in muscle progenitor populations will translate into physiological injury response mechanisms or medical condition treatment and prevention but nevertheless will provide further investigative avenues to our general understanding of adult stem cell multipotency. As mentioned previously, technological advances in lineage tracing, single cell mass cytometry, and transcriptome sequencing, combined with cell barcoding, could elucidate both the true osteogenic and brown adipose contribution from inherent and transplanted satellite and other skeletal muscle–derived cells.

8.5 CONCLUSION

This chapter has highlighted (i) the intricate developmental regulation of myogenesis and the establishment of a multipotential skeletal muscle reserve population in the form of satellite cells, (ii) the role these cells play in tissue regeneration upon activation, and (iii) the potential of satellite cells to differentiate into alternative tissue types.

Mounting evidence discussed herein highlights the conservation of satellite cells in skeletal muscle of many organisms, including humans. Furthermore, the developmental origin of these cells often is also conserved, as is the biological function of this deferred-use cell population in adults, along with the characteristic nodal regulatory transcription factor Pax7 expression as a defining identifier for satellite cells.

Whether the multipotency of skeletal muscle satellite cells, observed in some model organisms, is also a conserved characteristic of the population remains to be confirmed. Satellite cell preference to give rise to specific muscle fiber types can be considered a multipotential feature of the cells. Further studies using a range of model organisms will be required to establish whether this is either a cell-intrinsic predetermination of fiber type lineage or a combination of cell-specific features and the functional environment affecting activated satellite cell commitment to generating fibers with a distinct metabolic profile. Certainly, satellite cell regenerative and multipotential capacities offer a host of opportunities for clinical applications even beyond neuromuscular disorders. The potentiation of brown adipose tissue function is already a trending area of research for combating diabetes. Skeletal muscle satellite cells could therefore prove to be useful for restoring metabolic and wider homeostatic processes in the whole organism. We look forward to future research establishing clinically and therapeutically relevant roles for satellite cells in combating conditions also related to mesodermal tissues other than skeletal muscle.

REFERENCES

Abou-Khalil, R., Le Grand, F., Pallafacchina, G., Valable, S., Authier, F.-J., Rudnicki, M. A., Gherardi, R. K. et al. 2009. Autocrine and paracrine angiopoietin 1/Tie-2 signaling promotes muscle satellite cell self-renewal. *Cell Stem Cell* 5: 298–309. doi: 10.1016/j.stem.2009.06.001.

Anderson, J., and Pilipowicz, O. 2002. Activation of muscle satellite cells in single-fiber cultures. *Nitric Oxide: Biol. Chem.* 7: 36–41.

Asakura, A., Komaki, M., and Rudnicki, M. A. 2001. Muscle satellite cells are multipotential stem cells that exhibit myogenic, osteogenic, and adipogenic differentiation. *Differentiation* 68: 245–253.

Bachman, J. F., Klose, A., Liu, W., Paris, N. D., Blanc, R. S., Schmalz, M., Emma Knapp, E., and Chakkalakal, J. V. 2018. Prepubertal skeletal muscle growth requires Pax7-expressing satellite cell-derived myonuclear contribution. *Development* 145: dev167197. doi: 10.1242/dev.167197.

Baghdadi, M. B., and Tajbakhsh, S. 2018. Regulation and phylogeny of skeletal muscle regeneration. *Dev. Biol.* 433: 200–209. doi: 10.1016/j.ydbio.2017.07.026.

Bajard, L., Relaix, F., Lagha, M., Rocancourt, D., Daubas, P., and Buckingham, M. E. 2006. A novel genetic hierarchy functions during hypaxial myogenesis: Pax3 directly activates Myf5 in muscle progenitor cells in the limb. *Genes Dev.* 20: 2450–2464. doi: 10.1101/gad.382806.

Ben-Yair, R., and Kalcheim, C. 2005. Lineage analysis of the avian dermomyotome sheet reveals the existence of single cells with both dermal and muscle progenitor fates. *Development* 132: 689–701. doi: 10.1242/dev.01617.

Bentzinger, C. F., Wang, Y. X., and Rudnicki, M. A. 2012. Building muscle: Molecular regulation of myogenesis. *Cold Spring Harbor Perspect. Biol.* 4: doi: 10.1101/cshperspect.a008342.

Berberoglu, M. A., Gallagher, T. L., Morrow, Z. T., Talbot, J. C., Hromowyk, K. J., Tenente, I. M., Langenau, D. M., and Amacher, S. L. 2017. Satellite-like cells contribute to Pax7-dependent skeletal muscle repair in adult zebrafish. *Dev. Biol.* 424: 162–180. doi: 10.1016/j.ydbio.2017.03.004.

Bi, P., Ramirez-Martinez, A., Li, H., Cannavino, J., McAnally, J. R., Shelton, J. M., Sanchez-Ortiz, E., Bassel-Duby, R., and Olson, E. N. 2017. Control of muscle formation by the fusogenic micropeptide myomixer. *Science* 356: 323–327. doi: 10.1126/science.aam9361.

Biressi, S., and Rando, T. A. 2010. Heterogeneity in the muscle satellite cell population. *Semin. Cell Dev. Biol.* 21: 845–854. doi: 10.1016/j.semcdb.2010.09.003.

Biressi, S., Molinaro, M., and Cossu, G. 2007a. Cellular heterogeneity during vertebrate skeletal muscle development. *Dev. Biol.* 308: 281–293. doi: 10.1016/j.ydbio.2007.06.006.

Biressi, S., Tagliafico, E., Lamorte, G., Monteverde, S., Tenedini, E., Roncaglia, E., Ferrari, S. et al. 2007b. Intrinsic phenotypic diversity of embryonic and fetal myoblasts is revealed by genome-wide gene expression analysis on purified cells. *Dev. Biol.* 304: 633–651. doi: 10.1016/j.ydbio.2007.01.016.

Bjornson, C. R. R., Cheung, T. H., Liu, L., Tripathi, P. V., Steeper, K. M., and Rando, T. A. 2012. Notch signaling is necessary to maintain quiescence in adult muscle stem cells. *Stem Cells* 30: 232–242. doi: 10.1002/stem.773.

Bloemberg, D., and Quadrilatero, J. 2012. Rapid determination of myosin heavy chain expression in rat, mouse, and human skeletal muscle using multicolor immunofluorescence analysis. *PLoS One* 7(4): e35273. doi: 10.1371/journal.pone.0035273.

Bonavaud, S., Agbulut, O., Nizard, R., D'honneur, G., Mouly, V., and Butler-Browne, G. 2001. A discrepancy resolved: Human satellite cells are not preprogrammed to fast and slow lineages. *Neuromuscular Disord.* 11: 747–752.

Borycki, A. G., Brunk, B., Tajbakhsh, S., Buckingham, M., Chiang, C., and Emerson, C. P. 1999. Sonic hedgehog controls epaxial muscle determination through myf5 activation. *Development* 126: 4053–4063.

Boukhatmi, H., and Bray, S. 2018. A population of adult satellite-like cells in drosophila is maintained through a switch in rna-isoforms. *eLife* 7: 878. doi: 10.7554/eLife.35954.

Braun, T., Rudnicki, M. A., Arnold, H. H., and Jaenisch, R. 1992. Targeted inactivation of the muscle regulatory gene myf-5 results in abnormal rib development and perinatal death. *Cell* 71: 369–382.

Buckingham, M. 2001. Skeletal muscle formation in vertebrates. *Curr. Opin. Genet. Dev.* 11: 440–448. doi: 10.1016/S0959-437X(00)00215-X.

Buckingham, M., Bajard, L., Chang, T., Daubas, P., Hadchouel, J., Meilhac, S., Montarras, D., Rocancourt, D., and Relaix, F. 2003. The formation of skeletal muscle: From somite to limb. *J Anat.* 202: 59–68. doi: 10.1046/j.1469-7580.2003.00139.x.

Burgio, G. 2018. Redefining mouse transgenesis with CRISPR/Cas9 genome editing technology. *Genome Biol.* 19: 27. doi: 10.1186/s13059-018-1409-1.

Chal, J., and Pourquié, O. 2017. Making muscle: Skeletal myogenesis *in vivo* and *in vitro*. *Development* 144: 2104–2122. doi: 10.1242/dev.151035.

Chang, N. C., and Rudnicki, M. A. 2014. Satellite cells: The architects of skeletal muscle. *Curr. Top. Dev. Biol.* 107: 161–181. doi: 10.1016/B978-0-12-416022-4.00006-8.

Chang, N. C., Sincennes, M.-C., Chevalier, F. P., Brun, C. E., Lacaria, M., Segalés, J., Muñoz-Cánoves, P., Ming, H., and Rudnicki, M. A. 2018. The dystrophin glycoprotein complex regulates the epigenetic activation of muscle stem cell commitment. *Cell Stem Cell* 22: 755–756. doi: 10.1016/j.stem.2018.03.022.

Chaturvedi, D., Reichert, H., Gunage, R. D., and VijayRaghavan, K. 2017. Identification and functional characterization of muscle satellite cells in *Drosophila*. *eLife* 6: 641. doi: 10.7554/eLife.30107.

Chen, Y., Lin, G., and Slack, J. M. W. 2006. Control of muscle regeneration in the *Xenopus* tadpole tail by Pax7. *Development* 133: 2303–2313. doi: 10.1242/dev.02397.

Chen, Y., Ikeda, K., Yoneshiro, T., Scaramozza, A., Tajima, K., Wang, Q., Kim, K. et al. 2019. Thermal stress induces glycolytic beige fat formation via a myogenic state. *Nature* 565: 180–185. doi: 10.1038/s41586-018-0801-z.

Cheung, T. H., and Rando, T. A. 2013. Molecular regulation of stem cell quiescence. *Nat. Rev. Mol. Cell Biol.* 14: 329–340. doi: 10.1038/nrm3591.

Cornelison, D. D. W., Filla, M. S., Stanley, H. M., Rapraeger, A. C., and Olwin, B. B. 2001. Syndecan-3 and syndecan-4 specifically mark skeletal muscle satellite cells and are implicated in satellite cell maintenance and muscle regeneration. *Dev. Biol.* 239: 79–94. doi: 10.1006/dbio.2001.0416.

Cornelison, D. D. W., Wilcox-Adelman, S. A., Goetinck, P. F., Rauvala, H., Rapraeger, A. C., and Olwin, B. B. 2004. Essential and separable roles for syndecan-3 and syndecan-4 in skeletal muscle development and regeneration. *Genes Dev.* 18: 2231–2236. doi: 10.1101/gad.1214204.

Crist, C. G., Montarras, D., and Buckingham, M. 2012. Muscle satellite cells are primed for myogenesis but maintain quiescence with sequestration of Myf5 mRNA targeted by microRNA-31 in mRNP granules. *Cell Stem Cell* 11: 118–126. doi: 10.1016/j.stem.2012.03.011.

Csete, M., Walikonis, J., Slawny, N., Wei, Y., Korsnes, S., Doyle, J. C., and Wold, B. 2001. Oxygen-mediated regulation of skeletal muscle satellite cell proliferation and adipogenesis in culture. *J. Cell Physiol.* 189: 189–196. doi: 10.1002/jcp.10016.

Cypess, A. M., Lehman, S., Williams, G., Tal, I., Rodman, D., Goldfine, A. B., Kuo, F. C. et al. 2009. Identification and importance of brown adipose tissue in adult humans. *New Engl. J. Med.* 360: 1509–1517. doi: 10.1056/NEJMoa0810780.

de Morrée, A., van Velthoven, C. T. J., Gan, Q., Salvi, J. S., Klein, J. D. D., Akimenko, I., Quarta, M., Biressi, S., and Rando, T. A. 2017. Staufen1 inhibits MyoD translation to actively maintain muscle stem cell quiescence. *Proc. Natl. Acad. Sci. USA* 114: E8996–E9005. doi: 10.1073/pnas.1708725114.

Dumont, N. A., Wang, Y. X., von Maltzahn, J., Pasut, A., Bentzinger, C. F., Brun, C. E., and Rudnicki, M. A. 2015. Dystrophin expression in muscle stem cells regulates their polarity and asymmetric division. *Nat. Med.* 21: 1455–1463. doi: 10.1038/nm.3990.

Fei, J.-F., Schuez, M., Knapp, D., Taniguchi, Y., Drechsel, D. N., and Tanaka, E. M. 2017. Efficient gene knockin in axolotl and its use to test the role of satellite cells in limb regeneration. *Proc. Natl. Acad. Sci. USA* 114: 12501–12506. doi: 10.1073/pnas.1706855114.

Ferri, P., Barbieri, E., Burattini, S., Guescini, M., D'Emilio, A., Biagiotti, L., Del Grande, P. et al. 2009. Expression and subcellular localization of myogenic regulatory factors during the differentiation of skeletal muscle C2C12 myoblasts. *J. Cell. Biochem.* 108: 1302–1317. doi: 10.1002/jcb.22360.

Flanagan-Steet, H., Hannon, K., McAvoy, M. J., Hullinger, R., and Olwin, B. B. 2000. Loss of FGF receptor 1 signaling reduces skeletal muscle mass and disrupts myofiber organization in the developing limb. *Dev. Biol.* 218: 21–37. doi: 10.1006/dbio.1999.9535.

Fukada, S.-I., Uezumi, A., Ikemoto, M., Masuda, S., Segawa, M., Tanimura, N., Yamamoto, H., Miyagoe-Suzuki, Y., and Takeda, S. 2007. Molecular signature of quiescent satellite cells in adult skeletal muscle. *Stem Cells* 25: 2448–2459. doi: 10.1634/stemcells.2007-0019.

Fukada, S.-I., Yamaguchi, M., Kokubo, H., Ogawa, R., Uezumi, A., Yoneda, T., Matev, M. M. et al. 2011. Hesr1 and Hesr3 are essential to generate undifferentiated quiescent satellite cells and to maintain satellite cell numbers. *Development* 138: 4609–4619. doi: 10.1242/dev.067165.

Galli, L. M., Knight, S. R., Barnes, T. L., Doak, A. K., Kadzik, R. S., and Burrus, L. W. 2008. Identification and characterization of subpopulations of Pax3 and Pax7 expressing cells in developing chick somites and limb buds. *Dev. Dyn.* 237: 1862–1874. doi: 10.1002/dvdy.21585.

Gillespie, M. A., Le Grand, F., Scimè, A., Kuang, S., von Maltzahn, J., Seale, V., Cuenda, A., Ranish, J. A., and Rudnicki, M. A. 2009. P38-γ-dependent gene silencing restricts entry into the myogenic differentiation program. *J. Cell Biol.* 187: 991–1005. doi: 10.1083/jcb.200907037.

Giordani, L., He, G. J., Negroni, E., Sakai, H., Law, J. Y., Siu, M. M., Wan, R. et al. March 2019. High-dimensional single-cell cartography reveals novel skeletal muscle-resident cell populations. *Mol. Cell* 74: 609–621. doi: 10.1016/j.molcel.2019.02.026.

Grenier, G., Scimè, A., Le Grand, F., Asakura, A., Perez-Iratxeta, C., Andrade-Navarro, M. A., Labosky, P. A., and Rudnicki, M. A. 2007. Resident endothelial precursors in muscle, adipose, and dermis contribute to postnatal vasculogenesis. *Stem Cells* 25: 3101–3110. doi: 10.1634/stemcells.2006-0795.

Gros, J., Manceau, M., Thomé, V., and Marcelle, C. 2005. A common somitic origin for embryonic muscle progenitors and satellite cells. *Nature* 435: 954–958. doi: 10.1038/nature03572.

Gros, J., Scaal, M., and Marcelle, C. 2004. A two-step mechanism for myotome formation in chick. *Dev. Cell* 6: 875–882. doi: 10.1016/j.devcel.2004.05.006.

Gurevich, D. B., Nguyen, P. D., Siegel, A. L., Ehrlich, O. V., Sonntag, C., Phan, J. M. N., Berger, S. et al. 2016. Asymmetric division of clonal muscle stem cells coordinates muscle regeneration in vivo. *Science* 353: aad9969. doi: 10.1126/science.aad9969.

Halevy, O., Piestun, Y., Allouh, M. Z., Rosser, B. W. C., Rinkevich, Y., Reshef, R., Rozenboim, I., Wleklinski-Lee, M., and Yablonka-Reuveni, Z. 2004. Pattern of Pax7 expression during myogenesis in the posthatch chicken establishes a model for satellite cell differentiation and renewal. *Dev. Dyn.* 231: 489–502. doi: 10.1002/dvdy.20151.

Hernández-Hernández, J. M., García-González, E. G., Brun, C. E., and Rudnicki, M. A. 2017. The myogenic regulatory factors, determinants of muscle development, cell identity and regeneration. *Sem. Cell Dev. Biol.* 72: 10–18. doi: 10.1016/j.semcdb.2017.11.010.

Hirsinger, E., Jouve, C., Dubrulle, J., and Pourquié, O. 2000. Somite formation and patterning. *Int. Rev. Cytol.* 198: 1–65.

Horst, D., Ustanina, S., Sergi, C., Mikuz, G., Juergens, H., Braun, T., and Vorobyov, E. 2006. Comparative expression analysis of Pax3 and Pax7 during mouse myogenesis. *Int. J. Dev. Biol.* 50: 47–54. doi: 10.1387/ijdb.052111dh.

Hu, P., Geles, K. G., Paik, J.-H., Depinho, R. A., and Tjian, R. 2008. Codependent activators direct myoblast-specific MyoD transcription. *Dev. Cell* 15: 534–546. doi: 10.1016/j.devcel.2008.08.018.

Hutcheson, D. A., Zhao, J., Merrell, A., Haldar, M., and Kardon, G. 2009. Embryonic and fetal limb myogenic cells are derived from developmentally distinct progenitors and have different requirements for β-catenin. *Genes Dev.* 23: 997–1013. doi: 10.1101/gad.1769009.

Ishibashi, J., Perry, R. L., Asakura, A., and Rudnicki, M. A. 2005. MyoD induces myogenic differentiation through cooperation of its NH_2- and COOH-terminal regions. *J. Cell Biol.* 171: 471–482. doi: 10.1083/jcb.200502101.

Joe, A. W. B., Yi, L., Natarajan, A., Le Grand, F., So, L., Wang, J., Rudnicki, M. A., and Rossi, F. M. V. 2010. Muscle injury activates resident fibro/adipogenic progenitors that facilitate myogenesis. *Nat. Cell Biol.* 12: 153–163. doi: 10.1038/ncb2015.

Jones, N. C., Tyner, K. J., Nibarger, L., Stanley, H. M., Cornelison, D. D. W., Fedorov, Y. V., and Olwin, B. B. 2005. The P38α/βMAPK functions as a molecular switch to activate the quiescent satellite cell. *J. Cell Biol.* 169: 105–116. doi: 10.1083/jcb.200408066.

Juban, G., and Chazaud, B. 2017. Metabolic regulation of macrophages during tissue repair: Insights from skeletal muscle regeneration. *FEBS Lett.* 591: 3007–3021. doi: 10.1002/1873-3468.12703.

Kalhor, R., Kalhor, K., Mejia, L., Leeper, K., Graveline, A., Mali, P., and Church, G. M. 2018. Developmental barcoding of whole mouse via homing CRISPR. *Science* 361: eaat9804. doi: 10.1126/science.aat9804.

Kalhovde, J. M., Jerkovic, R., Sefland, I., Cordonnier, C., Calabria, E., Schiaffino, S., and Lømo, T. 2005. Fast and 'slow' muscle fibres in hind limb muscles of adult rats regenerate from intrinsically different satellite cells. *J. Physiol.* 562: 847–857. doi: 10.1113/jphysiol.2004.073684.

Kassar-Duchossoy, L., Giacone, E., Gayraud-Morel, B., Jory, A., Gomès, D., and Tajbakhsh, S. 2005. Pax3/Pax7 mark a novel population of primitive myogenic cells during development. *Genes Dev.* 19: 1426–1431. doi: 10.1101/gad.345505.

Katagiri, T., Yamaguchi, A., Komaki, M., Abe, E., Takahashi, N., Ikeda, T., Rosen, V., Wozney, J. M., Fujisawa-Sehara, A., and Suda, T. 1994. Bone morphogenetic protein-2 converts the differentiation pathway of C2C12 myoblasts into the osteoblast lineage. *J. Cell Biol.* 127: 1755–1766.

Kawabe, Y.-I., Wang, Y. X., McKinnell, I. W., Bedford, M. T., and Rudnicki, M. A. 2012. Carm1 regulates Pax7 transcriptional activity through MLL1/2 recruitment during asymmetric satellite stem cell divisions. *Cell Stem Cell* 11: 333–345. doi: 10.1016/j.stem.2012.07.001.

Kelly, A. M. 1978. Satellite cells and myofiber growth in the rat soleus and extensor digitorum longus muscles. *Dev. Biol.* 65: 1–10.

Kester, L., and van Oudenaarden, A. 2018. Single-cell transcriptomics meets lineage tracing. *Cell Stem Cell* 23: 166–179. doi: 10.1016/j.stem.2018.04.014.

Komaki, M., Asakura, A., Rudnicki, M. A., Sodek, J., and Cheifetz, S. 2004. MyoD enhances BMP7-induced osteogenic differentiation of myogenic cell cultures. *J. Cell Sci.* 117: 1457–1468. doi: 10.1242/jcs.00965.

Kuang, S., Kuroda, K., Le Grand, F., and Rudnicki, M. A. 2007. Asymmetric self-renewal and commitment of satellite stem cells in muscle. *Cell* 129: 999–1010. doi: 10.1016/j.cell.2007.03.044.

Kuang, S., Chargé, S. B., Seale, P., Huh, M., and Rudnicki, M. A. 2006. Distinct roles for Pax7 and Pax3 in adult regenerative myogenesis. *J. Cell Biol.* 172: 103–113. doi: 10.1083/jcb.200508001.

Latroche, C., Gitiaux, C., Chretien, F., Desguerre, I., Mounier, R., and Chazaud, B. 2015. Skeletal muscle microvasculature: A highly dynamic lifeline. *Physiology* 30: 417–427. doi: 10.1152/physiol.00026.2015.

Leclère, L., and Röttinger, E. 2016. Diversity of cnidarian muscles: Function, anatomy, development and regeneration. *Front. Cell Dev. Biol.* 4: 157. doi: 10.3389/fcell.2016.00157.

Lee, J. Y., Qu-Petersen, Z., Cao, B., Kimura, S., Jankowski, R., Cummins, J., Usas, A. et al. 2000. Clonal isolation of muscle-derived cells capable of enhancing muscle regeneration and bone healing. *J. Cell Biol.* 150: 1085–1100.

Le Grand, F., Jones, A. E., Seale, V., Scimè, A., and Rudnicki, M. A. 2009. Wnt7a activates the planar cell polarity pathway to drive the symmetric expansion of satellite stem cells. *Cell Stem Cell* 4: 535–547. doi: 10.1016/j.stem.2009.03.013.

Lepper, C., Conway, S. J., and Fan, C.-M. 2009. Adult satellite cells and embryonic muscle progenitors have distinct genetic requirements. *Nature* 460: 627–631. doi: 10.1038/nature08209.

Lepper, C., and Fan, C.-M. 2010. Inducible lineage tracing of Pax7-descendant cells reveals embryonic origin of adult satellite cells. *Genesis* 48: 424–436. doi: 10.1002/dvg.20630.

Liu, L., Cheung, T. H., Charville, G. W., Hurgo, B. M. C., Leavitt, T., Shih, J., Brunet, A., and Rando, T. A. 2013. Chromatin modifications as determinants of muscle stem cell quiescence and chronological aging. *Cell Rep.* 4: 189–204. doi: 10.1016/j.celrep.2013.05.043.

Livet, J., Weissman, T. A., Kang, H., Draft, R. W., Lu, J., Bennis, R. A., Sanes, J. R., and Lichtman, J. W. 2007. Transgenic strategies for combinatorial expression of fluorescent proteins in the nervous system. *Nature* 450: 56–62. doi: 10.1038/nature06293.

Matus, D. Q., Pang, K., Daly, M., and Martindale, M. Q. 2007. Expression of Pax gene family members in the anthozoan cnidarian, Nematostella vectensis. *Evol. Dev.* 9: 25–38. doi: 10.1111/j.1525-142X.2006.00135.x.

Mauro, A. 1961. Satellite cell of skeletal muscle fibers. *J. Biophys. Biochem. Cytol.* 9: 493–495.

McKenna, A., Findlay, G. M., Gagnon, J. A., Horwitz, M. S., Schier, A. F., and Shendure, J. 2016. Whole-organism lineage tracing by combinatorial and cumulative genome editing. *Science* 353: aaf7907. doi: 10.1126/science.aaf7907.

McKinnell, I. W., Ishibashi, J., Le Grand, F., Punch, V. G. J., Addicks, G. C., Greenblatt, J. F., Dilworth, F. J., and Rudnicki, M. A. 2008. Pax7 activates myogenic genes by recruitment of a histone methyltransferase complex. *Nat. Cell Biol.* 10: 77–84. doi: 10.1038/ncb1671.

Metzger, J. M., Scheidt, K. B., and Fitts, R. H. 1985. Histochemical and physiological characteristics of the rat diaphragm. *J. Appl. Physiol.* 58: 1085–1091. doi: 10.1152/jappl.1985.58.4.1085.

Millay, D. P., O'Rourke, J. R., Sutherland, L. B., Bezprozvannaya, S., Shelton, J. M., Bassel-Duby, J. R., and Olson, E. N. 2013. Myomaker is a membrane activator of myoblast fusion and muscle formation. *Nature* 499: 301–305. doi: 10.1038/nature12343.

Montarras, D., Morgan, J., Collins, C., Relaix, F., Zaffran, S., Cumano, A., Partridge, T., and Buckingham, M. 2005. Direct isolation of satellite cells for skeletal muscle regeneration. *Science* 309: 2064–2067. doi: 10.1126/science.1114758.

Motohashi, N., Uezumi, A., Asakura, A., Ikemoto-Uezumi, M., Mori, S., Mizunoe, Y., Takashima, R. et al. 2018. Tbx1 regulates inherited metabolic and myogenic abilities of progenitor cells derived from slow- and fast-type muscle. *Cell Death Differ.* 9: 493. doi: 10.1038/s41418-018-0186-4.

Mourikis, P., Gopalakrishnan, S., Sambasivan, R., and Tajbakhsh, S. 2012. Cell-autonomous Notch activity maintains the temporal specification potential of skeletal muscle stem cells. *Development* 139: 4536–4548. doi: 10.1242/dev.084756.

Murphy, M., and Kardon, G. 2011. Origin of vertebrate limb muscle: The role of progenitor and myoblast populations. *Curr. Top. Dev. Biol.* 96: 1–32. doi: 10.1016/B978-0-12-385940-2.00001-2.

Nowoshilow, S., Schloissnig, S., Fei, J.-F., Dahl, A., Pang, A. W. C., Pippel, M., Winkler, S. et al. 2018. The Axolotl genome and the evolution of key tissue formation regulators. *Nature* 554: 50–55. doi: 10.1038/nature25458.

Olguin, H. C., and Olwin, B. B. 2004. Pax-7 Up-regulation inhibits myogenesis and cell cycle progression in satellite cells: A potential mechanism for self-renewal. *Dev. Biol.* 275: 375–88. doi: 10.1016/j.ydbio.2004.08.015.

Oustanina, S., Hause, G., and Braun, T. 2004. Pax7 directs postnatal renewal and propagation of myogenic satellite cells but not their specification. *EMBO J.* 23: 3430–3439. doi: 10.1038/sj.emboj.7600346.

Owston, H., Giannoudis, P. V., and Jones, E. 2016. Do skeletal muscle MSCs in humans contribute to bone repair? A systematic review. *Injury* 47(Suppl 6): S3–S15. doi: 10.1016/S0020-1383(16)30834-8.

Ozeki, N., Lim, M., Yao, C.-C., Tolar, M., and Kramer, R. H. 2006. Alpha7 integrin expressing human fetal myogenic progenitors have stem cell-like properties and are capable of osteogenic differentiation. *Exp. Cell Res.* 312: 4162–4180. doi: 10.1016/j.yexcr.2006.09.017.

Pala, F., Di Girolamo, D., Mella, S., Yennek, S., Chatre, L., Ricchetti, M., and Tajbakhsh, S. 2018. Distinct metabolic states govern skeletal muscle stem cell fates during prenatal and postnatal myogenesis. *J. Cell Sci.* 131: jcs212977. doi: 10.1242/jcs.212977.

Park, S., Greco, V., and Cockburn, K. 2016. Live imaging of stem cells: Answering old questions and raising new ones. *Curr. Opin. Cell Biol.* 43: 30–37. doi: 10.1016/j.ceb.2016.07.004.

Pasut, A., Chang, N. C., Rodriguez, U. G., Faulkes, S., Yin, H., Lacaria, M., Ming, H., and Rudnicki, M. A. 2016. Notch signaling rescues loss of satellite cells lacking Pax7 and promotes brown adipogenic differentiation. *Cell Rep.* 16: 333–343. doi: 10.1016/j.celrep.2016.06.001.

Pipalia, T. G., Koth, J., Roy, S. D., Hammond, C. L., Kawakami, K., and Hughes, S. M. 2016. Cellular dynamics of regeneration reveals role of two distinct Pax7 stem cell populations in larval zebrafish muscle repair. *Dis. Models Mech.* 9: 671–684. doi: 10.1242/dmm.022251.

Pownall, M. E., Gustafsson, M. K., and Emerson, C. P. 2002. Myogenic regulatory factors and the specification of muscle progenitors in vertebrate embryos. *Annu. Rev. Cell Dev. Biol.* 18: 747–783. doi: 10.1146/annurev.cellbio.18.012502.105758.

Puri, P. L., Wu, Z., Zhang, P., Wood, L. D., Bhakta, K. S., Han, J., Feramisco, J. R., Karin, M., and Wang, J. Y. 2000. Induction of terminal differentiation by constitutive activation of p38 MAP kinase in human rhabdomyosarcoma cells. *Genes Dev.* 14: 574–584.

Quinn, M. E., Goh, Q., Kurosaka, M., Gamage, D. G., Petrany, M. J., Prasad, V., and Millay, D. P. 2017. Myomerger induces fusion of non-fusogenic cells and is required for skeletal muscle development. *Nat. Commun.* 8: 1–9. doi: 10.1038/ncomms15665.

Raj, B., Wagner, D. E., McKenna, A., Pandey, S., Klein, A. M., Shendure, J., Gagnon, J. A., and Schier, A. F. 2018. Simultaneous single-cell profiling of lineages and cell types in the vertebrate brain. *Nat. Biotechnol.* 36: 442–450. doi: 10.1038/nbt.4103.

Reimann, J., Brimah, K., Schröder, R., Wernig, A., Beauchamp, J. R., and Partridge, T. A. 2004. Pax7 distribution in human skeletal muscle biopsies and myogenic tissue cultures. *Cell Tissue Res.* 315: 233–242. doi: 10.1007/s00441-003-0833-y.

Relaix, F., Rocancourt, D., Mansouri, A., and Buckingham, M. 2004. Divergent functions of murine Pax3 and Pax7 in limb muscle development. *Genes Dev.* 18: 1088–1105. doi: 10.1101/gad.301004.

Relaix, F., Rocancourt, D., Mansouri, A., and Buckingham, M. 2005. A Pax3/Pax7-dependent population of skeletal muscle progenitor cells. *Nature* 435: 948–953. doi: 10.1038/nature03594.

Rudnicki, M. A., and Jaenisch, R. 1995. The MyoD family of transcription factors and skeletal myogenesis. *Bioessays* 17: 203–209. doi: 10.1002/bies.950170306.

Rudnicki, M. A., Braun, T., Hinuma, S., and Jaenisch, R. 1992. Inactivation of MyoD in mice leads to up-regulation of the myogenic HLH gene Myf-5 and results in apparently normal muscle development. *Cell* 71: 383–390.

Rudnicki, M. A., Le Grand, F., McKinnell, I., and Kuang, S. 2008. The molecular regulation of muscle stem cell function. *Cold Spring Harbor Symp. Quant. Biol.* 73: 323–331. doi: 10.1101/sqb.2008.73.064.

Rudnicki, M. A., Schnegelsberg, P. N., Stead, R. H., Braun, T., Arnold, H. H., and Jaenisch, R. 1993. MyoD or Myf-5 is required for the formation of skeletal muscle. *Cell* 75: 1351–1359.

Sabourin, L. A., Girgis-Gabardo, A., Seale, P., Asakura, A., and Rudnicki, M. A. 1999. Reduced differentiation potential of primary MyoD$^{-/-}$ myogenic cells derived from adult skeletal muscle. *J. Cell Biol.* 144: 631–643.
Sandoval-Guzmán, T., Wang, H., Khattak, S., Schuez, M., Roensch, K., Nacu, E., Tazaki, A., Joven, A. et al. 2014. Fundamental differences in dedifferentiation and stem cell recruitment during skeletal muscle regeneration in two salamander species. *Cell Stem Cell* 14: 174–187. doi: 10.1016/j.stem.2013.11.007.
Sánchez Alvarado, A., and Tsonis, P. A. 2006. Bridging the regeneration gap: Genetic insights from diverse animal models. *Nat. Rev. Gen.* 7: 873–884. doi: 10.1038/nrg1923.
Schiaffino, S. 2018. Muscle fiber type diversity revealed by anti-myosin heavy chain antibodies. *FEBS J.* 285: 3688–3694. doi: 10.1111/febs.14502.
Schiaffino, S., and Reggiani, C. 2011. Fiber types in mammalian skeletal muscles. *Physiol. Rev.* 91: 1447–1531. doi: 10.1152/physrev.00031.2010.
Schultz, E. 1976. Fine structure of satellite cells in growing skeletal muscle. *Am. J. Anat.* 147: 49–70. doi: 10.1002/aja.1001470105.
Seale, P., Ishibashi, J., Holterman, C., and Rudnicki, M. A. 2004. Muscle satellite cell-specific genes identified by genetic profiling of MyoD-deficient myogenic cells. *Dev. Biol.* 275: 287–300. doi: 10.1016/j.ydbio.2004.07.034.
Seale, P., Sabourin, L. A., Girgis-Gabardo, A., Mansouri, A., Gruss, P., and Rudnicki, M. A. 2000. Pax7 is required for the specification of myogenic satellite cells. *Cell* 102: 777–786.
Seale, P., Bjork, B., Yang, W., Kajimura, S., Chin, S., Kuang, S., Scimè, A. et al. 2008. PRDM16 controls a brown fat/skeletal muscle switch. *Nature* 454: 961–967. doi: 10.1038/nature07182.
Seger, C., Hargrave, M., Wang, X., Chai, R. J., Elworthy, S., and Ingham, P. W. 2011. Analysis of Pax7 expressing myogenic cells in zebrafish muscle development, injury, and models of disease. *Dev. Dyn.* 240: 2440–2451. doi: 10.1002/dvdy.22745.
Shefer, G., Wleklinski-Lee, M., and Yablonka-Reuveni, Z. 2004. Skeletal muscle satellite cells can spontaneously enter an alternative mesenchymal pathway. *J. Cell Sci.* 117: 5393–5404. doi: 10.1242/jcs.01419.
Siegel, A. L., Gurevich, D. B., and Currie, P. D. 2013. A myogenic precursor cell that could contribute to regeneration in zebrafish and its similarity to the satellite cell. *FEBS J.* 280: 4074–4088. doi: 10.1111/febs.12300.
Soleimani, V. D., Punch, V. G., Kawabe, Y.-I., Jones, A. E., Palidwor, G. A., Porter, C. J., Cross, J. W. et al. 2012a. Transcriptional dominance of Pax7 in adult myogenesis is due to high-affinity recognition of homeodomain motifs. *Dev. Cell* 22: 1208–1220. doi: 10.1016/j.devcel.2012.03.014.
Soleimani, V. D., Yin, H., Jahani-Asl, A., Ming, H., Kockx, C. E. M., van IJcken, W. F. J., Grosveld, F., and Rudnicki, M. A. 2012b. Snail regulates MyoD binding-site occupancy to direct enhancer switching and differentiation-specific transcription in myogenesis. *Mol. Cell* 47: 457–468. doi: 10.1016/j.molcel.2012.05.046.
Starkey, J. D., Yamamoto, M., Yamamoto, S., and Goldhamer, D. J. 2011. Skeletal muscle satellite cells are committed to myogenesis and do not spontaneously adopt non-myogenic fates. *J. Histochem. Cytochem.* 59: 33–46. doi: 10.1369/jhc.2010.956995.
Tapscott, S. J. 2005. The circuitry of a master switch: Myod and the regulation of skeletal muscle gene transcription. *Development* 132: 2685–2695. doi: 10.1242/dev.01874.
Tatsumi, R., Anderson, J. E., Nevoret, C. J., Halevy, O., and Allen, R. E. 1998. HGF/SF is present in normal adult skeletal muscle and is capable of activating satellite cells. *Dev. Biol.* 194: 114–128. doi: 10.1006/dbio.1997.8803.
Tatsumi, R., Hattori, A., Ikeuchi, Y., Anderson, J. E., and Allen, R. E. 2002. Release of hepatocyte growth factor from mechanically stretched skeletal muscle satellite cells and role of pH and nitric oxide. *Mol. Biol. Cell* 13: 2909–2918. doi: 10.1091/mbc.e02-01-0062.

Tatsumi, R., Liu, X., Pulido, A., Morales, M., Sakata, T., Dial, S., Hattori, A., Ikeuchi, Y., and Allen, R. E. 2006. Satellite cell activation in stretched skeletal muscle and the role of nitric oxide and hepatocyte growth factor. *Am. J. Physiol.* 290: C1487–C1494. doi: 10.1152/ajpcell.00513.2005.

Teboul, L., Gaillard, D., Staccini, L., Inadera, H., Amri, E. Z., and Grimaldi, P. A. 1995. Thiazolidinediones and fatty acids convert myogenic cells into adipose-like cells. *J. Biol. Chem.* 270: 28183–28187.

Tierney, M. T., and Sacco, A. 2016. Satellite cell heterogeneity in skeletal muscle homeostasis. *Trends Cell Biol.* 26: 434–444. doi: 10.1016/j.tcb.2016.02.004.

Tierney, M. T., Stec, M. J., Rulands, S., Simons, B. D., and Sacco, A. 2018. Muscle stem cells exhibit distinct clonal dynamics in response to tissue repair and homeostatic aging. *Cell Stem Cell* 22: 119–127. doi: 10.1016/j.stem.2017.11.009.

van Marken Lichtenbelt, W. D., Vanhommerig, J. W., Smulders, N. M., Drossaerts, J. M. A. F. L., Ger Kemerink, G. J., Nic Bouvy, N. D., Schrauwen, P., and Teule, G. J. J. 2009. Cold-activated brown adipose tissue in healthy men. *New Engl. J. Med.* 360: 1500–1508. doi: 10.1056/NEJMoa0808718.

Verma, M., Asakura, Y., Murakonda, B. S. R., Pengo, T., Latroche, C., Chazaud, B., McLoon, L. K., and Asakura, A. 2018. Muscle satellite cell cross-talk with a vascular niche maintains quiescence via VEGF and Notch signaling. *Cell Stem Cell* 23: 530–539. doi: 10.1016/j.stem.2018.09.007.

Virtanen, K. A., Lidell, M. E., Orava, J., Heglind, M., Westergren, R., Niemi, T., Taittonen, M. et al. 2009. Functional brown adipose tissue in healthy adults. *New Engl. J. Med.* 360: 1518–1525. doi: 10.1056/NEJMoa0808949.

von Maltzahn, J., Bentzinger, C. F., and Rudnicki, M. A. 2011. Wnt7a-Fzd7 signalling directly activates the Akt/mTOR anabolic growth pathway in skeletal muscle. *Nat. Cell Biol.* 14: 186–191. doi: 10.1038/ncb2404.

von Maltzahn, J., Jones, A. E., Parks, R. J., and Rudnicki, M. A. 2013. Pax7 is critical for the normal function of satellite cells in adult skeletal muscle. *Proc. Natl. Acad. Sci. USA* 110: 16474–16479. doi: 10.1073/pnas.1307680110.

Wang, Y. X., Feige, P., Brun, C. E., Hekmatnejad, B., Dumont, N. A., Renaud, J.-M., Faulkes, S., Guindon, D. E., and Rudnicki, M. A. 2019. EGFR-aurka signaling rescues polarity and regeneration defects in Dystrophin-deficient muscle stem cells by increasing asymmetric divisions. *Cell Stem Cell* 24: 419–432.e6. doi: 10.1016/j.stem.2019.01.002.

White, R. B., Biérinx, A.-S., Gnocchi, V. F., and Zammit, P. S. 2010. Dynamics of muscle fibre growth during postnatal mouse development. *BMC Dev. Biol.* 10: 21. doi: 10.1186/1471-213X-10-21.

Wozniak, A. C., and Anderson, J. E. 2007. Nitric oxide-dependence of satellite stem cell activation and quiescence on normal skeletal muscle fibers. *Dev. Dyn.* 236: 240–250. doi: 10.1002/dvdy.21012.

Wu, Z., Woodring, P. J., Bhakta, K. S., Tamura, K., Wen, F., Feramisco, J. R., Karin, M., Wang, J. Y., and Puri, P. L. 2000. P38 and extracellular signal-regulated kinases regulate the myogenic program at multiple steps. *Mol. Cell Biol.* 20: 3951–3964.

Wu, J., Boström, P., Sparks, L. M., Ye, L., Choi, J. H., Giang, A.-H., Khandekar, M. et al. 2012. Beige adipocytes are a distinct type of thermogenic fat cell in mouse and human. *Cell* 150: 366–376. doi: 10.1016/j.cell.2012.05.016.

Yablonka-Reuveni, Z., Rudnicki, M. A., Rivera, A. J., Primig, M., Anderson, J. E., and Natanson, P. 1999. The transition from proliferation to differentiation is delayed in satellite cells from mice lacking MyoD. *Dev. Biol.* 210: 440–455. doi: 10.1006/dbio.1999.9284.

Yada, E., Yamanouchi, K., and Nishihara, M. 2006. Adipogenic potential of satellite cells from distinct skeletal muscle origins in the rat. *J. Vet. Med. Sci.* 68: 479–486.

Yin, H., Pasut, A., Soleimani, V. D., Bentzinger, C. F., Antoun, G., Thorn, S., Seale, P. et al. 2013. MicroRNA-133 controls brown adipose determination in skeletal muscle satellite cells by targeting Prdm16. *Cell Metab.* 17: 210–224. doi: 10.1016/j.cmet.2013.01.004.

Yin, H., Price, F., and Rudnicki, M. A. 2013. Satellite cells and the muscle stem cell niche. *Physiol. Rev.* 93: 23–67. doi: 10.1152/physrev.00043.2011.

Zetser, A., Gredinger, E., and Bengal, E. 1999. P38 mitogen-activated protein kinase pathway promotes skeletal muscle differentiation participation of the Mef2c transcription factor. *J. Biol. Chem.* 274: 5193–5200.

Zhang, Q., Vashisht, A. A., O'Rourke, J., Corbel, S. Y., Moran, R., Romero, A., Miraglia, L. et al. 2017. The microprotein minion controls cell fusion and muscle formation. *Nat. Commun.* 8: 1–15. doi: 10.1038/ncomms15664.

Section III

Deferred-Use Cells in Development and Evolution

9 Sustained Pluripotency Underwrites Extreme Developmental and Reproductive Plasticity

Erin L Davies and Alejandro Sánchez Alvarado
Howard Hughes Medical Institute & Stowers Institute for Medical Research

CONTENTS

9.1 INTRODUCTION

Potency, or *developmental potential*, denotes the ability of a cell to assume alternate identities. As potency increases, so does the repertoire of developmental trajectories and attendant fates available to a cell. Potent cell populations underwrite many key aspects of animal biology. For instance, reproduction and the establishment of form and function during embryogenesis are driven by *totipotent* cells (e.g., the zygote), while *pluripotent* cells contribute to all lineages comprising the embryo proper. During adulthood, *multipotent* and *unipotent* adult stem cells frequently underwrite tissue homeostasis, producing cell type(s) comprising their host tissue. Potency dysregulation, or mutations arising within potent cell populations, can cause or contribute to infertility, early pregnancy loss, congenital defects, degenerative diseases, aging, or neoplasia. This immediate and impactful link to human health has galvanized efforts to elucidate the molecular mechanisms governing establishment, reprogramming, maintenance, and exit from pluripotency to inform applications in reproductive and regenerative medicine.

Here we explore the relationship between potency, regulation of potency, and their likely roles in driving the evolution of biological attributes such as regeneration. In mammals and most animals studied to date, pluripotency is fleeting, only existing for a short time early in embryogenesis prior to the construction of specialized anatomical systems. We consider planarian flatworms, emerging research organisms that naturally evolved strategies for postnatal perpetuation of pluripotency programs in somatic stem cells. We examine how *adult pluripotent stem cells* (PSCs) underwrite *developmental plasticity, regenerative ability, and reproductive strategies in planarians* and report on the developmental origin of adult PSCs. Finally, we consider the implications of sustained pluripotency on our understanding of regeneration and the regulation of cellular potency in other animals.

9.2 PLANARIAN PLURIPOTENT STEM CELLS FUEL TISSUE HOMEOSTASIS AND REGENERATION

Unique among well-established research organisms, planarian flatworms contain an abundant population of somatic adult PSCs and lineage-primed progenitors that constitute the cellular basis for tissue maintenance, regeneration, and reproduction in these bilaterally symmetrical, triploblastic animals (Elliott and Sánchez Alvarado 2013; Newmark and Sánchez Alvarado 2002; Reddien 2018; Rink 2013).

Planarians are free-living and long-lived carnivores indigenous to terrestrial, marine, and freshwater habitats. The adult body plan is mutable (liable to change) as the product of continuous development that is underwritten by resident PSCs. Planarians dynamically undergo proportional growth and degrowth according to nutrient availability (Baguñà and Romero 1981; Oviedo et al. 2003; Romero and Baguñà 1991). Most adult planarians are highly regenerative: small tissue fragments excised from virtually anywhere in the body are frequently capable of whole-body regeneration, seamlessly integrating new and preexisting tissues (Reddien and Sánchez Alvarado 2004). Indeed, freshwater-dwelling *Schmidtea mediterranea* remains the best-studied species among free-living Platyhelminths, due in large part to its stable diploid karyotype, existence of stable asexually and sexually reproducing strains, and seemingly inexhaustible whole-body regenerative abilities (Elliott and Sánchez Alvarado 2013; Newmark and Sánchez Alvarado 2002; Rink 2013). However, regeneration competence varies across species and may exhibit dependence on axial position, tissue composition, or life cycle stage within a species (Bardeen 1902; Liu et al. 2013; Sikes and Newmark 2013; Umesono et al. 2013).

While the remarkable plasticity of planarians has long been appreciated, more than a century of experimentation was required to molecularly and functionally characterize neoblasts, the potent cell population underlying tissue maintenance and regeneration. Neoblast, originally a term coined for purported set-aside embryonic cells required for new tissue formation in the segmented earthworm *Lumbriculus* (Randolph 1891), was later applied to cycling, undifferentiated mesenchymal cells in planarians (Randolph 1897), other Lophotrochozoans such as the marine flatworm

Macrostomum lignano (Grudniewska et al. 2016), the parasitic flatworm *Schistosoma mansoni* (Collins et al. 2013), and acoels such as the three-banded panther worm *Hofstenia miamia* (Gehrke and Srivastava 2016).

Prior to the introduction of modern molecular techniques, neoblast identification was largely based on morphological characteristics gleaned from light and transmission electron microscopy. Neoblasts comprise an abundant population of small, cycling mesenchymal cells with large nuclei housing decondensed chromatin and prominent nucleoli, and scant, RNA-rich cytoplasm containing perinuclear RNP granules called chromatoid bodies (Coward 1974; Hay and Coward 1975; Hori 1982; Pedersen 1959). Historical usage of the term and theories linking neoblasts to the regenerative capacity of planarians are nicely summarized by Baguñà (2012). Earlier studies showed that cycling cells could be completely and irreversibly eliminated by ionizing irradiation, compromising both long-term viability and regenerative and reproductive abilities (Bardeen and Baether 1904; Curtis and Hickman 1926). Subsequent reports corroborated that cells meeting the morphological criteria of neoblasts were irradiation sensitive, while experiments employing partial irradiation followed by amputation suggested that neoblasts were required for blastema formation, an early and necessary step in regeneration of missing tissue (Dubois and Wolff 1947; Dubois 1949; Wolff and Dubois 1948). BrdU pulse labeling experiments, coupled with immunostaining for the G2-M-specific modification H3S10p, later confirmed that neoblasts were both the sole cycling somatic cell population in adult asexually reproducing planarians and the source of new tissue in the regeneration blastema (Newmark and Sánchez Alvarado 2000). Moreover, continuous BrdU labeling experiments suggested that all neoblasts were actively cycling (Newmark and Sánchez Alvarado 2000), providing an explanation as to why neoblasts are acutely irradiation sensitive.

Premolecular era experiments, while correct in asserting broad requirements for the neoblast population in tissue dynamics, ultimately lacked sufficient granularity to address whether molecular and/or functional heterogeneity existed among cycling cells. Are neoblasts a homogeneous, equipotent population? Or rather, does the neoblast population encompass subgroups with unique contributions to tissue homeostasis and regeneration? Two key neoblast attributes, their cycling status and irradiation sensitivity, facilitated development of protocols that enabled molecular and functional characterization of neoblasts. Fluorescence-activated cell sorting protocols using Hoechst or vital DNA dyes were developed for prospective isolation of dividing (4n) neoblasts, early postmitotic progenitors and postmitotic differentiated cells from asexual adult animals (Reddien et al. 2005; Hayashi et al. 2006). Irradiation doses were established for complete, irreversible neoblast elimination (lethal irradiation) (Reddien et al. 2005) and for partial neoblast elimination (sublethal irradiation), which serves as the basis for colony formation and repopulation assays Lei et al. 2016; Wagner et al. 2011, 2012).

In combination with genomic tools (e.g., sequenced *S. mediterranea* genome assemblies (Robb et al. 2008, 2015; Grohme et al. 2018), customized *S. mediterranea* microarrays (Wagner et al. 2012; Eisenhoffer et al. 2008), comprehensive *S. mediterranea* transcriptomes (Adamidi et al. 2011; Abril et al. 2010; Blythe et al. 2010) and next-generation sequencing methodologies) and RNAi (Sánchez

Alvarado and Newmark 1999)), these techniques enabled detailed molecular characterization of neoblasts and their progeny (Reddien et al. 2005; Eisenhoffer et al. 2008; Wagner et al. 2012; Solana et al. 2012; Onal et al. 2012; Labbé et al. 2012; Shibata et al. 2012), analysis of clonal proliferative capacities (Wagner et al. 2012), and identification of genes regulating symmetric and asymmetric neoblast division (Lei et al. 2016). Key findings about the molecular logic of sustained pluripotency and lineage commitment have been reported by Dattani et al. (2018), Elliott and Sánchez Alvarado (2013), Krishna et al. (2019), Reddien (2013), Rink (2013), and Shibata et al. (2010).

9.2.1 Cell Transplantation Experiments Demonstrate Stem Cell Pluripotency

Cell transplantation experiments were key to determining whether pluripotency was a property of a collective or of individual cells in the neoblast population. Bulk cell transplantation experiments showed that neoblast-enriched donor fractions, but not differentiated donor cells, were capable of rescuing lethally irradiated hosts that completely lacked stem cells (Baguñà et al. 1989), providing strong evidence that the neoblast population, wholly or in part, was the cellular agent of homeostatic and regenerative potential in planarians. Tissue graft experiments between *S. mediterranea* sexual donors and lethally-irradiated asexual hosts similarly showed long-term rescue and donor-dependent reconstitution of the host, as determined by the development of secondary sex characteristics and the assumption of egg capsule laying (Guedelhoefer and Sánchez Alvarado 2012). Moreover, serial transplantation experiments showed that repopulating stem cells retained their collective pluripotency during colony expansion, migration, and homing to vacated niches in the irradiated host mesenchyme: donor tissue harvested anterior to the primary tissue graft at least 10 days post–primary transplant efficiently rescued lethally irradiated *S. mediterranea* asexual recipients (Guedelhoefer and Sánchez Alvarado 2012). Quantification of donor-derived, migratory neoblasts at the site of the secondary transplant suggested that a mean estimated 168 cells were required to rescue a lethally irradiated host (Guedelhoefer and Sánchez Alvarado 2012). Although elegantly executed, this experiment did not resolve putative differences in clonogenic potential or potency within the transplanted neoblast populations.

Wagner and colleagues experimentally demonstrated the existence of a PSC in asexual *S. mediterranea* (Wagner et al. 2011). Single asexual donor cells, collected using a modified cell sorting strategy that enriched for a heterogeneous cell population encompassing cycling neoblasts (X1(FS) cells), were injected into the mesenchyme of lethally irradiated *S. mediterranea* sexual hosts devoid of stem cells. A few injected donor cells were competent to engraft, undergo replicative expansion, produce differentiated progeny associated with several lineages, and rescue the host from lethality, a reproducible, albeit extremely low-frequency, occurrence. Long-term reconstitution of sexual hosts by asexual donor cells was confirmed via strain-specific SNP analysis on rescued hosts, as well as by resorption of the hermaphroditic reproductive system and assumption of asexual reproductive behavior (i.e., fission).

Titrated, sublethal doses of irradiation eliminate the vast majority of cycling neoblasts, facilitating analysis of cell division and differentiation potential for a handful of rare, spatially isolated cells in vivo (Wagner et al. 2011). Sublethal irradiation and single-cell transplantation are complementary assays enabling functional characterization of clonogenic neoblasts (cNeoblasts): highly proliferative, colony-forming cells with broad differentiation potential (Wagner et al. 2011). Notably, experiments that quantified and compared colony number, survivorship, and regeneration competence following sublethal irradiation estimated that three to five colonies were sufficient for rescue and regeneration (Wagner et al. 2011). Post hoc comparison of colony number, colony size, host survivorship, and regenerative potential across experiments suggests that small numbers of self-renewing, migratory adult PSCs robustly repopulate host animals and produce the suite of differentiating cell types necessary to rebuild and maintain the adult worm.

9.2.2 Unified Molecular and Functional Definition for Neoblasts

Cell transplantation and colony formation assays provide an operational, retrospective definition of cNeoblasts that conclusively demonstrate the existence of somatic adult PSCs in planarians. However, it remained unclear whether the low frequency of rescue in single-cell transplantation experiments was due to heterogeneity in the starting donor cell population—e.g., contamination by nonneoblast cells or inherent differences in potency, cell cycle, metabolic activity, or commitment among neoblasts)—or technical limitations of the assay. Moreover, whether cNeoblasts defined in different contexts (e.g., colony-forming units following sublethal or partial-body irradiation, repopulation following single cell transplantation, regenerative blastema formation) comprise one or many discrete neoblast subpopulations was unknown.

Single-cell RNA-Seq on sorted, cycling cells possessing high mRNA and protein levels for the neoblast-specific biomarker *piwi-1* identified a neoblast subpopulation (Nb2) that contained cNeoblasts (Zeng et al. 2018). Colony-founding cells that survived sublethal irradiation invariably expressed Nb2-enriched molecular markers, while knockdown of the Nb2-enriched gene, *tspan-1*, impeded neoblast repopulation in sublethally irradiated animals (Zeng et al. 2018). Nb2 neoblasts also appear to be critical players during regeneration. A specific increase in the Nb2 neoblast subpopulation was observed following wounding, suggesting that Nb2 neoblasts are responsive to amputation-induced cues that promote proliferation, migration, and construction of the regeneration blastema. Moreover, *tspan-1* knockdown impaired directed neoblast migration to the wound site in amputated, partially irradiated animals (Zeng et al. 2018).

Using antibodies raised against TSPAN-1, a cell surface transmembrane protein, Zeng and colleagues devised a cell sorting strategy to prospectively isolate Nb2 neoblasts from asexual *S. mediterranea*. When introduced into the mesenchyme of lethally irradiated sexual *S. mediterranea* adults, single TSPAN-1^{+} cells were capable of engrafting, proliferating, recolonizing, and rescuing hosts from lethality (Zeng et al. 2018). Notably, the rescue efficiency for single TSPAN-1^{+} donor cells was considerably higher than for X1(FS) cells (Zeng et al. 2018), likely reflecting the increased probability of selecting a cNeoblast from among the TSPAN-1^{+} sorted cells.

In sum, these studies suggest that the Nb2 neoblast subpopulation underlies tissue homeostasis, repopulation, and regeneration. Future studies will address whether additional neoblast subpopulations are multi- or pluripotent, whether specialized neoblast subpopulations contain long-term self-renewing cells or arise continuously from Nb2 cNeoblasts, whether interconversion occurs among neoblast states, and how the exit from pluripotency is executed during lineage commitment.

9.3 PLANARIAN PLURIPOTENT STEM CELLS FUEL REPRODUCTION

PSC-dependent plasticity also facilitates use of different reproductive strategies across and within planarian species. Cross-fertilizing hermaphrodites may reproduce sexually as in the freshwater triclad *S. mediterranea* (Guo et al. 2016; Hyman 1951) or parthenogenetically as in the freshwater flatworm *Schmidtea polychroa* (D'Souza and Michiels 2009)). Adults lacking a reproductive system reproduce asexually by fission, relying on neoblast-driven regeneration to form complete individuals from fission fragments (Newmark et al. 2008; Hyman 1951). Moreover, adults within a species may be competent to switch between sexual and asexual modes of reproduction as in the triclad *Dugesia ryukyuensis* (Hoshi et al. 2003). The reproductive plasticity of planarians in particular, and of the phylum Platyhelminthes in general (Laumer et al. 2015), may reflect an evolutionary novelty likely to have arisen by the ability of these organisms to retain adult somatic PSCs in their body plans.

Planarians, like mammals, undergo *inductive germ line specification* (Newmark et al. 2008). Unlike mammals, planaria retain the ability to specify new germ cells de novo throughout postnatal life. Existing data support the hypothesis that spermatogonial and oogonial stem cells are descendants of adult PSCs. A somatic origin for the male and female germ line was first suggested by Morgan, who demonstrated that head fragments devoid of gonads and somatic accessory reproductive organs regenerated the entire reproductive system during whole-body regeneration (Morgan 1902). Reproductive organ resorption occurs in response to amputation, either due to cessation of central nervous system–derived trophic support signals and/or to tissue remodeling and rescaling, i.e., reestablishing proper body allometric proportions after injury (Wang et al. 2007; Ghiradelli 1965). Sublethal irradiation (Fedecka-Bruner 1967, 1965) and starvation (Newmark et al. 2008) also lead to regression of the reproductive system. Notably, spermatogonial and oogonial stem cells frequently remain following resorption of the reproductive system (Wang et al. 2010; Tharp et al. 2014; Collins et al. 2010; Saberi et al. 2016; Zhang et al. 2018), allowing for regeneration of the reproductive system and the resumption of fertility.

Molecular analysis of reproductive system development suggests that spermatogonial stem cells expressing the RNA-binding protein *nanos* are absent early in *S. mediterranea* embryogenesis; they first arise as construction of the adult anatomy begins during Stage 6 (Davies et al. 2017). Upon hatching, *S. mediterranea* juveniles contain testes primordia (dorsolaterally positioned clusters of *nanos*+ spermatogonial stem cells); gonad and accessory reproductive system development proceeds postnatally (Wang et al. 2007; Handberg-Thorsager and Saló 2007). RNAi knockdown studies *suggest that spermatogonial and oogonial stem cells are specified via distinct genetic mechanisms* (Wang et al. 2010), though oogenesis remains an understudied, enigmatic process in planarians. Comparison of gene function during juvenile development and regeneration suggests that sex-specific germ cell specification mechanisms operate throughout life, and as a general rule, most genes required for germ line stem cell specification are similarly required for gonad maintenance (Wang et al. 2007, 2010; Tharp et al. 2014; Saberi et al. 2016; Iyer et al. 2016; Collins et al. 2010; Chong et al. 2013).

As expected, *nanos* function is required for both testis and ovary formation during *S. mediterranea* juvenile development, extending its evolutionarily conserved role in gametogenesis (Wang et al. 2007). Close-range, noncell autonomous cues from *dmd-1*$^{+}$ somatic gonadal niche cells, themselves adult PSC descendants, are required for specification of *nanos*$^{+}$ spermatogonial stem cells (Chong et al. 2013). During adulthood, *dmd-1* function is similarly required for maintenance of the testes and male accessory reproductive organs—seminal vesicles and sperm ducts (Chong et al. 2013). However, in direct contrast to its male-specific expression pattern and phenotype in sexually mature *S. mediterranea, dmd-1* function is required for ovary formation and development of the female reproductive system during juvenile development (Chong et al. 2013). These seemingly contradictory observations hint that the reproductive system may form in a stereotyped sequence, with ovary development being contingent on testes formation. Interestingly, *nanos* function was similarly required for development of the somatic reproductive system despite its early germ line–restricted expression pattern (Wang et al. 2007), suggesting that noncell autonomous, germ line–derived cues may likewise be required for specification of attendant reproductive organs. Downstream of germ cell specification, neuropeptide signaling within the central nervous system and soma-derived lipophilic hormone signaling are required for gonad formation and maintenance, underscoring the importance of noncell autonomous communication during gametogenesis (Saberi et al. 2016; Collins et al. 2010; Tharp et al. 2014; Zhang et al. 2018).

9.4 EMBRYONIC ORIGIN OF PLANARIAN PLURIPOTENT STEM CELLS

Our understanding of potency regulation and neoblast behavior is gleaned from studies performed using clonal, asexually reproducing *S. mediterranea* or *Dugesia japonica* adults, where neoblasts are ever-present.

Until recently, the developmental origin of neoblasts was completely obscure. Are neoblasts effectively persistent embryonic stem cells? Or rather, are neoblasts specified during embryogenesis? To determine the provenance of neoblasts and the molecular mechanisms that establish sustained postnatal pluripotency, Davies and colleagues transformed *S. mediterranea* into a bona fide developmental model system, establishing a molecular staging series informed by single embryo RNA-Seq and the technical toolkit necessary to study molecular processes and gene function during embryogenesis. Using these resources and heterochronic, heterotopic cell transplantation assays, they determined that neoblasts likely arise from undifferentiated, cycling blastomeres (undifferentiated embryonic cells) as adult organogenesis begins (Davies et al. 2017).

Bulk cell transplantation of staged embryonic cells into lethally irradiated *S. mediterranea* sexual adult hosts showed that embryos harvested during or after Stage 6 contain cells that collectively behave like cNeoblasts: they engraft into vacated niches, proliferate and, in some instances, are competent to rescue adult hosts from lethality (Davies et al. 2017). Embryonic cells from earlier stages were largely incapable of persisting in the adult microenvironment, even when comparable numbers of cells expressing the pan-neoblast marker *piwi-1* were injected. *In the future, single-cell transplantation experiments will show whether emergent adult pluripotency is a collective property of the embryonic donor cells or is activity harbored by single, clonogenic cells.* Additionally, transplantation of indelibly labeled embryonic cells will facilitate analysis of repopulation dynamics and fate determination in the adult microenvironment.

Surprisingly, many genes exhibiting neoblast-enriched expression in adult asexual animals were expressed throughout *S. mediterranea* embryogenesis (Davies et al. 2017), including chromatoid body-associated proteins, post-transcriptional regulators of gene expression, chromatin remodeling proteins, transcription factors required for stem cell maintenance, and Nb-2-enriched genes such as *tspan-1*. Whole-mount in situ hybridization experiments confirmed expression of adult neoblast-enriched genes throughout the undifferentiated, cycling blastomere population in early embryos (Davies et al. 2017). These observations suggest that neoblasts may sustain activity of pluripotency programs established early in embryogenesis. If true, this begs the question of why early embryonic cells were unresponsive to adult niche cues in bulk cell transplantation assays. Cell-intrinsic differences in gene expression between embryonic blastomeres and adult neoblasts, observed via single embryo RNA-Seq and corroborated by whole-mount in situ hybridization (Davies et al. 2017), provide molecular footholds that may hold the key to deciphering the molecular mechanisms underlying establishment of embryonic pluripotency and the transition to adult pluripotency.

9.5 OUTLOOK

In many animals, developmental programs execute their functions during embryogenesis and are silenced postnatally; inappropriate reactivation of embryonic programs later in life is frequently associated with malignancy. In contrast, *planarian development is a continuous, emergent, self-organizing process that is highly reliant on access to embryonic developmental programs throughout life.* Adult PSCs, specified during embryogenesis, underwrite de novo construction of the adult body plan during embryonic development and are similarly responsible for tissue maintenance and regeneration during adulthood (Davies et al. 2017). The abundance of adult PSCs, coupled with their unrivaled potency and high proliferation rates, suggests that novel tumor suppressive mechanisms must be in place in these long-lived animals. Future studies examining PSC clonal competition, lineage output, turnover, succession, and genetic drift will shed light on microevolution in planarians, with potential ramifications for understanding tumor progression in mammals. Finally, comparative studies of adult PSCs in planarians along with sponges, ctenophores, and other early branching bilaterians will inform theories of Metazoan evolution, highlighting core, conserved regulators of pluripotency across species.

ACKNOWLEDGMENTS

This work was supported by the Stowers Institute for Medical Research, the Howard Hughes Medical Institute, and NIH R37GM057260.

REFERENCES

Abril, J. F., F. Cebria, G. Rodriguez-Esteban, T. Horn, S. Fraguas, B. Calvo, K. Bartscherer, and E. Saló. 2010. S. med.454 dataset: unravelling the transcriptome of Schmidtea mediterranea. *BMC Genomics* 11: 731. doi: 10.1186/1471-2164-11-731.

Adamidi, C., Y. Wang, D. Gruen, G. Mastrobuoni, X. You, D. Tolle, M. Dodt, S. D. Mackowiak, et al. 2011. *De novo* assembly and validation of planaria transcriptome by massive parallel sequencing and shotgun proteomics. *Genome Res.* 21: 1193–1200. doi: 10.1101/gr.113779.110.

Baguñà, J. 2012. The planarian neoblast: the rambling history of its origin and some current black boxes. *Int. J. Dev. Biol.* 56: 19–37. doi: 10.1387/ijdb.113463jb.

Baguñà, J., and R. Romero. 1981. Quantitative analysis of cell types during growth, degrowth and regeneration in the planarians *Dugesia mediterranea* and *Dugesia tigrina*. *Hydrobiologia* 84: 181–194.

Baguñà, J., Saló, E., and C. Auladell. 1989. Regeneration and pattern formation in planarians. III. Evidence that neoblasts are totipotent stem cells and the source of blastema cells. *Development* 107: 77–86.

Bardeen, C. R. 1902. Embryonic and regenerative development in planarians. *Biol. Bull.* 3: 262–288.

Bardeen, C. R., and F. H. Baether. 1904. The inhibitive action of the roentgen rays on regeneration in planarians. *J. Exp. Zool.* 1: 191–195.

Blythe, M. J., D. Kao, S. Malla, J. Rowsell, R. Wilson, D. Evans, J. Jowett, A. Hall, et al. 2010. A dual platform approach to transcript discovery for the planarian *Schmidtea mediterranea* to establish RNAseq for stem cell and regeneration biology. *PLoS One* 5 (12): e15617. doi: 10.1371/journal.pone.0015617.

Chong, T., J. J. Collins, 3rd, J. L. Brubacher, D. Zarkower, and P. A. Newmark. 2013. A sex-specific transcription factor controls male identity in a simultaneous hermaphrodite. *Nat. Commun.* 4: 1814. doi: 10.1038/ncomms2811.

Collins, J. J., 3rd, X. Hou, E. V. Romanova, B. G. Lambrus, C. M. Miller, A. Saberi, J. V. Sweedler, and P. A. Newmark. 2010. Genome-wide analyses reveal a role for peptide hormones in planarian germline development. *PLoS Biol* 8 (10): e1000509. doi: 10.1371/journal.pbio.1000509.

Collins, J. J., 3rd, B. Wang, B. G. Lambrus, M. E. Tharp, H. Iyer, and P. A. Newmark. 2013. Adult somatic stem cells in the human parasite *Schistosoma mansoni*. *Nature* 494: 476–479. doi: 10.1038/nature11924.

Coward, S. J. 1974. Chromatoid bodies in somatic cells of the planarian: observations on their behavior during mitosis. *Anat. Rec.* 180: 533–545.

Curtis, W. C., and J. Hickman. 1926. Effects of X-rays and radium upon regeneration in planarians. *Anat. Rec.* 34: 145–146.

Dattani, A., D. Kao, Y. Mihaylova, P. Abnave, S. Hughes, A. Lai, S. Sahu, and A. A. Aboobaker. 2018. Epigenetic analyses of planarian stem cells demonstrate conservation of bivalent histone modifications in animal stem cells. *Genome Res.* 28: 1543–1554. doi: 10.1101/gr.239848.118.

Davies, E. L., K. Lei, C. W. Seidel, A. E. Kroesen, S. A. McKinney, L. Guo, S. M. Robb, E. J. Ross, et al. 2017. Embryonic origin of adult stem cells required for tissue homeostasis and regeneration. *Elife* 6: e21052. doi: 10.7554/eLife.21052.

D'Souza, T. G., and N. K. Michiels. 2009. Sex in parthenogenic planarians: phylogenetic relic or evolutionary resurrection? In *Lost Sex: The Evolutionary Biology of Parthenogenesis* (Eds) I. Schön, K. Martens, and P. Dijk, pp. 377–397. Springer, New York.

Dubois, F. 1949. Contribution a l'étude de la migration des cellules de régénération chez les planaires dulcicoles. *Bull. Biol. Fr. Belg.* 83: 213–283.

Dubois, F., and E. Wolff. 1947. Sur une méthode d'irradiation localisée permettant de mettre en évidence la migration des cellules de régenération chez les planaires. *C. R. Seances Soc. Biol. Ses Fil.* 141: 903–906.

Eisenhoffer, G. T., H. Kang, and A. Sánchez Alvarado. 2008. Molecular analysis of stem cells and their descendants during cell turnover and regeneration in the planarian *Schmidtea mediterranea. Cell Stem Cell* 3: 327–339. doi: 10.1016/j.stem.2008.07.002.

Elliott, S. A., and A. Sánchez Alvarado. 2013. The history and enduring contributions of planarians to the study of animal regeneration. *Wiley Interdiscip. Rev.:Dev. Biol.* 2: 301–326. doi: 10.1002/wdev.82.

Fedecka-Bruner, B. 1965. Régénération des testicules des planaires après destruction par les rayons X. In *Regeneration in Animals and Related Problems* (Eds) V. Kiortsis and H. A. L. Trampusch, pp. 185–192. North-Holland Publishing Company, Amsterdan.

Fedecka-Bruner, B. 1967. Études sur la régénération des organes genitaux chez la planaire Dugesia lugubris. I. Régénération des testicules après destruction. *Bull. Biol. Fr. Belg.* 101: 255–319.

Gehrke, A. R., and M. Srivastava. 2016. Neoblasts and the evolution of whole-body regeneration. *Curr. Opin. Genet. Dev.* 40: 131–137. doi: 10.1016/j.gde.2016.07.009.

Ghiradelli, E. 1965. Differentiation of the germ cells and regeneration of the gonads in planarians. In *Regeneration in Animals and Related Problems* (Eds) V. Kiortsis and H. A. L. Trampush, pp. 177–184. North-Holland Publishing Company, Amsterdam.

Grohme, M. A., S. Schloissnig, A. Rozanski, M. Pippel, G. R. Young, S. Winkler, H. Brandl, I. Henry, et al. 2018. The genome of *Schmidtea mediterranea* and the evolution of core cellular mechanisms. *Nature* 554: 56–61. doi: 10.1038/nature25473.

Grudniewska, M., S. Mouton, D. Simanov, F. Beltman, M. Grelling, K. de Mulder, W. Arindrarto, P. M. Weissert, et al. 2016. Transcriptional signatures of somatic neoblasts and germline cells in Macrostomum lignano. *Elife* 5: e20607. doi: 10.7554/eLife.20607.

Guedelhoefer, O. C., and A. S. Alvarado. 2012. Amputation induces stem cell mobilization to sites of injury during planarian regeneration. *Development* 139: 3510–3520. doi: 10.1242/dev.082099.

Guo, L., S. Zhang, B. Rubinstein, E. Ross, and A. Sánchez Alvarado. 2016. Widespread maintenance of genome heterozygosity in *Schmidtea mediterranea. Nat. Ecol. Evol.* 1: 19. doi: 10.1038/s41559-016-0019.

Handberg-Thorsager, M., and E. Saló. 2007. The planarian nanos-like gene *Smed-nos* is expressed in germline and eye precursor cells during development and regeneration. *Dev. Genes Evol.* 217: 403–411. doi: 10.1007/s00427-007-0146-3.

Hay, E. D., and S. J. Coward. 1975. Fine structure studies on the planarian, Dugesia: I. Nature of the neoblast and other cell types in non-injured worms. *J. Ultrastruct. Res.* 50: 1–21.

Hayashi, T., M. Asami, S. Higuchi, N. Shibata, and K. Agata. 2006. Isolation of planarian X-ray-sensitive stem cells by fluorescence-activated cell sorting. *Dev. Growth Differ.* 48: 371–380. doi: 10.1111/j.1440-169X.2006.00876.x.

Hori, I. 1982. An ultrastructural study of the chromatoid body in planarian regenerative cells. *J. Electron Microsc. Tech.* 31: 63–72.

Hoshi, M., K. Kobayashi, S. Arioka, S. Hase, and M. Matsumoto. 2003. Switch from asexual to sexual reproduction in the Planarian *Dugesia ryukyuensis. Integr. Comp. Biol.* 43: 242–246. doi: 10.1093/icb/43.2.242.

Hyman, L. H. 1951. *The Invertebrates: Platyhelminthes and Rhynchocoela. The Acoelomate Bilateria.* New York: McGraw Hill.

Iyer, H., J. J. Collins, 3rd, and P. A. Newmark. 2016. NF-YB regulates spermatogonial stem cell self-renewal and proliferation in the Planarian *Schmidtea mediterranea. PLoS Genet.* 12 (6): e1006109. doi: 10.1371/journal.pgen.1006109.

Krishna, S., D. Palakodeti, and J. Solana. 2019. Post-transcriptional regulation in planarian stem cells. *Semin. Cell Dev. Biol.* 87: 69–78. doi: 10.1016/j.semcdb.2018.05.013.

Labbé, R. M., M. Irimia, K. W. Currie, A. Lin, S. J. Zhu, D. D. Brown, E. J. Ross, V. Voisin, et al. 2012. A comparative transcriptomic analysis reveals conserved features of stem cell pluripotency in planarians and mammals. *Stem Cells* 30: 1734–1745. doi: 10.1002/stem.1144.

Laumer, C. E., A. Hejnol, and G. Giribet. 2015. Nuclear genomic signals of the 'microturbellarian' roots of platyhelminth evolutionary innovation. *eLife* 4: e05503. doi: 10.7554/eLife.05503.

Lei, K., H. T. K. Vu, R. D. Mohan, S. A. McKinney, C. W. Seidel, R. Alexander, K. Gotting, J. L. Workman, and A. Sánchez Alvarado. 2016. Egf signaling directs neoblast repopulation by regulating asymmetric cell division in Planarians. *Dev. Cell* 38: 413–429. doi: 10.1016/j.devcel.2016.07.012.

Liu, S. Y., C. Selck, B. Friedrich, R. Lutz, M. Vila-Farre, A. Dahl, H. Brandl, N. Lakshmanaperumal, et al. 2013. Reactivating head regrowth in a regeneration-deficient planarian species. *Nature* 500: 81–84. doi: 10.1038/nature12414.

Morgan, T. H. 1902. Growth and regeneration in *Planaria lugubris. Ann. Entomol. Soc. Am.* 13: 179–212.

Newmark, P. A., and A. Sánchez Alvarado. 2000. Bromodeoxyuridine specifically labels the regenerative stem cells of planarians. *Dev. Biol.* 220: 142–153. doi: 10.1006/dbio.2000.9645.

Newmark, P. A., and A. Sánchez Alvarado. 2002. Not your father's planarian: a classic model enters the era of functional genomics. *Nat. Rev. Genet.* 3: 210–219. doi: 10.1038/nrg759.

Newmark, P. A., Y. Wang, and T. Chong. 2008. Germ cell specification and regeneration in planarians. *Cold Spring Harbor Symp. Quant. Biol.* 73: 573–581. doi: 10.1101/sqb.2008.73.022.

Onal, P., D. Grun, C. Adamidi, A. Rybak, J. Solana, G. Mastrobuoni, Y. Wang, H. P. Rahn, et al. 2012. Gene expression of pluripotency determinants is conserved between mammalian and planarian stem cells. *EMBO J.* 31: 2755–2769. doi: 10.1038/emboj.2012.110.

Oviedo, N. J., P. A. Newmark, and A. Sánchez Alvarado. 2003. Allometric scaling and proportion regulation in the freshwater planarian *Schmidtea mediterranea. Dev. Dyn.* 226: 326–333. doi: 10.1002/dvdy.10228.

Pedersen, K. J. 1959. Cytological studies on the planarian neoblast. *Z. Zellforsch. Mikrosk. Anat.* 50: 799–817.

Randolph, H. 1891. Regeneration of the tail in *Lumbriculus. Zool. Anzr.* 14: 154–156.

Randolph, H. 1897. Observations and experiments on regeneration in planarians. *Arch. Entw. Mech. Org.* 5: 352–372.

Reddien, P. W. 2013. Specialized progenitors and regeneration. *Development* 140: 951–957. doi: 10.1242/dev.080499.

Reddien, P. W. 2018. The cellular and molecular basis for planarian regeneration. *Cell* 175: 327–345. doi: 10.1016/j.cell.2018.09.021.

Reddien, P. W., and A. Sánchez Alvarado. 2004. Fundamentals of planarian regeneration. *Annu. Rev. Cell Dev. Biol.* 20: 725–757. doi: 10.1146/annurev.cellbio.20.010403.095114.

Reddien, P. W., N. J. Oviedo, J. R. Jennings, J. C. Jenkin, and A. Sánchez Alvarado. 2005. S. MED.WI-2 is a PIWI-like protein that regulates planarian stem cells. *Science* 310: 1327–1330. doi: 10.1126/science.1116110.

Rink, J. C. 2013. Stem cell systems and regeneration in planaria. *Dev. Genes Evol.* 223: 67–84. doi: 10.1007/s00427-012-0426-4.

Robb, S. M., K. Gotting, E. Ross, and A. Sánchez Alvarado. 2015. S. med.GD 2.0: The Schmidtea mediterranea genome database. *Genesis* 53: 535–546. doi: 10.1002/dvg.22872.

Robb, S. M., E. Ross, and A. Sánchez Alvarado. 2008. S. med.GD: the Schmidtea mediterranea genome database. *Nucleic Acids Res.* 36 (Database issue): D599–D606. doi: 10.1093/nar/gkm684.

Romero, R., and J. Baguñà. 1991. Quantitative cellular analysis of growth and reproduction in freshwater planarians (Turbellaria; Tricladida). I. A cellular description of the intact organism. *Invertebr. Reprod. Dev.* 19: 157–165.

Saberi, A., A. Jamal, I. Beets, L. Schoofs, and P. A. Newmark. 2016. GPCRs direct germline development and somatic gonad function in Planarians. *PLoS Biol.* 14 (5): e1002457. doi: 10.1371/journal.pbio.1002457.

Sánchez Alvarado, A., and P. A. Newmark. 1999. Double-stranded RNA specifically disrupts gene expression during planarian regeneration. *Proc. Natl. Acad. Sci. U S A* 96: 5049–5054.

Shibata, N., L. Rouhana, and K. Agata. 2010. Cellular and molecular dissection of pluripotent adult somatic stem cells in planarians. *Dev. Growth Differ.* 52: 27–41. doi: 10.1111/j.1440-169X.2009.01155.x.

Shibata, N., T. Hayashi, R. Fukumura, J. Fujii, T. Kudome-Takamatsu, O. Nishimura, S. Sano, F. Son, et al. 2012. Comprehensive gene expression analyses in pluripotent stem cells of a planarian, Dugesia japonica. *Int. J. Dev. Biol.* 56: 93–102. doi: 10.1387/ijdb.113434ns.

Sikes, J. M., and P. A. Newmark. 2013. Restoration of anterior regeneration in a planarian with limited regenerative ability. *Nature* 500: 77–80. doi: 10.1038/nature12403.

Solana, J., D. Kao, Y. Mihaylova, F. Jaber-Hijazi, S. Malla, R. Wilson, and A. Aboobaker. 2012. Defining the molecular profile of planarian pluripotent stem cells using a combinatorial RNAseq, RNA interference and irradiation approach. *Genome Biol.* 13: R19. doi: 10.1186/gb-2012-13-3–r19.

Tharp, M. E., J. J. Collins, 3rd, and P. A. Newmark. 2014. A lophotrochozoan-specific nuclear hormone receptor is required for reproductive system development in the planarian. *Dev. Biol.* 396: 150–157. doi: 10.1016/j.ydbio.2014.09.024.

Umesono, Y., J. Tasaki, Y. Nishimura, M. Hrouda, E. Kawaguchi, S. Yazawa, O. Nishimura, K. Hosoda, et al. 2013. The molecular logic for planarian regeneration along the anterior-posterior axis. *Nature* 500: 73–76. doi: 10.1038/nature12359.

Wagner, D. E., J. J. Ho, and P. W. Reddien. 2012. Genetic regulators of a pluripotent adult stem cell system in planarians identified by RNAi and clonal analysis. *Cell Stem Cell* 10: 299–311. doi: 10.1016/j.stem.2012.01.016.

Wagner, D. E., I. E. Wang, and P. W. Reddien. 2011. Clonogenic neoblasts are pluripotent adult stem cells that underlie planarian regeneration. *Science* 332: 811–816. doi: 10.1126/science.1203983.

Wang, Y., J. M. Stary, J. E. Wilhelm, and P. A. Newmark. 2010. A functional genomic screen in planarians identifies novel regulators of germ cell development. *Genes Dev.* 24: 2081–2092. doi: 10.1101/gad.1951010.

Wang, Y., R. M. Zayas, T. Guo, and P. A. Newmark. 2007. *Nanos* function is essential for development and regeneration of planarian germ cells. *Proc. Natl. Acad. Sci. U S A* 104: 5901–5906. doi: 10.1073/pnas.0609708104.

Wolff, E., and F. Dubois. 1948. Sur la migration des cellules de régénération chez les planaires. *Rev. Suisse Zool.* 55: 218–227.

Zeng, A., H. Li, L. Guo, X. Gao, S. McKinney, Y. Wang, Z. Yu, J. Park, et al. 2018. Prospectively isolated tetraspanin(+) neoblasts are adult pluripotent stem cells underlying planaria regeneration. *Cell* 17: 1593–1608 e20. doi: 10.1016/j.cell.2018.05.006.

Zhang, S., L. Guo, C. Guerrero-Hernández, E. J. Ross, K. Gotting, S. A. McKinney, W. Wang, Y. Xiang, et al. 2018. A nuclear hormone receptor and lipid metabolism axis are required for the maintenance and regeneration of reproductive organs. *bioRxiv* 279364. doi: 10.1101/279364.

10 The Coordination of Insect Imaginal Discs and the Regulation and Evolution of Complex Worker Caste Systems of Ants

Sophie Koch and Ehab Abouheif
McGill University

CONTENTS

10.1 INTRODUCTION: IMAGINAL DISCS AS SET-ASIDE DEVELOPMENTAL MODULES

Traits originate from *developmental modules*, each of which is comprised of a highly integrated gene regulatory network (West-Eberhard 2003).

10.1.1 Developmental Modules and Evolution

Developmental modules are quasi-independent entities that are dissociable from one another and can be decoupled and recombined, making each a potential target of selection (Abouheif et al. 2014; Lewontin 1970; West-Eberhard 2003, 2005). For example, eyespots on butterfly wings are each a developmental module specified by a gene network that can be decoupled from other components on the wing through development and evolution to generate diversity of butterfly wing patterns (Carroll et al. 1994; Beldade et al. 2002; Monteiro et al. 2003). Genetic and environmental perturbations have the potential to modify the module, the connections that link modules, and/or the strength of those connections (Wagner 1996; West-Eberhard 2003; Yang 2001; Klingenberg 2008). Such modifications generate variation, and if positively selected for, this variation can lead to specialization in morphology, behavior, or life history and even to the evolution of novel developmental programs (Wagner 1996; West-Eberhard 2003).

At the same time, developmental modules must coordinate with one another to ensure development of a viable and reproducible phenotype. The coordination between developmental modules facilitates homeostasis at the organismal level in response to environmental or genetic perturbations (Klingenberg 2008; Shingleton and Frankino 2018; Mirth and Shingleton 2019). Homeostatic forces confer developmental robustness and coordinate a canalized developmental system; however, if perturbed outside the limits of homeostasis, these forces can also facilitate the establishment of a new developmental equilibrium. Therefore, perturbations to *developmental modules can produce variation, which can then be stabilized through existing coordination mechanisms and be operated on by selection* (Moczek 2008; Waddington 1956; Wilkins 2003; Abouheif et al. 2014).

The evolutionary significance of developmental modules can be readily observed in holometabolous insects, which make up ~85% of insect diversity, and in which ontogenetic life-stage modularity is thought to have facilitated astounding evolutionary success and ecological dominance (Engel 2015; Misof et al. 2014; Yang 2001). Ametabolous insects have direct development without metamorphosis, and hemimetabolous insects have incomplete metamorphosis between juvenile nymphal and adult stages. This means that juvenile and adult stages typically show few phenotypic and life history differences—adult development is marked only by increases in size, reproductive maturity, and, in the Hemimetabola, wing development. In contrast, holometabolous insects develop from the egg into a series of larval instars before undergoing complete metamorphosis to develop into the pupal and, subsequently, the adult stages (Truman and Riddiford 1999; Yang 2001). The separation and increase in ontogenetic modularity in the Holometabola is suggested to have encouraged their evolutionary diversification by facilitating ecological niche partitioning between larval and adult stages and allowing independent selection on each stage (Yang 2001).

10.1.2 Imaginal Discs as Developmental Modules in Holometabolous Insects

The high level of modularity in holometabolous insect ontogeny when compared with hemimetabolous insects is in part due to cells that are set aside during development called *imaginal discs*. These populations of cells grow during the larval stage and give rise to adult appendages and organs during metamorphosis (West-Eberhard 2003; Yang 2001).

Imaginal disc development begins during embryogenesis, when clusters of cells are specified and set aside. During the larval stage, these cells proliferate to form imaginal discs, which are composed of two epithelial layers – the disc proper and a peripodial membrane that encases the disc (Held 2002). Imaginal disc growth and patterning is regulated by largely conserved gene regulatory networks that generate intrinsic morphogen gradients and by systemic endocrine factors. During the larval-pupal transition, the discs undergo morphogenesis and differentiate into the adult appendages (Cohen 1993; Held 2002). Their high degree of modularity coupled with modifications to their growth and patterning throughout insect evolution has led to an astounding degree of morphological diversity in trait shape, size, and scaling within the Holometabola. Examples include the development of enlarged beetle horns (Moczek and Rose 2009), butterfly wing patterns and eyespots (Brakefield et al. 1996; Beldade et al. 2002), enlarged eyes in stalk-eyed flies (Hurley et al. 2002), and big-headed soldiers in ant colonies (Wilson 1953; Huang and Wheeler 2011). Although imaginal discs are quasi-independent developmental modules that can be modified to generate morphological diversity, these modules are simultaneously regulated through mechanisms that synchronize the disc growth and coordinate them with larval ontogeny to ensure robust development (Andersen et al. 2013; Shingleton and Frankino 2018; Mirth and Shingleton 2019). Therefore, coordination of imaginal discs can canalize developmental programs and, in the face of environmental or genetic perturbations, may play a role in stabilizing novel developmental equilibria.

In part due to the developmental properties discussed above, insect imaginal discs have been widely used for over a century to investigate organ autonomy, tissue patterning, morphogenesis, biomechanical forces, regeneration, and tumorogenesis (Beira and Paro 2016). The majority of these studies focused on the model fruit fly *Drosophila melanogaster* by using its remarkable array of genetic and molecular tools to make major contributions to the fields of developmental and cell biology, genetics, and medicine. Studies in holometabolous insects have demonstrated that imaginal disc growth is synchronized with other discs and is coordinated with larval development. Furthermore, studies in Drosophila have begun to identify the molecular mechanisms through which this synchronization and coordination occurs (Andersen et al. 2013; Colombani et al. 2012, 2015; Droujinine and Perrimon 2016; Garelli et al. 2012, 2015). Here, we focus on several elegant and pioneering experiments carried out during the twentieth century (from 1933 to 1985) that were some of the first to demonstrate: (i) synchronization of growth between imaginal discs and (ii) their coordination with larval development and metamorphosis across holometabolous insects. We then examine this older literature in the context of how the environment regulates development of *worker castes in ants*, where environmental and social regulation of imaginal disc and larval development has facilitated the evolution of novel morphological castes (Rajakumar et al. 2012, 2018; Wheeler and Nijhout, 1981a; Wheeler 1986). Imaginal discs in ants provide us with a case study on how the development and evolution of set-aside cells may have facilitated the elaboration of biological complexity.

10.2 PIONEERS IN IMAGINAL DISC ABLATION AND TRANSPLANTATION

> With the imaginal disks, [Ernst Hadorn] provided biologists with the object which allowed them to study development with some hope of understanding its principles in genetic terms.
>
> **(Nöthiger 2002, p. 25)**

Prior to the development of molecular techniques allowing for gene and cell-specific manipulations in insects, researchers studied insect development using techniques to remove, introduce, or cultivate imaginal discs. Specifically, ablation, transplantation, and tissue cultivation methods served to marry the accumulating resources of *Drosophila* genetics and insect physiology with experimental embryological approaches that had previously been used in organisms for which genetic techniques were difficult (Kohler 1994; Katsuyama and Paro 2016).

One technique was transplantation of imaginal discs, which was initiated in the Mediterranean flour moth *Ephestia kuehniella* in Alfred Kuhn's lab and also in the fruit fly *D. melanogaster* through a collaboration between George W. Beadle and Boris Ephrussi (Beadle and Ephrussi 1935, Ephrussi and Beadle 1936; Caspari 1933; Kohler 1994). Based on the well-characterized *Drosophila* eye color mutant phenotypes, Beadle and Ephrussi used transplantation to test the genetic interactions of eye pigment genes and identified the interactions that specify eye pigment autonomously and nonautonomously through diffusible factors (Beadle et al. 1937; Beadle and Ephrussi 1936). Ablation of imaginal discs provided insight into their patterning, growth, and function within the developing larvae. Transplantation of imaginal discs furthered this and permitted investigation of the genetic interactions and biochemistry that had previously been only possible through rare gynandromorphs and artificially induced genetic mosaics (Ephrussi and Beadle 1936; Kohler 1994; Sturtevant 1929; Xu and

Rubin 2012). As a result of this work, transplantation spread quickly throughout the developmental genetics and insect physiology communities (Beadle and Ephrussi 1935; Katsuyama and Paro 2016; Beira and Paro 2016; Kohler 1994; Nöthiger 2002; Rhinsberger 2000). Imaginal discs served as a tangible and tractable model—as stated by Dietrich Bodenstein in 1972 "the choice of the object was perfect" (cited in Nöthiger 1998, p. 519), and its study facilitated the examination of questions that have and continue to inform numerous fields of biology. Here, we focus on experiments using disc ablation and transplantation in holometabolous insects, as these techniques reveal the shared developmental and physiological synchronization of disc growth and coordination with larval development. Our goal is to discuss the relationships between the synchronization of organ development and coordination of overall ontogeny with the evolution of novel developmental programs to apply these findings to the evolution of castes in ants.

10.3 ABLATION AND TRANSPLANTATION OF IMAGINAL DISCS REVEALS MECHANISMS OF COORDINATION

10.3.1 Synchronization and Competition between Imaginal Discs

Pioneers in insect developmental biology showed that when imaginal discs are transplanted into an adult fly's abdomen, they grow reliably to the final size typically observed in the donor larvae (Bryant and Levinson 1985; Hadorn 1963). This demonstrated that intrinsic properties of the disc regulate its final size in an autonomous manner. Yet despite this autonomous development, a transplanted disc is not able to undergo complete morphogenesis in the adult abdomen (Hojyo and Fujiwara 1997; Oberlander 1972; Postlethwait and Schneiderman 1968). Furthermore, transplantation of eye-antennal discs of certain genotypes develop autonomously in wild-type host larvae, while *vermillion* and *cinnabar* mutants develop nonautonomously and take on a wild-type phenotype (Beadle and Ephrussi 1936; Ephrussi and Beadle 1937). Phenotype and morphogenesis can be regulated by diffusible signals within the larval environment.

Extrinsic regulation of imaginal disc growth is also observed when a disc is damaged or ectopically introduced in a larval host, which shows that other discs adjust their growth to synchronize with the affected organ (Figure 10.1). This regulation provides feedback mechanisms that ensure that all presumptive adult structures, including wings, legs, and genital structures, develop correctly and symmetrically prior to the larval-pupal transition and commitment to the adult form (Garelli et al. 2012; Jaszczak et al. 2015; Parker and Shingleton 2011; Vallejo et al. 2015).

Synchronization of appendage and organ growth during development is widespread across animals as seen in vertebrates and insects (Busse et al. 2018; Gontijo and Garelli 2018; Rosello-Diez et al. 2018; Tanner 1963). These processes are elegantly elucidated further in insects through disc transplantation. Larval development is dependent on nutrition; when a more developed disc is transplanted into young or starved larvae, in which the host's discs are less developed, the transplant slows its growth to synchronize with the discs of the host larvae (Bodenstein 1939, 1940). Synchronization also occurs when a less well-developed disc is transplanted

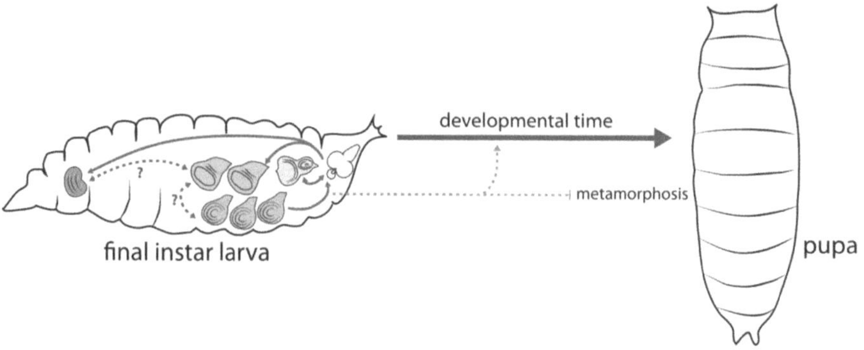

FIGURE 10.1 A summary of synchronization and coordination mechanisms regulating imaginal discs shown as schematics in a final instar *Drosophila* spp. larva (left) and pupa (right). Schematics within the larva indicate larval wing, leg, eye-antennal, and genital imaginal discs and larval brain and ring gland. Solid lines within larva indicate signaling mechanisms between imaginal discs and the larval brain that regulate developmental time and affect metamorphosis to the pupal stage (dashed lines) (Colombani et al. 2012, 2015; Garelli et al. 2012, 2015; Boone et al. 2016; Jaszczak et al. 2015, 2016; Vallejo et al. 2015). Dashed lines within larva indicate potential positive or negative regulatory mechanisms directly between imaginal discs.

into an older or well-fed larva, resulting in the transplant accelerating its growth while the growth of the host's discs slows (Bodenstein 1940). Strikingly, synchronization of growth between a transplanted disc and the discs of the host larvae is even observed in interspecies transplants between *D. melanogaster* and *Drosophila hydei*. However, transplantations between *D. melanogaster* and *D. pseudoobscura* do not synchronize with their host, a difference that may be explained by differences in the timing of metamorphosis or potentially due to evolutionary divergence (Bodenstein 1939). These experiments and others showed that while final size may rely in part on intrinsic properties, the growth rate of the disc can be regulated by extrinsic systemic mechanisms that act to synchronize the growth of all imaginal discs within the larva (Figure 10.1) (Gontijo and Garelli 2018; Andersen et al. 2013). Future work is required to tease apart these processes and explore how final size varies as a function of growth rate in inter-species transplants.

In addition to mechanisms that synchronize disc growth, perturbations to imaginal disc growth also uncovered competitive or differential resource acquisition mechanisms (Emlen and Nijhout 2000). An example that reveals this complexity is the development of the large elaborated wings in butterflies. While growth of lepidopteran wing discs is dependent on larval nutrition and hormone signals, competitive interactions between discs also occur (Nijhout and Callier 2015; Nijhout and Grunert 2010). Surgical removal of both hind wing discs in the butterfly *Precis coenia* results in an increase in the dry weight of the adult forewing, mid- and hind legs, while removal of only one hind wing results in individuals with heavier but similarly sized forewings on the side of removal (Klingenberg and Nijhout 1998; Nijhout and Emlen 1998). This suggests that in the absence of hind wing disc(s), the forewing disc(s) acquire(s) more resources and accumulate(s) more mass (Klingenberg and Nijhout 1998). More recently, these findings have been discussed in the context of coordination between imaginal discs to suggest that differential resource acquisition may be a product of the disruption of endocrine mechanisms that synchronize the discs, resulting in asymmetrical development (Shingleton and Frankino 2018). This may be possible as the signaling mechanisms that regulate disc regeneration and the delay of metamorphosis in *D. melanogaster* also ensure that the appendages are bilaterally symmetric during normal development (Vallejo et al. 2015). Future experiments testing these process in different holometabolous lineages will be required to identify how mechanisms of homeostatic coordination and resource competition interact and have been modified by selection (Figure 10.1).

Imaginal discs can also be targeted by selection to favor morphs with distinct life histories. In scarab beetles, while large males develop disproportionately large horns for sexual display, defense, and combat, smaller males lack horns and perform alternative mating tactics (Emlen and Nijhout 2000; Moczek and Nijhout 2004). The disproportionate growth of horns develops through differential tissue-specific sensitivity to nutrition, yet experimental evidence also reveals that growth of the horn is correlated with other organs (Casasa and Moczek 2018; Emlen 1997; Emlen et al. 2012). Selection for large horns in the dung beetles *Onthophagus taurus* and *Onthophagus acuminatus* results in disproportionate reduction in eye size, which, as with the compensation observed in the butterfly experiments, is suggestive of differential resource acquisition (Nijhout and Emlen 1998).

Furthermore, manipulations of the horn and genital precursors in beetles have shown that regulation of imaginal disc growth can generate distinct morphs with different life history traits. Cauterization of the horn precursor in *Onthophagus* spp. results in larger individuals that develop larger genital discs, whereas cauterization of the genital precursor results in larger horns (Moczek and Nijhout 2004; Simmons and Emlen 2006). This suggests a developmental and reproductive strategy trade-off between investment in horns for competition, mate acquisition, and potentially multiple matings, versus investment in the genital precursor for sperm production and mating efficiency (Emlen and Nijhout 2000; Moczek and Nijhout 2004; Shingleton and Frankino 2018; Simmons and Emlen 2006). Therefore, while mechanisms exist during insect development to synchronize organ growth, they may be modified through selective pressures or may exist in tandem with additional developmental trade-offs to produce novel and highly adaptive morphologies.

10.3.2 Developmental and Physiological Coordination between Imaginal Disc Growth and Developmental Time

In addition to mechanisms that synchronize the development and growth of imaginal discs, perturbation of imaginal discs also results in the extension of larval development and the delay of the larval-pupal transition (Figure 10.1).

The extension of larval development is widespread in hemimetabolous and holometabolous insects; it has been observed in cockroaches (Kunkel 1977; Pohley 1959; Stock and O'Farrell 1954), crickets (Lakes and Mucke 1988), beetles (Wu et al. 2019), moths (Dewes 1973; Madhavan and Schneiderman 1969; Muth 1961; Pohley 1960), and flies (Parker and Shingleton 2011; Poodry and Woods 1990; Simpson et al. 1980; Simpson and Schneiderman 1975; Stieper et al. 2008). Extension of developmental time is proportional to the amount of damaged tissue—surgical removal of wing tissue from all four wing discs through extirpation in the greater wax moth *Galleria mellonella* extends larval development for 14 days, while surgical extirpation of tissue from a single wing disc extends it by only 4 days (Madhavan and Schneiderman 1969). During this developmental extension, the damaged discs regenerate and undamaged discs exhibit slowed growth (Madhavan and Schneiderman 1969).

In contrast to surgical extirpation, which leaves behind a small portion of the imaginal disc thus permitting regeneration, cauterization of the wound site following extirpation prevents imaginal disc regeneration. When the site of extirpation is cauterized, it eliminates the extension of larval development showing that regeneration of the damaged tissue is necessary for the extension of larval development (Madhavan and Schneiderman 1969). In *D. melanogaster*, imaginal disc damage through a heat-sensitive cell lethal mutation extends larval developmental time proportional to the observed damage. However, in cases where individuals withstand minimal damage or elimination of the entire disc, no developmental delay is observed (Simpson et al. 1980). Similarly, in the ladybug, *Coccinella septempunctata,* a higher level of damage results in a more significant developmental delay (Wu et al. 2019). In contrast, in cockroaches, damage to presumptive appendages extends developmental time to allow for regeneration but is not proportional to the amount of regenerating tissue (Kunkel 1977). This difference is likely because cockroaches are hemimetabolous insects that undergo incomplete metamorphosis through a series of nymphal stages and lack imaginal discs (Truman and Riddiford 1999; Yang 2001). Therefore, while *mechanisms to ensure organ regeneration through coordination with developmental time may be ancestral in insects*, coordination mechanisms may differ in sensitivity or robustness in holometabolous insects to ensure correct repair and regeneration prior to the larval-pupal transition.

Delay of larval-pupal transition occurs also when imaginal tissues are introduced ectopically during larval development. Transplantation of fragmented genital discs into larvae of the moth *E. kuehniella* results in regeneration of the disc and extension of host development (Dewes 1973). Furthermore, in *D. melanogaster,* larval development is extended when dissociated wing disc cells are injected. However, unlike transplantation of whole discs, these cells proliferate rapidly to form large, disorganized cell masses causing the host larvae to grow very large and often halt development in the larval stage (Ursprung and Hadorn 1962). This demonstrates that larval resources may serve as a limiting factor to the amount of damage that is repairable during disc regeneration. Collectively, this shows that general and robust mechanisms link imaginal disc development with the transition to the pupal stage in order to ensure that presumptive appendages grow and, if they are damaged, that the larval to pupal transition is delayed allowing for tissue regeneration.

The extension of the larval stage by regenerating imaginal discs requires the larval nervous system and the ring gland that acts as the endocrine signaling center to regulate growth, molting, and metamorphosis (Hadorn 1937; Kunkel 1977). Imaginal discs exist in a feedback circuit with the larval brain and ring gland. Larval endocrine centers regulate disc growth through secreted hormones and neuropeptides, and discs regulate larval growth and developmental time through secreted signals during their growth or regeneration (Figure 10.1) (Colombani et al. 2012, 2015; Edgar 2006; Garelli et al. 2012, 2015; Jaszczak et al. 2016). This signaling circuit ensures both interorgan synchronization and coordination with larval developmental time.

The larval ring gland includes the prothoracic gland, which is necessary and sufficient to induce the transition to metamorphosis (Faltus and Oberlander 1970; Fukuda 1940; Horikawa and Sugahara 1960; Oberlander 1972; Vogt 1942). The prothoracic gland secretes the molting hormone *ecdysone*, which, at high levels, triggers larval molts and transition to the pupal stage (Bodenstein 1938; Fukuda 1940; Truman and Riddiford 1999; Yamanaka et al. 2013). Perturbation of ecdysone biosynthesis, degradation, or temporal signaling can produce small larvae that undergo pupation prematurely or that grow larger and have delayed pupation (Caldwell et al. 2005; Mirth and Shingleton 2014; Schwartz et al. 1984). In addition to molting and metamorphosis, ecdysone is also necessary for organ growth and differentiation in a dose-dependent manner and can induce differentiation in environments in which discs do not normally differentiate, including in the adult fly abdomen and in in vitro cultivation (Bodenstein 1938, 1943; Herboso et al. 2015; Howland et al. 1937; Oberlander 1972; Tobler and Nijhout 2010).

Therefore, ecdysone signaling appears to have a central role in regulating imaginal disc growth and the timing of metamorphosis and has been shown to link these processes through systemic coordination. Low levels of ecdysone are necessary for synchronization of growth within and between discs (Gokhale et al. 2016). The removal of wing disc tissue through surgical extirpation in the moth *G. mellonella* causes regeneration in the damaged disc and slows growth of unaffected discs. However, extirpation in larvae that are ligated to block endogenous ecdysone signaling eliminates this synchronization. Furthermore, injection of these ligated larvae with ecdysone rescues interdisc synchronization showing that ecdysone is necessary and sufficient for interorgan synchronized growth (Madhavan and Schneiderman 1969). Similarly, slowing of wing imaginal disc growth in *D. melanogaster* carrying the damage-inducing *Minute* mutation ($RpS3^{GD4577}$) results in the slowing and synchronization of other imaginal discs. Yet systemic applications of ectopic ecdysone override these existing mechanisms and restore the normal growth trajectories of both imaginal discs and larval developmental time (Parker and Shingleton 2011).

Therefore, disc growth and synchronization is regulated by the prothoracic gland and its dose-dependent secretion of ecdysone, which is also necessary to generate the extension of larval development that occurs to permit disc regeneration (Figure 10.1) (Madhavan and Schneiderman 1969; Oberlander 1972; Pohley 1960, 1970; Sehnal and Bryant 1993). Together, these experiments reveal the coordinated and phylogenetically widespread regulation that exists between the synchronization of disc growth and developmental timing and place ecdysone signaling as a key regulator of this process.

Recent work in *D. melanogaster* has uncovered the molecular signals that underlie the coordination of imaginal disc growth with larval developmental timing and ecdysone regulation of the larval-pupal transition (Colombani et al. 2012; Garelli et al. 2012). The identification of an insulin-like peptide *(Dilp8)* and its neural-expressed relaxin receptor Lgr3 has facilitated further identification of the molecular mechanisms that regulate the intrinsic properties of imaginal disc growth, developmental time, symmetry, and regeneration (Boone et al. 2016; Colombani et al. 2015; Garelli et al. 2015; Hariharan 2012; Jaszczak et al. 2015, 2016; Vallejo et al. 2015). Furthermore, the secreted imaginal disc morphogen *decapentaplegic* (*dpp*) also regulates ecdysone biosynthesis showing that it may interact with insulin signaling to coordinate developmental time (Setiawan et al. 2018). Ecdysone signaling interacts with insulin and juvenile hormone (JH) signaling to regulate disc growth, larval growth, and developmental time to surpass size checkpoints and permit the transition to metamorphosis (Edgar 2006; Nijhout et al. 2014; Nijhout 1994). These and other studies, which are revealing the signals that operate to regulate these molecular mechanisms in *D. melanogaster*, are the focus of recent reviews (Droujinine and Perrimon 2016; Gontijo and Garelli 2018; Hariharan 2012; Hariharan and Serras 2017). While a review of the molecular literature is beyond the scope of this chapter, future work can examine these signals and be synthesized with the literature examining organ-level manipulations to further our understanding of the shared and modified regulation of imaginal disc growth and larval development across holometabolous insect lineages.

10.4 ANTS: MODIFICATION OF IMAGINAL DISC GROWTH AND COORDINATION IN THE WINGLESS WORKER CASTE

Social insects (order: Hymenoptera) are an ecologically dominant and evolutionarily successful lineage within the holometabolous insects (Engel 2015; Misof et al. 2014). Ants belong to family Formicidae, which represents one of the most diverse and biologically complex hymenopteran groups with over ~16,000 described species (AntWeb 2019; Ward 2014). All extant ants are eusocial with a reproductive division of labor and colonies typically consisting of two female castes: a queen caste and a worker caste (Wilson and Hölldobler 2005; Hölldobler and Wilson 2009). While the queen and male castes develop functional wings that are used for seasonal mating flights, the worker caste is wingless and performs foraging, brood-care, and defense tasks (Hölldobler and Wilson 1990, 2009). The potential for a queen-laid embryo to develop into a winged queen or a wingless worker based on environmental cues is called *wing polyphenism* and is shared by all extant ants (Wilson 1953; Abouheif and Wray 2002). Although workers are wingless, their larvae still develop rudiments of wing imaginal discs that degenerate prior to the adult stage, and the morphology of these rudiments varies between species and between worker sub castes (Abouheif and Wray 2002; Behague et al. 2018; Rajakumar et al. 2012; Sameshima et al. 2004; Wheeler and Nijhout 1981a; Shbailat and Abouheif 2013; Nahmad et al. 2008). Therefore, in ant colonies, environmental and social cues regulate larval and imaginal disc development to produce distinct morphological castes (Abouheif and Wray 2002; Rajakumar et al. 2012, 2018; Wheeler and Nijhout 1981a, 1984). Here, we explore and discuss the possibility that the *major transitions in ant evolution may have been facilitated by modification of mechanisms that synchronize imaginal disc growth and coordination larval development* and have subsequently facilitated their evolutionary success and ecological dominance (Wilson 1953; Hölldobler and Wilson 2009).

10.5 WING IMAGINAL DISC DEVELOPMENT AND THE ORIGIN OF WING POLYPHENISM

In ants, queen and worker caste determination is generally regulated by environmental, nutritional, and social cues that activate juvenile hormone (JH), which, if it surpasses a threshold, activates the queen developmental program and, if it fails to surpass the threshold, activates the worker developmental program (Figure 10.2) (Brian 1963; Passera and Suzzoni 1978a, 1978b; Penick et al. 2012; Schwander et al. 2010). JH is a sesquiterpenoid hormone that is secreted by a pair of larval endocrine glands near the brain called the corpora allata (Nijhout 1994; Truman et al. 2006). JH is highly conserved within the insects, promoting larval development, and interacts with insulin and ecdysone signaling to regulate larval growth and developmental time (Nijhout 1994; Mirth and Shingleton 2014; Nijhout et al. 2014; Truman et al. 2006).

The timing of queen-worker caste determination differs between ant species—in the phylogenetically basal ponerine *Harpegnathos saltator*, caste determination can occur as late as the fourth larval instar, whereas in the phylogenetically derived species *Pheidole pallidula*, caste determination occurs during embryogenesis (Figure 10.2A–C) (Passera and Suzzoni 1978a, 1978b; Penick et al. 2012). In basal branching lineages, the most notable difference between queen and worker castes are the traits of wing polyphenism, where workers lack wings and associated thoracic musculature required for reproductive mating flights (Abouheif and Wray

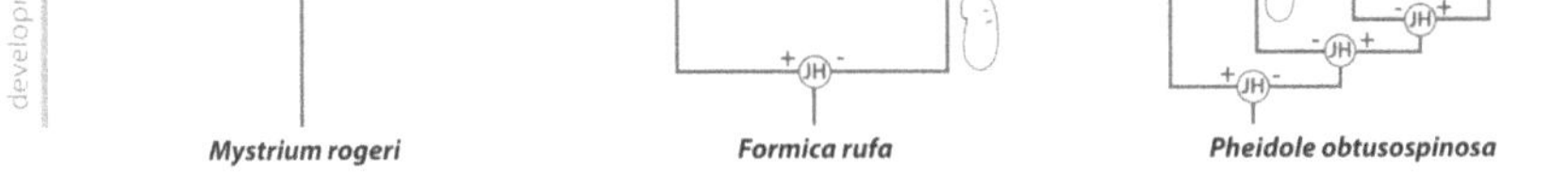

FIGURE 10.2 Larval and rudimentary wing imaginal disc development in different caste systems in ants shown as schematics of developmental switches (dark gray lines) mediated by juvenile hormone (JH) in three representative ant species. In (A–C), the length of the switches indicates approximate developmental time. Activation (+) or lack of activation (–) by JH regulates development of queen (Q) and worker (W) castes and minor worker (MW), soldier (SD), and supersoldier (XSD) subcastes. Terminal larvae are indicated by outlines: one side of a larval (rudimentary) wing disc(s) is represented as gray ovals within the outline of a larva; asterisks indicate the absence of the wing disc. Larval schematics and developmental time are to approximate scale within a species. (A) *Mystrium rogeri* (Behague et al. 2018; M. Molet personal communications); (B) *Formica rufa* (Dewitz 1878; Gosswald and Bier 1954), and (C) *Pheidole obtusospinosa* (Rajakumar et al. 2012). *Pheidole* spp. that lack the XSD subcaste have as single subcaste JH-mediated switch between MW and SD (Wheeler and Nijhout 1981b; Rajakumar et al. 2012, 2018).

2002; Keller et al. 2014). Despite lacking wings, worker larvae develop rudiments of wing discs (Figure 10.2A). In the basal genus *Mystrium*, these rudiments are slightly smaller compared to queen wing discs yet no differences in the expression of the wing gene regulatory network are observed (Behague et al. 2018). This suggests that worker rudimentary wing discs degenerate through either interruption of peripheral network genes or activation of apoptosis; however, the mechanisms and exact developmental timing of their degeneration remains to be determined (Abouheif and Wray 2002; Behague et al. 2018).

In other holometabolous insects, disc growth is synchronized to ensure robust and symmetrical development. While the worker rudiments develop extensively in basal ant species, they degenerate at some point during the late larval-pupal stages. This suggests that the origin of the queen-worker developmental switch may have initiated the "developmental character release" of the worker wing discs. Character release describes the increasing independence or modularity of a trait, following the origin of a switch, that reduces links with other traits (West-Eberhard 2003). Wing disc degeneration in the worker program and release of the wing discs from their wing function led to the origin of wing polyphenism, which is suggested to have been a driver of the colonization of the subterrestrial niche (Behague et al. 2018; Hölldobler and Wilson 2009; Rabeling et al. 2008; Keller et al. 2014). Therefore, modification of larval development and wing imaginal discs in ants may have been a necessary step in early ant evolution through one of two possibilities: (i) wing polyphenism originated prior to and facilitated the origin of eusociality in ants in the subterrestrial niche, or (ii) wing polyphenism may have evolved after the origin of eusociality to reinforce the reproductive division of labor between winged queens and wingless workers in a colony because wingless workers cannot engage in mating flights.

10.6 WING IMAGINAL DISC DEVELOPMENT AND THE POINT OF NO RETURN: REPRODUCTIVE AND MORPHOLOGICAL DIFFERENTIATION BETWEEN QUEEN AND WORKER CASTES

Derived ant genera represent lineages past the "point of no return"—a major transition in ant evolution in which workers become increasingly integrated into the colony, which is characterized by the notable increase in queen-worker dimorphism, reduced worker reproductive potential, and a marked increase in colony sizes (Fjerdingstad and Crozier 2006; Wilson 1953, 1954).

Earlier timing of the queen-worker switch in derived species past the point of no return generates significant size dimorphism—and is suggested to have evolved through changes in nutrition and insulin signaling—to produce worker larvae that are generally smaller and have a shorter developmental time compared to queen larvae (Figure 10.2B) (Abouheif and Wray 2002; Brian 1955; Chandra et al. 2018; Gosswald and Bier 1954; Mutti et al. 2011; West-Eberhard 2003; Wheeler 1986; Wilson 1954). These larvae typically possess rudimentary wing discs that are significantly reduced in size (Abouheif and Wray 2002; Bowsher et al. 2007; Rajakumar et al. 2018; Shbailat and Abouheif 2013). In the genus *Monomorium*, worker larvae lack wing rudiments completely revealing that their growth is not necessary for larval development or to the larval-pupal transition (Fave et al. 2015; Rajakumar et al. 2018). Reduced growth or lack of growth of worker rudimentary wing discs suggests that their growth has become increasingly modular and further dissociated from the development of other imaginal discs, raiseing the question of whether these rudiments participate in any interorgan coordination or have been fully disconnected from mechanisms of coordination due to their suppressed growth.

Worker larvae in species past the point of no return generally show interruption points upstream in the wing patterning gene network suggesting that, compared to basal ants, the interruption points have moved up in the network to limit the growth of wing rudiments (Abouheif and Wray 2002; Behague et al. 2018; Shbailat and Abouheif 2013). Nonetheless, different ant genera show striking variation in the expression of the wing gene network showing that these interruption points are a mosaic of conserved and labile interruptions (Abouheif and Wray 2002; Bowsher et al. 2007; Nahmad et al. 2008; Shbailat and Abouheif 2013; Fave et al. 2015). While it is unclear whether this is a result of neutral drift, selection, or a mix of both, it also raises the question of whether these rudiments are neutral or may have a specific function during larval development and/or whether they are still coordinated with larval developmental time (Abouheif and Wray 2002; Hall 2003; Nahmad et al. 2008). This could be investigated through ablation of the worker rudimentary wing discs across different ant genera. *Therefore the evolution of an early queen-worker switch, including the reduction of rudimentary wing disc growth, may have contributed to the elaboration of queen-worker dimorphism before or after the "point of no return," thereby facilitating this major transition in ant evolution.*

10.7 WING IMAGINAL DISC DEVELOPMENT AND THE EVOLUTION OF COMPLEX WORKER CASTE SYSTEMS IN ANTS

10.7.1 Repurposing Wing Rudiments to Generate the Novel Soldier Subcaste

Past the point of no return, several derived ant lineages independently underwent a major evolutionary transition, in which a unimodally distributed and uniformly sized monomorphic worker caste split into morphologically distinct subcastes called "*minor workers*" and "*major workers*" or "*soldiers*" (Blanchard and Moreau 2017; Ward 2014; Wilson 1953). These represent some of the most ecologically dominant and evolutionary successful lineages in ants, including genera such as the carpenter ants (*Camponotus* spp.), army ants (*Eciton* spp.), leaf-cutter ants (*Atta* spp.), and big-headed ants (*Pheidole* spp.) (Economo et al. 2015; Hölldobler and Wilson 2009; Ward et al. 2015, 2016). The novel soldier subcaste differs from minor workers in both size and scaling ("*allometry*") of the head relative to body size—their head is disproportionately large relative to their body size, facilitating the soldier's ability to efficiently perform defense, prey capture, or seed milling (Figure 10.3) (Hölldobler and Wilson 1990, 2009; Huang and Wheeler 2011; Wilson 1953).

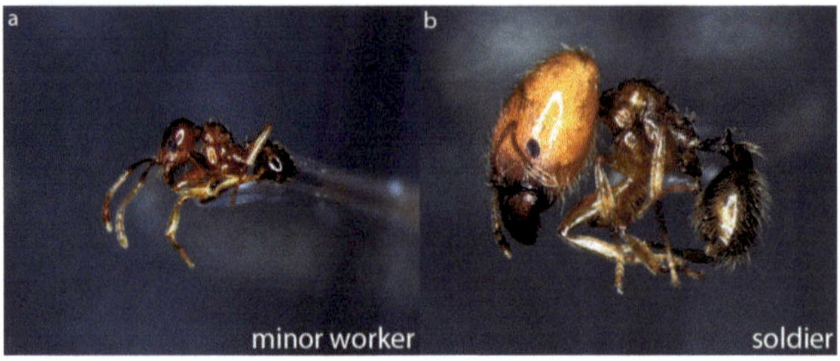

FIGURE 10.3 *Pheidole spadonia* minor worker (a) and soldier (b) adult showing disproportionate head-to-body size scaling between worker subcastes. Images to scale.

In the hyperdiverse genus *Pheidole,* environmental cues regulate soldier-minor worker subcaste determination through an additional developmental switch during the final larval instar, after which the larvae become distinguishable based on size and duration of development (Figure 10.2C) (Lillico-Ouachour and Abouheif 2017; Wheeler and Nijhout 1981a, 1981b, 1983; Passera 1974). A larva is specified as a minor worker or soldier during a critical window of larval development: a high-protein diet induces high levels of JH and activates the soldier developmental program, while low levels of JH activate the minor worker developmental program (Metzl et al. 2018; Rajakumar et al. 2012; Wheeler and Nijhout, 1981b, 1983).

In addition to changes in larval size and morphology, *Pheidole* soldier larvae are also characterized by JH-dependent growth of a single pair of rudimentary forewing discs (Figure 10.2C) (Rajakumar et al. 2012, 2018; Wheeler and Nijhout 1981a). While in *Pheidole* minor worker larvae wing rudiments are never activated, the rudimentary forewing discs in soldier larvae proliferate rapidly until they undergo apoptosis during the larval-prepupal transition (Sameshima et al. 2004; Shbailat et al. 2010; Wheeler and Nijhout 1981a). These forewing rudiments show interrupted expression of genes, such as *spalt*, downstream in the wing gene network (Abouheif and Wray 2002; Rajakumar et al. 2012; Shbailat et al. 2010). The soldier larval phenotype is characterized by an increase in larval size and extension of developmental time, which is strikingly similar to the phenotype observed when ectopic imaginal tissue is transplanted into host larvae or to the phenotypes observed following mutations in the ecdysone biosynthesis pathway in *Drosophila*. This suggests that similar signals may underlie soldier development (Bridges and Gabritschevsky 1928; Dewes 1973; Gabritschevsky and Bridges 1928; Hadorn and Buck 1962; Muth 1961; Simpson et al. 1980).

The elaborated growth of rudimentary wing discs in soldier larvae is not limited to the genus *Pheidole*. Rajakumar et al. (2018) detected a phylogenetic correlation between the presence of a soldier subcaste and the presence of discrete intersubcaste variation in rudimentary wing discs. For example, in the carpenter ant *Camponotus floridanus* and in the fire ant *Solenopsis geminata*—two species that also have smaller minor workers and large-headed soldiers—minor worker larvae develop two pairs of small rudimentary wing discs while larger soldier larvae develop two pairs of large rudimentary wing discs (Tschinkel 2013; Alvarado et al. 2015; Rajakumar et al. 2018). The development of these rudiments has long been used as a marker of the *soldier developmental program*, but they were thought to be a pleiotropic by-product of JH signaling (Wheeler and Nijhout 1981a). More recently, however, Rajakumar et al. (2018) demonstrated that these rudimentary discs acquired a novel function during development to generate disproportionate head-to-body scaling in the soldier subcaste in *Pheidole*.

Using imaginal disc ablation and RNA interference, Rajakumar et al. (2018) tested the function of the soldier rudimentary forewing discs in *Pheidole hyatti*. Experimentally reducing growth of the soldier rudimentary forewing disc disproportionately reduced head and body size, producing smaller individuals with altered head-to-body scaling with a more substantial effect on head size (Figure 10.4). This generated intermediate individuals that range between the distributions of typical soldiers and minor workers (Rajakumar et al. 2018). These intermediates are not typically observed in nature or in lab colonies; all *Pheidole* species described have a discrete dimorphic worker caste system (Wilson 2003). These phenotypes differ from those suggesting a resource competition model discussed above, where perturbations to disc development result in an increase in either mass or size of another appendage and also differ from those supporting a homeostatic model, where perturbations result in regeneration and interorgan synchronization or complete elimination and no reported change to adult morphology (Madhavan and Schniederman 1969; Simpson et al. 1980; Nijhout and Emlen 1998; Parker and Shingleton 2011; Simmons and Emlen 2006). Instead, perturbations to the soldier rudimentary forewing disc show that it positively regulates disproportionate head-to-body scaling in *P. hyatti*. While it remains to be determined if the soldier rudiment regulates head and body size directly or regulates head size through regulation of body size, these results show that these rudiments have acquired a novel function during development and evolution of the soldier subcaste (Rajakumar et al. 2018; Nijhout 2019).

In addition to the effects on disproportionate head-to-body scaling, experimental knockdown of the soldier rudimentary forewing discs using RNAi reduces larval developmental time by ~1 day (unpublished observations; Rajakumar et al. 2018). In contrast, disc elimination that does not allow for regeneration has no significant effect on developmental time (Madhavan and Schniederman 1969; Simpson et al. 1980). Therefore, in *P. hyatti*, the extension of larval developmental time typically observed during soldier development, as compared to minor worker development, may be a downstream product of growth of the rudimentary forewing disc (Figure 10.4). While future experiments are required to quantify and further investigate this process, we predict that the rapid proliferation of these rudiments functions to extend the fourth larval instar, perhaps using the same mechanism as regenerating or transplanted discs do to extend developmental time in *Drosophila*. In other holometabolous insects, an extension of larval development occurs when an imaginal disc is damaged and facilitates regeneration. In soldier development, the rapid proliferation of the rudimentary forewing disc may signal through ancestral

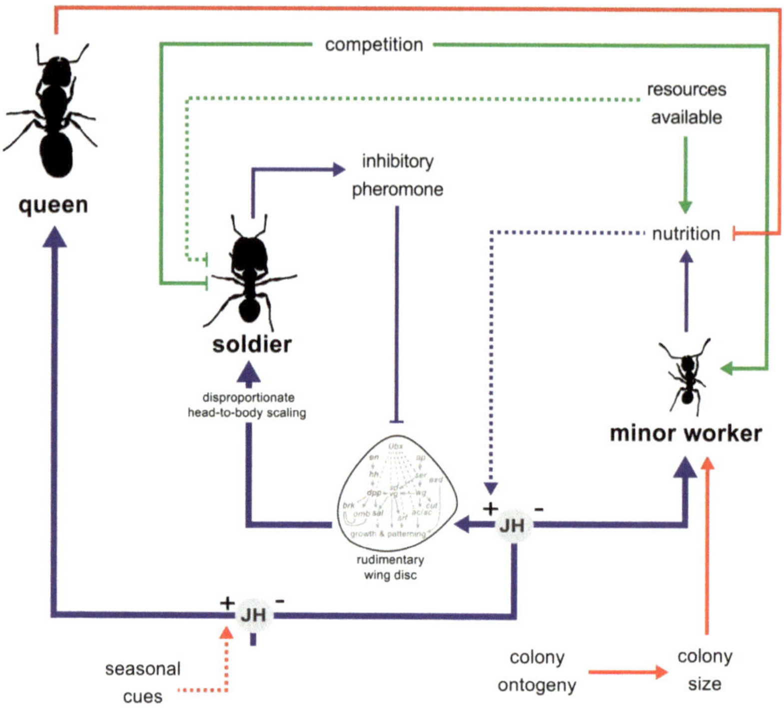

FIGURE 10.4 Model of environmental and social regulation of larval development and rudimentary wing discs in *Pheidole.* Solid lines indicate regulatory mechanisms supported by sufficient evidence, while dashed lines indicate regulatory mechanisms supported by evidence but requiring further study. All lines may indicate direct or indirect interactions. Lines with arrowheads (—>) indicate activation; lines with perpendicular bars (——|) indicate repression. Three major classes of interactions are internal and external influences (blue and red lines, respectively), colony development, and life cycle (red lines). **Internal influences** (blue lines). A first JH-mediated developmental switch during embryogenesis determines queen or worker fate; a second during late larval development determines soldier or minor worker fate. Nutrition activates JH, which activates growth of soldier rudimentary wing discs (gray outline with wing gene network inside) (Rajakumar et al. 2012, 2018; Wheeler and Nijhout 1981a). The soldier rudimentary wing disc is necessary to generate disproportionate head-to-body scaling and soldier subcaste determination. Adult soldiers suppress soldier production though the soldier inhibitory pheromone, which reduces growth of the soldier rudimentary wing disc (Rajakumar et al. 2018). **External influences** (green lines). Resources available in the external environment promote soldier development while reducing their numbers in the colony due to foraging needs. Competition reduces the adult soldier force and thus promotes soldier production through release of the soldier inhibitory pheromone. **Colony development and life cycle** (red lines). Seasonal cues increase the production of virgin queens by promoting JH, which reduces resources available for soldier development. Colony ontogeny promotes overall colony development and thus production of the worker force. External and colony-level influences are discussed in detail in Lillico-Ouachour and Abouheif (2017). Schematic modified from Lillico-Ouachour and Abouheif (2017) and Rajakumar et al. (2018).

systems of coordination that link imaginal discs to endocrine systems to extend developmental time and ultimately generate disproportionate head-to-body scaling. This may occur through the exploitation of ancestral mechanisms, including ecdysone signaling, or through tissue co-option and the evolution of novel signals (Rajakumar et al. 2018; Nijhout 2019).

The molecular mechanisms underlying the generation of disproportionate head-to-body scaling and its relationship with developmental timing remain unexplored but could be further investigated using transplantation. We would predict that transplantation of the soldier rudimentary forewing disc into bipotential larvae will extend developmental time and bias a larger proportion of larvae to develop into soldiers with disproportionate head-to-body scaling. These experiments are necessary to investigate if the soldier rudimentary forewing disc is sufficient to generate soldier allometry and if it regulates head and body size directly or indirectly through regulation of neuroendocrine centers and developmental time. While difficult to tease apart, we propose that coupling classic embryological techniques with transcriptomics and manipulations of endocrine signaling pathways could begin to identify the mechanisms through which disproportionate head-to-body scaling is generated during soldier development in *Pheidole*.

In addition to their developmental role in generating disproportionate head-to-body scaling, these soldier rudimentary forewing discs are necessary for determination of the soldier subcaste (Rajakumar et al. 2018). *Pheidole* colonies typically consist of approximately 95% minor workers and 5% soldiers (Passera 1977). Soldier development is activated by nutrition and JH and is negatively regulated by a contact-based inhibitory pheromone that is produced by adult soldiers (Lillico-Ouachour and Abouheif 2017; Metzl et al. 2018; Rajakumar et al. 2012, 2018; Wheeler and Nijhout 1983, 1984). The inhibitory pheromone suppresses growth of the soldier rudimentary forewing disc and subsequently regulates worker subcaste determination (Figure 10.4) (Rajakumar et al. 2018; Wheeler and Nijhout 1984).

Regulation of soldier subcaste determination through the rudimentary wing disc by adult soldiers facilitates dynamic social regulation of subcaste ratios within the colony (Passera et al. 1996; Wheeler and Nijhout 1984; Rajakumar et al. 2018). Internal colony properties are further regulated by external environmental cues, including seasonal temperature, available resources, competition as well as by colony ontogeny and size as illustrated and discussed further in Figure 10.4 (reviewed in detail in Lillico-Ouachour and Abouheif 2017). We propose that this dynamic regulation may have evolved through integration of environmental cues and social pheromones with the coordination mechanisms between imaginal discs and larval development that generate disproportionate head-to-body scaling and the soldier subcaste. Collectively, this shows that rudimentary forewing discs exist within a regulatory network that extends across levels of biological organization, from genes and cells to societies and ecological communities (Figure 10.4) (Rajakumar et al. 2018). Future experiments should investigate the mechanisms through which pheromones regulate rudimentary wing disc growth and will provide insight into how this mechanism of regulation may have evolved.

10.7.2 Further Elaboration of Rudimentary Wing Disc Growth May Have Facilitated Parallel Evolution of a Novel Supersoldier Subcaste

In addition to the evolution of the soldier subcaste, at least eight species in the genus *Pheidole* evolved a third worker subcaste called the "*supersoldier*" subcaste (Moreau 2008; Rajakumar et al. 2012; Wilson 2003). Supersoldiers have disproportionately even larger heads relative to their body sizes. Together, the three subcastes—minor workers, soldiers, and supersoldiers—produce a trimorphic worker caste system (Huang and Wheeler 2011; Rajakumar et al. 2012; Wilson 1953).

Supersoldier development differentiates from soldiers through a third nutritionally activated JH-mediated switch that produces individuals that use their large heads for seed milling and to block the nest entrance (Huang 2010; Rajakumar et al. 2012). Supersoldier larvae develop for a longer period of time and are characterized by the development of rudimentary fore- and hind wing discs, which degenerate prior to metamorphosis (Figure 10.2C) (Rajakumar et al. 2012). The supersoldier subcaste is ancestral in the genus *Pheidole.* Although it has been lost in the majority of *Pheidole* species, Rajakumar et al. (2012) showed that the developmental potential to produce this subcaste is retained across the genus and can be experimentally induced with JH. Induced supersoldier larvae grow larger and also develop two pairs of rudimentary wing discs showing that supersoldier development may be induced by environmental cues to produce variation in natural colonies that can then be genetically fixed (Rajakumar et al. 2012). Rudimentary forewing discs function to generate disproportionate scaling in soldiers of *P. hyatti*, which suggests that they may function to retain the ancestral developmental potential to produce supersoldiers and may generate the disproportionate head-to-body allometry of supersoldiers (Rajakumar et al. 2012, 2018).

We return now to the observations in *Galleria*, *Drosophila*, and *Coccinella* that the extension of larval developmental time is proportional to the amount of regenerating imaginal tissue (see Section 10.3B) (Madhavan and Schneiderman 1969; Simpson et al. 1980, Wu et al. 2019). This could suggest that the elaborated growth of rudimentary wing discs in supersoldiers may function additively or synergistically to generate disproportionate head-to-body scaling. While naturally occurring supersoldier larvae in *Pheidole rhea* and *Pheidole obtusospinosa* consistently develop large and symmetric rudimentary fore- and hind wing discs, induced supersoldier larvae in other *Pheidole* species that lack a supersoldier caste show a high degree of variability in the number and size of rudimentary wing discs (Rajakumar et al. 2012). This suggests that elaboration of rudimentary wing disc growth is an integral part of the supersoldier developmental program and that selection has stabilized their variation to generate discrete subcastes (Abouheif et al. 2014).

Investigating the role of these rudiments in generating disproportionate head-to-body scaling using a series of ablation experiments during natural and induced supersoldier development will test if they are necessary to generate supersoldier head-to-body scaling and if there is any difference in function between fore- and hind wing rudiments. As shown in other insects, one would predict that they function additively or synergistically and removal of multiple rudimentary wing discs will have a larger effect on disproportionate head-to-body scaling compared to removal of a single disc. Furthermore, transplanting a soldier rudimentary wing disc into soldier-destined larvae would test if these tissues are collectively sufficient to activate the supersoldier developmental program and further our understanding of how the developmental potential to produce this third novel subcaste is retained across the genus.

While *Pheidole* worker subcastes are discrete in size and allometry and typically show discrete elaboration in the size and number of rudimentary forewing discs in soldiers and supersoldiers, other worker caste systems are continuous in body size with proportional variation, such as *Solenopsis invicta,* while others have disproportionate head-to-body allometry, such as *C. floridanus* and *S. geminata* (Alvarado et al. 2015; Rajakumar et al. 2018; Tschinkel 2013; Tschinkel et al. 2003; Wilson 1953). Rajakumar et al. (2018) showed that perturbations to rudimentary forewing disc growth in *P. hyatti* transform the dimorphic worker caste system into a continuous system that mimics other species. We predict that in continuous worker caste systems, elaboration of rudimentary wing disc growth will also be continuous and that, in species with disproportionate head-to-body scaling, will show discrete differential growth trajectories that correspond with those in allometry. This would suggest that rudimentary wing discs are developmental modules that can be regulated by colony properties to release variation, which can be targeted by selection to generate novel worker caste systems and may have played an essential role in the evolution of discrete and continuous complex worker caste systems in ants.

10.8 CONCLUSIONS

In holometabolous insects, signaling mechanisms synchronize imaginal disc growth and coordinate them with developmental time, acting as a buffer against environmental perturbations to ensure proportional and symmetrical development (Gontijo and Garelli 2018; Shingleton and Frankino 2018; Vallejo et al. 2015; Andersen et al. 2013). The existence of physiological, developmental, and genetic mechanisms that underlie these processes have been shown to exist in many holometabolous lineages and have been studied for the past century through numerous techniques, including imaginal disc ablation and transplantation. Here, we have examined how these synchronization and coordination mechanisms may have been modified or released through environmental and social regulation to generate novel morphologies and developmental programs in ants.

In ants, environmental cues regulate imaginal disc and larval development to generate an astounding range of morphologies that produce functionally and morphologically distinct castes and subcastes. We propose that a change that was necessary for the origin of wing polyphenism—the evolution of degeneration of wing discs in the worker caste—may have been a necessary step that preceded the reduction and eventual repurposing of the wing rudiments to generate the soldier subcaste. This may have occurred through modification of ancestral coordination mechanisms between imaginal discs, larval developmental time, and the larval-pupal transition (Rajakumar et al. 2018). Furthermore, this may be why novel and distinct subcastes with disproportionate head-to-body scaling have evolved multiple times independently in ants but have not been characterized in bees and wasps, which maintain development of functional wings across castes (Branstetter et al. 2017).

More generally, set-aside cells provide a major source of modularity and thus facilitate the evolution of complexity by providing raw material for selection to act upon (Lewontin 1970; West-Eberhard 2003, 2005). Despite their quasi-independence, modules must be coordinated during development and may provide a framework to support and stabilize novel developmental programs (Klingenberg 2008). In holometabolous insects, imaginal discs are set aside during larval development, which has facilitated modification of their growth and the diversification of insect lineages (Shingleton and Frankino 2018; Yang 2001). In ants, wing imaginal discs have been further set aside as rudiments during the evolution of the worker caste following the origin of wing polyphenism. These rudiments exist within a regulatory network that extends across levels of biological organization (Abouheif and Wray 2002; Rajakumar et al. 2018). Therefore, the elaboration of biological complexity may be rooted in the setting aside, modification, and repurposing of rudimentary organs as developmental modules that can be tinkered with to generate novel morphology and developmental programs. Finally, revisiting older experiments and literature that provides links between experimental and natural phenotypes across phylogenetic groups may reveal common threads in the origin, regulation, and evolution of complex systems.

ACKNOWLEDGMENTS

We thank Brian Hall and Cory Bishop for the invitation to write and for their comments on this chapter; R. Rajakumar and members of the Abouheif Lab for discussion and comments on the manuscript; M. Morris at the Schulich Library at McGill University for assistance with literature queries; and C. Metzl for invaluable help with translations. This work was supported by an NSERC Discovery Grant to E.A. and an FRQNT graduate scholarship to S.K.

REFERENCES

Abouheif, E., Fave, M.-J., Ibarraran-Viniegra, A. S., Lesoway, M. P., Rafiqi, A. M., and Rajakumar, R. 2014. Eco evo devo: the time has come. *Adv. Exp. Med. Biol.* 781: 107–125.

Abouheif, E., and Wray, G. A. 2002. Evolution of the gene network underlying wing polyphenism in ants. *Science* 297(5579): 249–252. doi: 10.1126/science.1071468.

Alvarado, S., Rajakumar, R., Abouheif, E., and Szyf, M. 2015. Epigenetic variation in the Egfr gene generates quantitative variation in a complex trait in ants. *Nat. Commun.* 6: 6513. doi: 10.1038/ncomms7513.

Andersen, D. S., Colombani, J., and Leopold, P. 2013. Coordination of organ growth: principles and outstanding questions from the world of insects. *Trends Cell Biol.* 23: 336–344. doi: 10.1016/j.tcb.2013.03.005.

AntWeb v8.4. Available from http://www.antweb.org. Accessed 4 March 2019.

Beadle, G. W., Clancy, C. W., and Ephrussi, B. 1937. Development of eye colours in *Drosophila* pupal transplants and the influence of body fluid on vermilion. *Proc. R. Soc. London, Ser. B* 122: 98–105. doi: 10.1098/rspb.1937.0012.

Beadle, G. W., and Ephrussi, B. 1935. Transplantation in *Drosophila. Proc. Natl. Acad. Sci. U.S.A.* 21: 642–646. doi: 10.1073/pnas.21.12.642.

Beadle, G. W., and Ephrussi, B. 1936. The differentiation of eye pigments in *Drosophila* as studied by transplantation. *Genetics* 21: 225–247.

Behague, J., Fisher, B. L., Peronnet, R., Rajakumar, R., Abouheif, E., and Molet, M. 2018. General ladder-like evolution followed by reversals of the gene regulatory network underlying wing polyphenism in ants. *J. Exp. Zool. Part B* 330: 109–117.

Beira, J. V., and Paro, R. 2016. The legacy of *Drosophila* imaginal discs. *Chromosoma* 125: 573–592. doi: 10.1007/s00412-016-0595-4.

Beldade, P., Koops, K., and Brakefield, P. M. 2002. Modularity, individuality, and evo-devo in butterfly wings. *Proc. Natl. Acad. Sci. U.S.A.* 99: 14262–14267. doi: 10.1073/pnas.222236199.

Bier, K. 1954. Ueber den Saisondimorphismus des Oogenese von *Formica rufa* rufo-pratensis minor Gössw und dessen Bedeutung für die Kastendetermination. *Biol. Zbl.* 73: 170–190.

Blanchard, B. D., and Moreau, C. S. 2017. Defensive traits exhibit an evolutionary trade–off and drive diversification in ants. *Evolution* 71: 315–328. doi: 10.1111/evo.13117.

Bodenstein, D. 1938. Investigations on the problem of metamorphosis I combined experiments in clipping and transplanting on *Drosophila. Wilhelm Roux Arch. Entwicklungsmech. Org.* 137: 474–505.

Bodenstein, D. 1939. Investigations on the problem of metamorphosis IV. Developmental relations of interspecific organ transplants in Drosophila. *J. Exp. Zool.* 82: 1–30.

Bodenstein, D. 1940. Growth regulation of transplanted eye and leg discs in *Drosophila. J. Exp. Zool.* 84: 23–37. doi: 10.1002/jez.1400840104.

Bodenstein, D. 1943. Hormones and tissue competence in the development of *Drosophila. Biol. Bull.* 84: 34–58.

Boone, E., Colombani, J., Andersen, D. S., and Leopold, P. 2016. The hippo signalling pathway coordinates organ growth and limits developmental variability by controlling dilp8 expression. *Nat. Commun.* 7: 13505. doi: 10.1038/ncomms13505.

Bowsher, J. H., Wray, G. A., and Abouheif, E. 2007. Growth and patterning are evolutionarily dissociated in the vestigial wing discs of workers of the red imported fire ant, Solenopsis invicta. *J. Exp. Zool., Part B.* 308: 769–776. doi: 10.1002/jez.b.21200.

Brakefield, P. M., Gates, J., Keys, D., Kesbeke, F., Wijngaarden, P. J., Monteiro, A., French V., and Carroll, S. B. 1996. Development, plasticity and evolution of butterfly eyespot patterns. *Nature* 384: 236–242. doi: 10.1038/384236a0.

Branstetter, M. G., Danforth, B. N., Pitts, J. P., Gates, M. W., Kula, R. R., and Brady, S. G. 2017. Phylogenomic insights into the evolution of stinging wasps and the origins of ants and Bees. *Curr. Biol.* 27: 1019–1025. doi: 10.1016/j.cub.2017.03.027.

Brian, M. V. 1955. Studies of caste differentiation in *Myrmica rubra*: 2. The growth of workers and intercastes. *Insectes Soc.* 2: 1–34.

Brian, M. V. 1963. Studies of caste differentiation in *Myrmica rubra* L. 6. Factors influencing the course of female development in the early third instar. *Insectes Soc.* 10: 91–102.

Bridges, C. B., and Gabritschevsky, E. 1928. The giant mutation in *Drosophila melanogaster. Z. Indukt. Abstamm. Vererbungsl.* 46: 231–247.

Bryant, P. J., and Levinson, P. 1985. Intrinsic growth control in the imaginal primordia of Drosophila, and the autonomous action of a lethal mutation causing overgrowth. *Dev. Biol.* 107: 355–363. doi: 10.1016/0012-1606(85)90317-3.

Busse, S. M., McMillen, P. T., and Levin, M. 2018. Cross-limb communication during Xenopus hindlimb regenerative response: non-local bioelectric injury signals. *Development* 145: dev164210. doi: 10.1242/dev.164210.

Caldwell, P. E., Walkiewicz, M., and Stern, M. 2005. Ras activity in the *Drosophila* prothoracic gland regulates body size and developmental rate via ecdysone release. *Curr. Biol.* 15: 1785–1795. doi: 10.1016/j.cub.2005.09.011.

Carroll, S. B., Gates, J., Keyes, D. N., Paddock, S. W., Panganiban, G. E., Selegue, J. E., and Williams, J. A. 1994. Pattern formation and eyespot determination in butterfly wings. *Science* 265(5168): 109–114.

Casasa, S., and Moczek, A. P. 2018. Insulin signalling's role in mediating tissue-specific nutritional plasticity and robustness in the horn-polyphenic beetle Onthophagus taurus. *Proc. R. Soc. B* 285: 20181631. doi: 10.1098/rspb.2018.163110.6084/m9.

Caspari, E. 1933. Uber die Wirkung eines pleiotropen Gens bei der Mehlmotte *Ephestia kuhniella* Zeller. *Wilhelm Roux' Arch. Entwicklungsmech. Org.* 130: 353–381.

Chandra, V., Fetter-Pruneda, I., Oxley, P. R., Ritger, A. L., McKenzie, S. K., Libbrecht, R., and Kronauer, D. J. C. 2018. Social regulation of insulin signaling and the evolution of eusociality in ants. *Science* 361: 398–402. doi: 10.1126/science.aar5723.

Cohen, S. M. 1993. Imaginal Disc Development. In *The Development of Drosophila melanogaster* (Bate, M. and Martinez-Arias, A., Eds.) Vol. II, pp. 747–842. Cold Spring Harbor, NY: Cold Spring Harbor Laboratory Press.

Colombani, J., Andersen, D. S., Boulan, L., Boone, E., Romero, N., Virolle, V., Texada, M., and Leopold, P. 2015. *Drosophila* Lgr3 couples organ growth with maturation and ensures Ddevelopmental stability. *Curr. Biol.* 25: 2723–2729. doi: 10.1016/j.cub.2015.09.020.

Colombani, J., Andersen, D. S., and Leopold, P. 2012. Secreted peptide Dilp8 coordinates *Drosophila* tissue growth with developmental timing. *Science* 336: 582–585.

Dewes, E. 1973. Regeneration in transplanted halves of male genital disks and its influence upon duration of development in *Ephestia kuhniella* Z. *Wilhelm Roux' Arch. Entwicklungsmech. Org.* 172: 349–354.

Dewitz, H. 1878. Beitrage zur postembryonalen Gliedmassenbildung bei den lnsekten. *Z. Wiss. Zool.* 30: 78–105.

Droujinine, I. A., and Perrimon, N. 2016. Interorgan communication pathways in physiology: focus on *Drosophila. Annu. Rev. Genet.* 50: 539–570. doi: 10.1146/annurev-genet-121415-122024.

Economo, E. P., Klimov, P., Sarnat, E. M., Guenard, B., Weiser, M. D., Lecroq, B., and Knowles, L. L. 2015. Global phylogenetic structure of the hyperdiverse ant genus *Pheidole* reveals the repeated evolution of macroecological patterns. *Proc. Biol. Sci.* 282: 20141416. doi: 10.1098/rspb.2014.1416.

Edgar, B. A. 2006. How flies get their size: genetics meets physiology. *Nat. Rev. Genet.* 7: 907–916. doi: 10.1038/nrg1989.

Emlen, D. J. 1997. Diet alters male horn allometry in the beetle *Onthophagus acuminatus* (Coleoptera: Scarabaeidae). *Proc. R. Soc. London, Ser. B* 264: 567–574.

Emlen, D. J., and Nijhout, H. F. 2000. The development and evolution of exaggerated morphologies in insects. *Annu. Rev. Entomol.* 45: 661–708.

Emlen, D. J., Warren, I. A., Johns, A., Dworkin, I., and Lavine, L. C. 2012. A mechanism of extreme growth and reliable signaling in sexually selected ornaments and weapons. *Science* 337: 860–864. doi: 10.1126/science.1224286.

Engel, M. S. 2015. Insect evolution. *Curr. Biol.* 25: R868–R872. doi: 10.1016/j.cub.2015.07.059.

Ephrussi, B., and Beadle, G. W. 1936. A technique of transplantation for *Drosophila. Am. Nat.* 70: 218–225.

Ephrussi, B., and Beadle, G. W. 1937. Development of eye colors in *Drosophila*: transplantation experiments on the interaction of *Vermilion* with other eye colors. *Genetics* 22: 65–75.

Faltus, F., and Oberlander, H. 1970. Ecdysone induced differentiation of pulsating regions in the genital imaginal disks after culture in vivo. *Dros. Inf. Serv.* 45: 155.

Fave, M. J., Johnson, R. A., Cover, S., Handschuh, S., Metscher, B. D., Muller, G. B., Gopalan, S., and Abouheif, E. 2015. Past climate change on Sky Islands drives novelty in a core developmental gene network and its phenotype. *BMC Evol. Biol.* 15: 183. doi: 10.1186/s12862-015-0448-4.

Fjerdingstad, E. J., and Crozier, R. H. 2006. The evolution of worker caste diversity in social insects. *Am. Nat.* 167: 390–400. doi: 10.1086/499545.
Fukuda, S. 1940. Induction of pupation in silkworm by transplanting the prothoracic gland. *Proc. Imp. Acad. Japan* 16: 414–416.
Gabritschevsky, E., and Bridges, C. B. 1928. Part II: physiological aspects of the giant race the giant "caste". *Z. Indukt. Abstamm. Vererbungsl.* 46: 248–284.
Garelli, A., Gontijo, A. M., Miguela, V., Caparros, E., and Dominguez, M. 2012. Imaginal discs secrete insulin-like peptide 8 to mediate plasticity of growth and maturation. *Science* 336: 579–582. doi: 10.1126/science.1216735.
Garelli, A., Heredia, F., Casimiro, A. P., Macedo, A., Nunes, C., Garcez, M., Dias, A. R., Volonte, Y. A. et al. 2015. Dilp8 requires the neuronal relaxin receptor Lgr3 to couple growth to developmental timing. *Nat. Commun.* 6: 8732. doi: 10.1038/ncomms9732.
Gokhale, R. H., Hayashi, T., Mirque, C. D., and Shingleton, A. W. 2016. Intra-organ growth coordination in *Drosophila* is mediated by systemic ecdysone signaling. *Dev. Biol.* 418: 135–145. doi: 10.1016/j.ydbio.2016.07.016.
Gontijo, A. M., and Garelli, A. 2018. The biology and evolution of the Dilp8-Lgr3 pathway: A relaxin–like pathway coupling tissue growth and developmental timing control. *Mech. Dev.* 154: 44–50. doi: 10.1016/j.mod.2018.04.005.
Gosswald, K., and Bier, K. 1954. Untersuchungen zur Kastendetermination in der Gattung Formica. Die Kastendetermination von Formica rufa. *Insectes Soc.* 1: 229–246.
Hadorn, E. 1937. An accelerating effect of normal "ring-glands" on puparium-formation in lethal larvae of *Drosophila melanogaster. Proc. Natl. Acad. Sci. U.S.A.* 23: 478–484.
Hadorn, E. 1963. Differenzierungsleistungen wiederholt fragmentierter Teilstücke männlicher Genitalscheiben von *Drosophila melanogaster* nach Kultur *in vivo. Dev. Biol.* 7: 617–629.
Hadorn, E., and Buck, D. 1962. On the differentiation of transplanted wing imaginal disc fragments of *Drosophila melanogaster. Rev. Suisse Zool.* 69: 302–310.
Hall, B. K. 2003. Descent with modification: the unity underlying homology and homoplasy as seen through an analysis of development and evolution. *Biol. Rev.* 78(3): 409–433.
Hariharan, I. K. 2012. How growth abnormalities delay "puberty" in *Drosophila. Sci. Signal.* 5: pe27. doi: 10.1126/scisignal.200323S.
Hariharan, I. K., and Serras, F. 2017. Imaginal disc regeneration takes flight. *Curr. Opin. Cell Biol.* 48: 10–16. doi: 10.1016/j.ceb.2017.03.005.
Held, L. I. J. 2002. *Imaginal Discs: The Genetic and Cellular Logic of Pattern Formation.* Cambridge: Cambridge University Press.
Herboso, L., Oliveira, M. M., Talamillo, A., Perez, C., Gonzalez, M., Martin, D., Sutherland, J. D., Shingleton, A. W. et al. 2015. Ecdysone promotes growth of imaginal discs through the regulation of Thor in D. melanogaster. *Sci. Rep.* 5: 12383. doi: 10.1038/Srep12383.
Hojyo, T., and Fujiwara, H. 1997. Reciprocal transplantation of wing discs between a wing deficient mutant (fl) and wild type of the silkworm, *Bombyx mori. Dev. Growth Differ.* 39: 599–606.
Hölldobler, B., and Wilson, E. O. 1990. *The Ants.* Cambridge: The Belknap Press of Harvard University Press.
Hölldobler, B., and Wilson, E. O. 2009. *The Superorganism.* New York: W.W Norton and Company Inc.
Horikawa, M., and Sugahara, T. 1960. Studies on the effects of radiation on living cells in tissue culture: I. Radiosensitivity of various imaginal discs and organs in larvae of Drosophila melanogaster. *Rad. Res.* 12: 266–275.
Howland, R. B., Sonnenblick, B. P., and Glancy, E. A. 1937. Transplantation of wing–thoracic primordia in *Drosophila melanogaster. Am. Nat.* 71: 158–166. doi: 10.1086/280716.
Huang, M. H. 2010. Multi-phase defense by the big-headed ant, Pheidole obtusospinosa, against raiding army ants. *J. Insect Sci.* 10: 1. doi: 10.1673/031.010.0101.

Huang, M. H., and Wheeler, D. E. 2011. Colony demographics of rare soldier-polymorphic worker caste systems in Pheidole ants (Hymenoptera, Formicidae). *Insectes Soc.* 58: 539–549. doi: 10.1007/s00040-011-0176-8.

Hurley, I., Pomiankowski, A., Fowler, K., and Smith, H. 2002. Fate map of the eye-antennal imaginal disc in the stalk–eyed fly *Cyrtodiopsis dalmanni. Dev. Genes Evol.* 212: 38–42.

Jaszczak, J. S., Wolpe, J. B., Bhandari, R., Jaszczak, R. G., and Halme, A. 2016. Growth coordination during *Drosophila melanogaster* imaginal disc regeneration is mediated by signaling through the relaxin receptor lgr3 in the prothoracic gland. *Genetics* 204: 703–709. doi: 10.1534/genetics.116.193706.

Jaszczak, J. S., Wolpe, J. B., Dao, A. Q., and Halme, A. 2015. Nitric oxide synthase regulates growth coordination during *Drosophila melanogaster* imaginal disc regeneration. *Genetics* 200: 1219–1228. doi: 10.1534/genetics.115.178053.

Katsuyama, T., and Paro, R. 2016. Imaginal Disc Transplantation in Drosophila. In *Polycomb Group Proteins: Methods and Protocols. C. Lenzuolo and B. Bodega* Vol. 1480, pp. 301–310. Totowa: Humana Press Inc.

Keller, R. A., Peeters, C., and Beldade, P. 2014. Evolution of thorax architecture in ant castes highlights trade-off between flight and ground behaviors. *eLife* 3:e01539.

Klingenberg, C. P. 2008. Morphological integration and developmental modularity. *Annu. Rev. Ecol. Evol. Sys.* 39(1): 115–132.

Klingenberg, C. P., and Nijhout, H. F. 1998. Competition among growing organs and developmental control of morphological asymmetry. *Proc. R. Soc. London, Ser. B* 265: 1135–1139. doi: 10.1098/rspb.1998.0409.

Kohler, R. E. 1994. *Lords of the Fly: Drosophila Genetics and the Experimental Life.* Chicago: The University of Chicago Press.

Kunkel, J. G. 1977. Cockroach molting. II. The nature of regeneration induced delay of molting hormone secretion. *Biol. Bull.* 153: 145–162.

Lakes, R., and Mucke, A. 1988. Regeneration of the foreleg tibia and tarsi of *Ephippiger ephippiger* (Orthoptera:Tettigoniidae). *J. Exp. Zool.* 250: 176–187.

Lewontin, R. C. 1970. The units of selection. *Annu. Rev. Eco. Syst.* 1: 1–18.

Lillico-Ouachour, A., and Abouheif, E. 2017. Regulation, development, and evolution of caste ratios in the hyperdiverse ant genus *Pheidole. Curr. Opin. Insect Sci.* 19: 43–51. doi: 10.1016/j.cois.2016.11.003.

Madhavan, K., and Schneiderman, H. A. 1969. Hormonal control of imaginal disc regeneration in *Galleria mellonella* (Lepidoptera). *Biol. Bull.* 137: 321–331. doi: 10.2307/1540104.

Metzl, C., Wheeler, D. E., and Abouheif, E. 2018. Wilhelm Goetsch (1887–1960): pioneering studies on the development and evolution of the soldier caste in social insects. *Myrmecol. News* 26: 81–96.

Mirth, C. K., and Shingleton, A. W. 2014. The roles of juvenile hormone, insulin/target of rapamycin, and ecydsone signaling in regulating body size in Drosophila. *Commun. Integr. Biol.* 7(5): e971568. doi: 10.4161/cib.29240.

Mirth, C. K. and Shingleton, A.W. 2019. Coordinating development: how do animals integrate plastic and robust developmental processes? *Front. Cell Dev. Biol.* 7: 8. doi: 10.3389/fcell.2019.00008.

Misof, B., Liu, S., Meusemann, K., Peters, R. S., Donath, A., Mayer, C., Frandsen, P. B., Ware, J. et al. 2014. Phylogenomics resolves the timing and pattern of insect evolution. *Science* 346: 763–767. doi: 10.1126/science.1257570.

Moczek, A. P. 2008. On the origins of novelty in development and evolution. *Bioessays* 30: 432–447. doi: 10.1002/bies.20754.

Moczek, A. P. and Nijhout, H. F. 2004. Trade-offs during the development of primary and secondary sexual traits in a horned beetle. *Am. Nat.* 163(2): 184–191.

Moczek, A. P., and Rose, D. J. 2009. Differential recruitment of limb patterning genes during development and diversification of beetle horns. *Proc. Natl. Acad. Sci. U S A.* 106: 8992–8997. doi: 10.1073/pnas.0809668106.

Moreau, C. S. 2008. Unraveling the evolutionary history of the hyperdiverse ant genus *Pheidole* (Hymenoptera: Formicidae). *Mol. Phylogenet. Evol.* 48: 224–239. doi: 10.1016/j.ympev.2008.02.020.

Monteiro, A., Prijs, J., Bax, M., Hakkaart, T., and Brakefield, P. M. 2003. Mutants highlight the modular control of butterfly eyespot patterns. *Evol. Dev.* 5(2): 180–187.

Muth, F. W. 1961. Untersuchungen zur wirkungsweise der mutatne 'kfl' bei der methlmotte *Ephestia kuhniella. Wilhelm Roux' Arch. Entwicklungsmech. Org.* 153: 370–418.

Mutti, N. S., Dolezal, A. G., Wolschin, F., Mutti, J. S., Gill, K. S., and Amdam, G. V. 2011. IRS and TOR nutrient-signaling pathways act via juvenile hormone to influence honey bee caste fate. *J. Exp. Biol.* 214: 3977–3984. doi: 10.1242/jeb.061499.

Nahmad, M., Glass, L., and Abouheif, E. 2008. The dynamics of developmental system drift in the gene network underlying wing polyphenism in ants: a mathematical model. *Evol. Dev.* 10: 360–374. doi: 10.1111/j.1525-142X.2008.00244.x.

Nijhout, H. F. 1994. *Insect Hormones.* Princeton: Princeton University Press.

Nijhout, H. F. 2019. Larval development: making ants into soldiers. *Curr. Biol.* 29(1): R32–R34.

Nijhout, H. F., and Callier, V. 2015. Developmental mechanisms of body size and wing–body scaling in insects. *Annu. Rev. Entomol.* 60: 141–156. doi: 10.1146/annurev-ento-010814-020841.

Nijhout, H. F., and Emlen, D. J. 1998. Competition among body parts in the development and evolution of insect morphology. *Proc. Natl. Acad. Sci. U S A.* 95: 3685–3689.

Nijhout, H. F., and Grunert, L. W. 2010. The cellular and physiological mechanism of wing-body scaling in *Manduca sexta. Science* 330: 1693–1695. doi: 10.1126/science.1197292.

Nijhout, H. F., Riddiford, L. M., Mirth, C., Shingleton, A. W., Suzuki, Y., and Callier, V. 2014. The developmental control of size in insects. *Wiley Interdiscip Rev. Dev. Biol.* 3: 113–134. doi: 10.1002/wdev.124.

Nöthiger, R. 1998. Antonio García-Bellido at Hadorn's laboratory in Zurich. *Int. J. Dev. Biol.* 42: 519–521.

Nöthiger, R. 2002. Ernst Hadorn, a pioneer of developmental genetics. *Int. J. Dev. Biol.* 46: 23–27.

Oberlander, H. 1972. The Hormonal Control of Development of Imaginal Disks. In *The Biology of Imaginal Disks* (Ursprung, H. and Nöthiger, R., Eds.), pp. 155–172. Germany: Springer-Verlag Berlin.

Parker, N. F., and Shingleton, A. W. 2011. The coordination of growth among *Drosophila* organs in response to localized growth-perturbation. *Dev. Biol.* 357: 318–325. doi: 10.1016/j.ydbio.2011.07.002.

Passera, L. 1977. Production of soldiers in colonies coming out of hibernation in ant *Pheidole pallidula* (Nyl). *Insectes Soc.* 24: 131–146.

Passera, L., Roncin, E., Kaufmann, B., and Keller, L. 1996. Increased soldier production in ant colonies exposed to intraspecific competition. *Nature* 379: 630–631.

Passera, L., and Suzzoni, J. P. 1978a. JH-treatment of queens and sexualization of the brood in *Pheidole pallidula* (NYL) (Hymenoptera, Formicidae). *C. R. Hebd. Seances Acad. Sci. Ser D* 287: 1231–1233.

Passera, L., and Suzzoni, J. P. 1978b. Obtaining of sexual brood after JH treatment in ant *Pheidole pallidula*-(Nyl) (Hymenoptera, Formicidae). *C. R. Hebd. Seances Acad. Sci. Ser D* 286: 615–618.

Penick, C. A., Prager, S. S., and Liebig, J. 2012. Juvenile hormone induces queen development in late-stage larvae of the ant *Harpegnathos saltator. J. Insect Physiol.* 58: 1643–1649. doi: 10.1016/j.jinsphys.2012.10.004.

Pohley, H.-J. 1959. Experimentelle Beitrage zur Lenkung der Organentwicklung, des Hautungsrhythmus und der Metamorphose bei der Schabe *Periplaneta americana. Wilhelm Roux' Arch. Entwicklungsmech. Org.* 151: 323–380.

Pohley, H.-J. 1960. Experimentelle Untersuchungen uber die Steuerung des Hautungsrhythmus bei dei Mehlmotte Ephestia kuhniella Zeller. *Roux's Arch. Dev. Biol.* 152: 183–203.

Pohley, H.-J. 1970. Interactions between the endocrine system and the developing tissue in *Ephestia kuhniella. Behav. Sci.* 15: 46–56.

Poodry, C. A., and Woods, D. F. 1990. Control of the developmental timer for *Drosophila* pupariation. *Roux's Arch. Dev. Biol.* 199: 219–227.

Postlethwait, J. H., and Schneiderman, H. A. 1968. Effects of an ecdysone on growth and cuticle formation of *Drosophila* imaginal discs cultured *in vivo. Biol. Bull.* 135: 431.

Rabeling, C., Brown, J. M., and Verhaagh, M. 2008. Newly discovered sister lineage sheds light on early ant evolution. *Proc. Natl. Acad. Sci. U S A.* 105: 14913–14917. doi: 10.1073/pnas.0806187105.

Rajakumar, R., Koch, S., Couture, M., Fave, M.-J., Lilico-Ouachour, A., Chen, T., De Blasis, G., Rajakumar, J. et al. 2018. Social regulation of a rudimentary organ generates complex worker caste systems in ants. *Nature* 562: 574–577.

Rajakumar, R., Mauro, D. S., Dijkstra, M. B., Huang, M. H., Wheeler, D. E., Hiou–Tim, F., Khila, A. et al. 2012. Ancestral developmental potential facilitates parallel evolution in ants. *Science* 335: 79–82. doi: 10.1126/science.1211451.

Rhinsberger, H.-J. 2000. The experimental design of Alfred Kuhn's physiological developmental genetics. *J. Hist. Biol.* 33: 535–576.

Rosello-Diez, A., Madisen, L., Bastide, S., Zeng, H., and Joyner, A. L. 2018. Cell-nonautonomous local and systemic responses to cell arrest enable long-bone catch-up growth in developing mice. *PLoS Biol.* 16: e2005086. doi: 10.1371/journal.pbio.2005086.

Sameshima, S., Miura, T., and Matsumoto, T. 2004. Wing disc development during caste differentiation in the ant *Pheidole megacephala* (Hymenoptera : Formicidae). *Evol. Dev.* 6: 336–341.

Schwander, T., Lo, N., Beekman, M., Oldroyd, B. P., and Keller, L. 2010. Nature versus nurture in social insect caste differentiation. *Trends Ecol. Evol.* 25: 275–282. doi: 10.1016/j.tree.2009.12.001.

Schwartz, M. B., Imberski, R. B., and Kelly, T. J. 1984. Analysis of metamorphosis in *Drosophila melanogaster*: characterization of *giant*, an ecdysteroid–deficient mutant. *Dev. Biol.* 103: 85–95.

Sehnal, F., and Bryant, P. J. 1993. Delayed pupariation in *Drosophila* imaginal disc overgrowth mutants is associated with reduced ecdysteroid titer. *J. Ins. Physiol.* 39: 1051–1059. doi: 10.1016/0022-1910(93)90129-F.

Setiawan, L., Woods, A. L., and Hariharan, I. K. 2018. The BMP2/4 ortholog Dpp can function as an inter-organ signal that regulates developmental timing. *Life Sci. Alliance* 1: e201800216. doi: 10.1101/180562.

Shbailat, S. J., and Abouheif, E. 2013. The wing-patterning network in the wingless castes of Myrmicine and Formicine ant species is a mix of evolutionarily labile and non-labile genes. *J. Exp. Zool. B Mol. Dev. Evol.* 320: 74–83. doi: 10.1002/jez.b.22482.

Shbailat, S. J., Khila, A., and Abouheif, E. 2010. Correlations between spatiotemporal changes in gene expression and apoptosis underlie wing polyphenism in the ant *Pheidole morrisi. Evol. Dev.* 12: 580–591. doi: 10.1111/j.1525-142X.2010.00443.x.

Shingleton, A. W., and Frankino, W. A. 2018. The (ongoing) problem of relative growth. *Curr. Opin. Insect Sci.* 25: 9–19. doi: 10.1016/j.cois.2017.10.001.

Simmons, L. W., and Emlen, D. 2006. Evolutionary trade-off between weapons and testes. *Proc. Natl. Acad. Sci. U.S.A.* 103: 16346–16351.

Simpson, P., Berreur, P., and Berreur-Bonnenfant, J. 1980. The initation of pupariation in *Drosophila*: dependence on growth of the imaginal discs. *J. Embryol. Exp. Morphol.* 57: 155–165.

Simpson, P., and Schneiderman, H. A. 1975. Isolation of temperature sensitive mutations blocking clone development in *Drosophila melanogaster*, and the effects of a temperature sensitive cell lethal mutation of pattern formation in imaginal discs. *Wilhelm Roux's Arch. Dev. Biol.* 178: 247–275.

Stieper, B. C., Kupershtok, M., Driscoll, M. V., and Shingleton, A. W. 2008. Imaginal discs regulate developmental timing in *Drosophila melanogaster*. *Dev. Biol.* 321: 18–26. doi: 10.1016/j.ydbio.2008.05.556.

Stock, A., and O'Farrell, A. F. 1954. Regeneration and the moulting cycle in *Blattella germanica*: II. Simultaneous regeneration of both metathoracic legs. *Aust. J. Biol. Sci.* 7: 302–307.

Sturtevant, A. H. 1929. The claret mutant type of *Drosophila* simulans: a study of chromosome elimination and cell lineage. *Z. Wiss. Zool.* 135: 323–356.

Tanner, J. M. 1963. Regulation of growth in size in mammals. *Nature* 199: 845–850.

Tobler, A., and Nijhout, H. F. 2010. A switch in the control of growth of the wing imaginal disks of *Manduca sexta*. *PLoS One* 5: e10723. doi: 10.1371/journal.pone.0010723.

Truman, J. W., Hiruma, K., Allee, J. P., MacWhinnie, S. G. B., Champlin, D. T., and Riddiford, L. M. 2006. Juvenile hormone is required to couple imaginal disc formation with nutrition in insects. *Science* 312: 1385–1388. doi: 10.1126/science.1123652.

Truman, J. W., and Riddiford, L. M. 1999. The origins of insect metamorphosis. *Nature* 401: 447–452.

Tschinkel, W. R. 2013. The morphometry of Solenopsis fire ants. *PLoS One* 8: e79559. doi: 10.1371/journal.pone.0079559.

Tschinkel, W. R., Mikheyev, A. S., and Storz, S. R. 2003. Allometry of workers of the fire ant, *Solenopsis invicta*. *J. Insect Sci.* 3: 11.

Ursprung, H., and Hadorn, E. 1962. Further investigations of pattern formation in combinations of partly dissociated imaginal wing discs of *Drosophila melanogaster*. *Dev. Biol.* 4: 40–66.

Vallejo, D. M., Juarez-Carreno, S., Bolivar, J., Morante, J., and Dominguez, M. 2015. A brain circuit that synchronizes growth and maturation revealed through Dilp8 binding to Lgr3. *Science* 350: aac6767. doi: 10.1126/10.1126/science.aac6767.

Vogt, M. 1942. Induction of metamorphotic processes by implanted ring glands in *Drosophila*. *Wilhelm Roux' Arch. Entwicklungsmech. Org.* 142: 131–182.

Waddington, C. H. 1956. Genetic assimilation of the Bithorax phenotype. *Evolution* 10: 1–13.

Wagner, G. P. 1996. Homologues, natural kinds and the evolution of modularity. *Am. Zool.* 36: 36–43.

Ward, P. S. 2014. The phylogeny and evolution of ants. *Annu. Rev. Ecol. Evol. Syst.* 45: 23–43. doi: 10.1146/annurev-ecolsys-120213-091824.

Ward, P. S., Blaimer, B. B., and Fisher, B. L. 2016. A revised phylogenetic classification of the ant subfamily Formicinae (Hymenoptera: Formicidae), with resurrection of the genera Colobopsis and Dinomyrmex. *Zootaxa* 4072: 343–357. doi: 10.11646/zootaxa.4072.3.4.

Ward, P. S., Brady, S. G., Fisher, B. L., and Schultz, T. R. 2015. The evolution of myrmicine ants: phylogeny and biogeography of a hyperdiverse ant clade (Hymenoptera: Formicidae). *Syst. Entomol.* 40: 61–81. doi: 10.1111/syen.12090.

West-Eberhard, M. J. 2003. *Developmental Plasticity and Evolution*. Oxford, New York: Oxford University Press.

West-Eberhard, M. J. 2005. Developmental plasticity and the origin of species differences. *Proc. Natl. Acad. Sci. U.S.A.* 102: 6543–6549. doi: 10.1073/pnas.0501844102.

Wheeler, D. E. 1986. Developmental and physiological determinants of caste in social hymenoptera - evolutionary implications. *Am. Nat.* 128: 13–34.

Wheeler, D. E., and Nijhout, H. F. 1981a. Imaginal wing discs in larvae of the soldier caste of *Pheidole bicarinata vinelandica forel* (Hymenoptera: Formicidae). *Int. J. Insect Morphol. Embryol.* 10: 131–139.

Wheeler, D. E., and Nijhout, H. F. 1981b. Soldier determination in ants: new role for juvenile–hormone. *Science* 213: 361–363.

Wheeler, D. E., and Nijhout, H. F. 1983. Soldier determination in *Pheidole bicarinata*: effect of methoprene on caste and size within castes. *J. Insect Physiol.* 29: 847–854.

Wheeler, D. E., and Nijhout, H. F. 1984. Soldier determination in *Pheidole bicarinata*: Inhibition by adult soldiers. *J. Insect Physiol.* 30: 127–135.

Wilkins, A. 2003. Canalization and Genetic Assimilation. In *Keywords and Concepts in Evolutionary Developmental Biology* (Hall, B. K. and Olson, W. M., Eds.), pp. 23–30. Cambridge, Massachusetts: Harvard University Press.

Wilson, E. O. 1953. The origin and evolution of polymorphism in ants. *Q. Rev. Biol.* 28: 136–156.

Wilson, E. O. 1954. A new interpretation of the frequency curves associated with ant polymorphism. *Insectes Soc.* 1: 75–80.

Wilson, E. O. 2003. *Pheidole in the New World.* Cambridge, Massachusetts: Harvard University Press.

Wilson, E. O., and Hölldobler, B. 2005. Eusociality: origin and consequences. *Proc. Natl. Acad. Sci. U.S.A.* 102: 13367–13371. doi: 10.1073/pnas.0505858102.

Wu, P., Wu, F., Yan, S., Liu, C., Shen, Z., Xiong, X., Li, Z., Zhang, Q., and Liu, X. 2019. Developmental cost of leg-regenerated *Coccinella septempunctata* (Coleoptera: Coccinellidae). *PLoS One* 14(1): e0210615.

Xu, T., and Rubin, G. M. 2012. The effort to make mosaic analysis a household tool. *Development* 139: 4501–4503. doi: 10.1242/dev.085183.

Yamanaka, N., Rewitz, K. F., and O'Connor, M. B. 2013. Ecdysone control of developmental transitions: lessons from Drosophila research. *Annu. Rev. Entomol.* 58: 497–516. doi: 10.1146/annurev-ento-120811-153608.

Yang, A. S. 2001. Modularity, evolvability, and adaptive radiations: a comparison of the hemi- and holometabolous insects. *Evol. Dev.* 3: 59–72.

11 Evolution of Adaptive Immunity through Set-Aside Cells

Kurt Buchmann
University of Copenhagen

CONTENTS

11.1 ADAPTIVE VERSUS INNATE IMMUNITY

Any organism on the earth will face the danger of being colonized or invaded by more or less opportunistic or specialized commensals or parasites. Immune mechanisms have evolved in the hosts to neutralize the invader, reduce the impact, and secure survival.

The word "*immunity*" is derived from Latin *immunis*, which means "released from a burden." This highly suitable term illustrates an ability to combat invading pathogens that has been crucial for survival of even the most primitive organisms since their first appearance more than a billion years ago. During evolution, a plethora of protective and defensive mechanisms in host organisms have appeared (Dzik 2010). Some of these variants have probably been lost together with major animal groups during mass extinction events and major geological changes as occurred, for example, 65 and 250 MYA.

Still, effective immune mechanisms are associated with extant animal groups. It is convenient to group these into *innate* and *adaptive responses*. One part of the immunological armament with associated sentinels can recognize potentially harmful elements and subsequently activate the needed counterattack immediately upon challenge. This system is under one umbrella—the innate immune system—and covers the concept of *innate immunity*. In contrast to these basic reactions, the term "*adaptive immunity*" is generally used to characterize the sophisticated ability of vertebrates to (i) differentiate *self* from *nonself*, (ii) memorize the specific shape, structure, and charge of environmental antigens, and (iii) induce clonal expansion of antigen-specific lymphocytes, some of which are set aside as memory cells.

The immunity raised in vertebrates by the use of the classic adaptive immune mechanisms involves the major histocompatibility complex (MHC), B- and T lymphocytes associated with RAG1 and RAG2 (recombination-activating genes 1 and 2 securing variability of antigen-binding sites), and somatic hypermutations. This system has demonstrated its evolutionary functionality as judged from the successful expansion of all classes of vertebrates. It is noteworthy that animals, including earthworms and shrimps, also are able to distinguish self from nonself but, most importantly, only by applying innate recognition mechanisms.

The specificity and memory associated with MHC and memory lymphocytes in vertebrates (Kulski et al. 2002) endows the organism with an ability to survive continuous colonization and invasion by pathogens from the ecosystem—especially when regarded over an extended time span. The efficacy of these immune elements in vertebrates is without doubt a major asset when combatting invading pathogens such as viruses, bacteria, and parasites and has probably been crucial for survival of all types of vertebrates. However, vertebrates comprise merely 3% of all described animal species, which suggests that other immune reaction mechanisms may possess a rather high efficacy. Similarly, the least developed adaptive responses within vertebrates are found in more than 30,000 described species of fish, which comprise more than 50% of living vertebrate species, indicating that highly sophisticated adaptive immune mechanisms—as found in mammals—do not necessarily lead to superior survival and expansion. Invertebrates lacking a classic adaptive immune system have enormous numbers of species. Molluscs (snails, slugs, mussels, and oysters) count far more than 120,000 extant species, insects more than a million, crustaceans more than 70,000, and echinoderms (sea urchins, sea stars) more than 8,000.

Although the basic elements in the adaptive immune system (lymphocytes, thymus) became visible in chordates (Kishikita and Nagawa 2014), it is noteworthy that some regulating cytokines applied in the adaptive responses can be traced to invertebrates (Buchmann 2014). Although still in its infancy, the search for MHC-like molecules, immunoglobulin-related proteins, and signal molecules such as cytokines and chemokines has achieved some success (Danchin et al. 2003; Suurväli et al. 2014). Furthermore, alternative mechanisms determining long-lasting immunity among invertebrates have been revealed as comprising epigenetic reprogramming of immune cells and regulation of cellular behavior (phagocytosis, encapsulation) to increased release of immune molecules (Melillo et al. 2018).

11.2 SET-ASIDE CELLS: AN EVOLUTIONARY PERSPECTIVE

Recognition molecules in invertebrates, as part of the innate immune system, comprise pattern recognition receptors (PRRs), which bind molecules on various pathogens carrying pathogen-associated molecular patterns (PAMPs) (Chettri et al. 2011) or danger-associated molecular patters (DAMPs) (Cooper 2010). Some of these recognition molecules—members of the innate system—also are highly involved in reaction patterns associated with some memory and specificity. PRRs include, among others, complement factors, Toll-like receptors (TLRs), and lectins and execute crucial tasks using cytokines for communication securing cellular integrity and survival. Eukaryotic organisms, including ciliates, produce these molecules using their cellular apparatus and carry a wide range of receptors on their surface membrane that endow them with the ability to recognize self from nonself and initiate an appropriate response (Vallesi et al. 2016).

Recent studies point to abilities of even unicellular organisms to communicate with conspecifics to establish an integrated reaction to combat external pathogens. Social amoebae can establish colonies consisting of specialized subpopulations of cells. The reactions in these intermediate organisms on the threshold between unicellular and simple multicellular organisms are associated with sets and subsets of cell types. Sponges, the most basal multicellular organisms, exhibit further development of this specialization and compartmentalization, providing the first model for sophisticated specialization of cell populations in an organism. Therefore, when discussing adaptive immunity in a narrow sense, sensu stricto, we restrict the talk to vertebrates. When addressing this concept in a broad sense, sensu lato, it is relevant to look into the evolution of immune cells across the animal kingdom and consider a variety of alternative innate elements that confer lasting protection and immune memory (Melillo et al. 2018).

11.3 IMMUNE CELLS IN VARIOUS EVOLUTIONARY GROUPS AND THEIR LINKS TO THE ADAPTIVE RESPONSES

11.3.1 Amoebae

The most basic eukaryotic organisms, amoebae, are able to recognize external antigens, a process termed *allorecognition*, by applying various receptors on the plasma membrane and subsequently performing phagocytosis by engulfing foreign elements (particles and microorganisms) (Desjardins et al. 2005). Other basic cytological functions among amoebae include intercellular interactions and the ability to mount a coordinated destruction of potential pathogens in their surroundings.

The social amoebae *Dictyostelium discoideum* form a primitive multicellular organism but with some specialization between compartments (Figure 11.1). While some differentiating amoebae loose phagocytic ability when establishing the

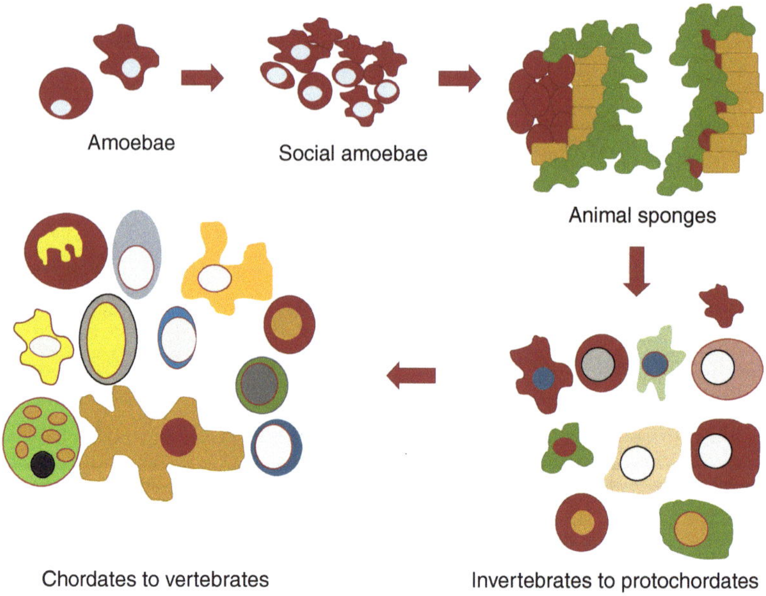

FIGURE 11.1 Diagrammatic overview of cells involved in immune responses from a basic unicellular organism to social amoebae, sponges, classes of invertebrates, protochordates, and chordates to vertebrates. Cell types have expanded from the first phagocytic cells (red) to a plethora of lymphocyte lineages and myeloid cell lines, shown here as cells of different colors and shapes.

multicellular phase of the life cycle, other cell types specialize in detoxification and removal of external organisms, including bacteria. Sentinel cells (S-cells) possess a trapping and killing function by applying extracellular DNA traps, which capture extracellular pathogens (Zhang et al. 2015). They are in principle functioning as corresponding to DNA and protein-composed webs from mammalian neutrophilic and eosinophilic granulocytes.

11.3.2 Sponges (Porifera)

This group represents the most basal metazoans, and already at the end of the nineteenth century, Metchnikoff demonstrated that cells (archaeocytes) in sponges performed phagocytosis and engulfed bacteria. The cells are equipped with receptors recognizing various antigens, both intracellular such as Nod-like receptors (NLRs) (Yuen et al. 2013) and extracellular binding fungi and Gram-negative bacteria. Following ligand binding, signal transduction processes (Müller and Müller 2003) lead to the export of fibrinogen-like proteins and epidermal growth factors (Perovic-Ottstadt et al. 2004). No trace of molecules directly related to adaptive immunity is known in sponges, but it is noteworthy that the clearly innate response involves the TLR adaptor molecule MyD88 and a perforin-like molecule (Wiens et al. 2005).

11.3.3 Platyhelminths (Platyhelminthes)

Platyhelminth, a large group of animals generally called flatworms due to their dorsoventrally flattened body, contains both free-living and parasitic representatives. A classic and well-known group are the free-living planarians, which have been applied in research and teaching for extended periods. Serving as models for regeneration and stem cell biology, planarians occupy a prominent position in research. Parasitic flatworm forms include monogeneans, cestodes, and trematodes occupying various niches and microhabitats within vertebrate hosts. As do all organisms, flatworms face continuous exposure to opportunistic and specialized pathogens. They apply a number of different defense mechanisms, which all are considered parts of the innate immune system. Planarians have been experimentally induced to respond faster to a second pathogen challenge by simple priming by neoblasts (stem cell–like), which may differentiate into reticular cells with immune functions (Torre et al. 2017).

11.3.4 Cnidaria (Jellyfish, Sea Anemones, and their Allies)

The cnidarians comprise—among others—jellyfish and sea anemones and are equipped with different types of immune-related cell types, which are able to communicate with vertebrate-like cytokines (Detournay et al. 2012). The basic cell type is still an amoebocyte, but more specialized cell types, equipped with various functions and receptors enabling the cells to react to a wide spectrum of ligands, have been identified in cnidarians. Binding initiates various pathways, leading to production and export of antimicrobial molecules. TLRs on cells interact with ligands and perform basic functions that do not always correspond to pathways reported in more derived animal lineages (Franzenburg et al. 2012). All the reactions are considered part of innate response patterns; no adaptive-like processes have been described. Although studies on immunological memory in this group are lacking, it is suggested that it exists and plays a prominent role; one major group of cnidarians, the myxozoans, are obligate parasitic in vertebrates and invertebrates and therefore dependent on long-lasting protections from host attacks.

11.3.5 Annelids (Segmented Worms)

Already in the 1960s, it was recognized that oligochaetes have the capability to differentiate self from nonself. Transplantation experiments performed by Cooper (1968, 1969) demonstrated that earthworms differentiate self from nonself. The reactive immune cells, coelomocytes, are actively engaged in interactions with both auto- and allografts but are far more aggressive in response to the latter. Elimination and rejection of xenografts is accelerated if transplantation is performed twice, which indicates that at least a short-term memory exists in annelids.

The cellular machinery of earthworms has been extensively studied, and several cell types described. Prominent hyaline and granular amoebocytes are capable of phagocytosis and encapsulation of foreign elements (Fuller-Espie 2010). Another cell type in earthworms is the eleocyte, which does not perform phagocytosis, but produces lysenin in response to exposure to antigens such as Gram-positive bacteria

(Opper et al. 2013). The coelomocytes may communicate with vertebrate-like cytokines (Fuller-Espie 2010), display cytotoxic abilities (Engelmann et al. 2004), and may, as do other annelid amoebocytes, specifically recognize antigens based on their possession of PRRs, including TLRs (Skanta et al. 2013) and the coelomic cytolytic factor isolated from the coelomic fluid of earthworms (Beschin et al. 1999). The latter molecule binds PAMPs such as O-antigen of lipopolysaccharides from Gram-negative bacteria and binds peptidoglycans from Gram-positive bacteria.

TLRs are membrane glycoproteins consisting of extracellular domains that interact with ligands and initiate a downstream signaling pathway when exposed to bacteria, which may lead to production of various antimicrobial peptides (Bilej et al. 2000). Also NLRs have been described in annelids such as the medical leech *Hirudo medicinalis* (Cuvillier-Hot et al. 2011). Annelids display self/nonself-recognition, recognition of foreign antigens, and short-term memory, although the cellular tools involved are far from a classic adaptive response mechanisms.

11.3.6 Nematodes (Round Worms)

These worms occupy an isolated position among invertebrates, but due to extensive research programs targeting *Caenorhabditis elegans*, round worms have become one of the most well-studied animal groups.

The *C. elegans* immune system based on nonspecialized coelomocytes is relatively simple and far from vertebrate adaptive immune immunity. Several of the PRRs recognizing PAMPs and DAMPs, well known in most invertebrate groups, are missing in nematodes. A DAMP recognition molecule described is DCAR-1, which is a G protein–coupled receptor protein probably involved in chemosensing (Zugasti et al. 2014). It is associated with the worm surface and may detect a tyrosine derivative (hydroxyphenyllactic acid), appearing during wounding or fungal infection of the nematode. Genes encoding a TLR, TOL-1, have been described, although their functions may not be entirely immunologically (Pujol et al. 2001; Tenor and Aballay 2007). The nematode genome encodes a range of leucine-rich repeat sequences. Although similar genes are building stones in cyclostomes, variable lymphocyte receptors—presenting an alternative adaptive immune system—it has not been established whether nematodes apply these elements for a similar purpose.

11.3.7 Molluscs

Bivalves (mussels and oysters) apply hemocytes for their immune responses, and it is particularly noteworthy that long-lasting immunity toward viral infections has been described in oysters (Lafont et al. 2017). Due to lack of specific markers, characterization of an immune response relies on morphometric and morphological characters such as the presence of cytoplasmic granules and nuclear morphology.

Various types of hemocytes have been described from different molluscs, and the cells apply an extensive signaling system for communication (Venier et al. 2016). The blue mussel *Mytilus* carries granular hemocytes, agranular hemocytes, and blast-like cells that communicate with vertebrate-related cytokines (Betti et al. 2006). The Pacific oyster *Crassostrea gigas* is equipped with granulocytes, semigranulocytes, and agranulocytes, also interacting via vertebrate-like cytokines (Roberts et al. 2008; Zhang et al. 2013). The cuttlefish (cephalopod) *Sepia officinalis* applies granulocytes, and the octopus *Octopus vulgaris* displays differentiated cell types such as large granulocytes, small granulocytes, granulocytes, hyalinocytes, and hemoblast-like cells. Granulocytes are the main players in the phagocytic response that produces radical oxygen/nitrogen species (Schultz and Adema 2017). Also, complement factors (Nonaka and Yoshizaki 2004), e.g., in razor clams (Peng et al. 2016) and antimicrobial peptides, provide the cells with the ability to kill foreign cells, demonstrating the ability of molluscs to differentiate self/nonself and to eliminate pathogens. This corresponds to studies in snails such as the air-breathing freshwater snail *Biomphalaria*, which develop lasting protection, mediated by vertebrate-like cytokines (Garcia et al. 2010) against parasite reinfection associated with intense production of reactive oxygen species (Hahn et al. 2000, 2001) and a series of humoral immune components (Pinaud et al. 2016).

11.3.8 Insects

A certain level of specificity and memory in the immune response of insects has been known for almost a century (Cytrynska et al. 2016).

Injection of inactivated bacteria into the greater wax moth *Galleria mellonella* induces protection against live pathogens and even elicits transfer immunity in larvae (Metalnikow 1920; Chigasaki 1925). Insect immune responses were confirmed through numerous studies on several species of fruit flies (*Drosophila* spp., Salazar-Jaramillo et al. 2014) and the American cockroach *Periplaneta americana* exposed to various species of bacteria (Faulhaber and Karp 1992). Cytokines, including IL-8-like peptides (Malagoli et al. 2008), occupy pivotal roles in immune regulation. The mechanisms involved in immune memory may be many, but the yellow fever mosquito *Aedes aegypti* reacts to dengue virus by intensive DNA synthesis and activation of several biochemical pathways (Schonhofer et al. 2016). Responses in another mosquito, *Anopheles gambiae*, are associated with differentiation of hemocytes into forms with different functions (Rodrigues et al. 2010). In addition, RNA methylation has been described in larvae of the mealworm beetle *Tenebrio molitor*, where primed adults and larvae showed lower levels of methylation than the control groups (Castro-Vargas et al. 2017). Nevertheless, the lasting protection conferred by various pathogens is not directly comparable to adaptive immunity in vertebrates.

11.3.9 Crustaceans

The cellular equipment of crustaceans allows representatives of this group to sense and phagocytose foreign elements and potential pathogens, leading to protection against reexposure to pathogen (Johansson and Söderhäll 1989; Roth and Kurtz 2009). A wide series of immune functions of crustacean cells allows them to differentiate self from nonself and execute killing of invading pathogens (Hauton 2012).

Some of the cytokines included in the cellular interactions in crayfish are ancient types (Lin et al. 2010; Lin and Söderhäll 2011), and the cells carry various PRRs that recognize ligands on the surface of pathogens and initiate various response cascades.

First of all, the production of reactive oxygen species such as hydrogen peroxide, nitric oxide, superoxide anion, hydrochloric acid, and hydroxyls may kill the intruder. The prophenoloxidase system adds to the immediate response armament raised in crustaceans as it leads to relatively fast melanization of the pathogen (and its surroundings). Both reaction types can be followed by the prophenoloxidase system (Cerenius and Söderhäll 2004) and subsequent encapsulation of the pathogen by various cell types.

These clearly innate responses, conferring extended protection, may participate in the extended protection described as *immune priming*: Giant tiger prawns (*Penaeus monodon*) experimentally exposed to viral proteins (white spot syndrome virus proteins) obtained protection against reexposure (Witteveldt et al. 2004); the copepod *Macrocyclops albidus* established lasting immunity against larvae of the tapeworm *Schistocephalus solidus* (Kurtz and Franz 2003). Although a series of immune mechanisms, mediated by vertebrate-like cytokines in king prawns (*Litopenaeus vannamei*; Zeng et al. 2013), could be involved in these long-lasting reactions, no evidence for classic adaptive elements has been presented.

11.3.10 Echinoderms

This animal group represents an important bridge between invertebrates and chordates, reflected in the term "protochordates" applied to sea stars and sea urchins. It has been thought that clues to parts of the classic adaptive immune system in vertebrates might be found in reaction patterns or prototype molecules in echinoderms. Nonetheless, innate immune systems are dominated in echinoderms (Arizza and Schillaci 2016), and as discussed in Chapter 1, the concept of set-aside cells was developed for echinoderms.

Investigated species show characteristic gene pools encoding numerous PRRs with a wide range of specificities, but only few genes that may have a function associated with the classic adaptive immune system (Hibino et al. 2006; Rast et al. 2006). Gene families such as the TLR gene family have been expanded to more than 221 TLR genes in the purple sea urchin *Strongylocentrotus purpuratus*, 238 in *Strongylocentrotus fragilis*, and 276 in the red sea urchin *Mesocentrotus franciscanus* (Buckley and Rast 2012). Evidently, the rich variety secures a strong potential for microbial recognition capacity.

In addition, the complement system is relatively well developed in echinoderms. Phagocytes in the green sea urchin, *Strongylocentrotus droebachiensis*, improve their phagocytic capacity following opsonization of foreign cells with complement of human origin, an effect that can be blocked by mammalian complement inhibitors (Bertheussen 1983). Although the classic activation pathway of the complement system in vertebrates includes binding complement factor C1 with an antigen/antibody complex, other activation pathways—alternative and lectin activation pathways—work fully without an adaptive immune player.

Therefore, the echinoderm immune system must be characterized as innate, although certain elements involved and interacting with vertebrate adaptive immune elements play important roles in echinoderm immunity. This reflects the finding that invertebrates are equipped with numerous genes encoding receptors for a range of PAMPs and DAMPs as part of the innate response system. The most promising clue for echinoderm molecules reminiscent of an ancient adaptive pathway is the gene pair *SpRAG1L* and *SpRAG2L* detected in the genome of *Strongylocentrotus purpuratus, which* resemble the RAGs of vertebrate (Fugmann et al. 2006). RAGs encode enzymes that function to control recombination of the genes of both immunoglobulin and T cell receptors in vertebrates. Together, these observations add to the notion that a few molecules of the adaptive immune system may find a very early molecular basis in the echinoderms.

11.3.11 Chordates

The three major branches of chordates are urochordates, cephalochordates, and chordates. Urochordates and cephalochordates represent the stepping stone from the most derived invertebrates, the echinoderms, toward the chordates (vertebrates). We should therefore expect to find in ascidians (urochordates) and/or in amphioxus (cephalochordate) some more or less well-developed prototypes of the adaptive immune mechanisms seen in vertebrates.

A Cephalochordates (Lancelets)

The cephalochordates, comprising amphioxus (*Branchiostoma lanceolatum*)[1] as a classic and prominent representative, carry lymphocyte-like cells in the pharynx (Huang et al. 2007), suggesting the presence of a primordial thymus-like organs with functional lymphocytes. The amphioxus genome contains genes encoding vertebrate-like cytokines (Jin et al. 2012) and complement C1-like factors (Gao et al. 2014). Supporting evidence for elements of an immune system that is more fully developed in vertebrates comes from the description of a primitive gene region of the human MHC in this model animal (Abi-Rached et al. 2002).

The RAGs *RAG1* and *RAG2* are pivotal elements of classic adaptive responses, indispensable for antigen receptor gene assembly (Zhang et al. 2014). Recall from the discussion above that the green sea urchin has both *RAG1L and RAG2L (SpRAG1L* and *SpRAG2L* (Fugmann et al. 2006). Gnathostomes (jawed fish) have both *RAG1* and *RAG2*, two genes. It is noteworthy that a homolog of the *RAG1* core gene has been detected in amphioxus (Kapitonov and Jurka 2005; Zhang et al. 2014), implicating RAGs as early evolving components of a complex immune system.

B Urochordates (Ascidians, Tunicates)

Discrimination of self from nonself is a central part of the vertebrate adaptive immune system using MHC. It has therefore been highly relevant to unravel the *fusibility histocompatibility complex*—termed Fu/HC—in urochordates (Weissman et al. 1990), which is responsible for rejection or fusion. It is now known that histocompatibility in the colonial star ascidian *Botryllus schlosseri* is based on at least two genes that code for highly polymorphic proteins (Nydam et al. 2013). The cellular components display considerable variability. Immune cells of urochordates include a range of cell types from basic hemocytes to specialized cytotoxic cells termed morula cells

[1] Amphioxus is the common name for lancelets, *Branchiostoma lanceolatum* is the most common and well-studied species of lancelets.

exhibiting various types of molecules, including various complement factors used for protection (Franchi and Ballarin 2016; Nicola and Loriano 2017).

C Vertebrates

I Agnathans

Lymphocyte-like cells with genes associated with lymphocyte functions in higher vertebrates have been characterized in lampreys and hagfish, the extant agnathans known as cyclostomes, which are the most basal extant vertebrates (Kishishita and Nagawa 2014). Still, these animals occupy a special position in evolution, and although they carry variable lymphocyte receptors, the nature of the receptors differs totally from the system in gnathostomes (Litman et al. 2010); lamprey T cell–like and B cell–like lymphocytes express variable lymphocyte receptors based on highly variable leucine-rich repeat modules. This is an alternate form of adaptive immunity to that seen in gnathostomes.

II Gnathostomes

In contrast to cyclostomes (jawless vertebrates, agnathans), gnathostomes (jawed vertebrates) use immunoglobulin as B cell receptors, but in common with cyclostomes, gnathostomes produce T cell receptors on their T cells (Abelli 2016). A defining characteristic of the adaptive immune system is V(D)J recombination of gene segments, which occurs only in the early stages of T- and B cell development. Variable (V), diversity (D), and joining (J) gene segments recombine randomly to determine the antigen-binding regions of T- and B cells (Coscia et al. 2016). During this process, the RAGs *RAG1* and *RAG2* facilitate recombination by encoding enzymes that mediate *V(D)J* rearrangement.

Within vertebrates, the adaptive immune response system arose in its basic form in the first fish. It was further developed in amphibians, reptiles, birds, and mammals. Immunoglobulin M (IgM) is the basic immunoglobulin class of antibodies found in all major vertebrate groups and the first to be expressed upon exposure to an antigen. Where teleost fishes possess immunoglobulin classes IgM, IgD, IgT (IgZ), other classes occur in tetrapods (Zhang et al. 2010); except for birds, all jawed vertebrates have IgD, while IgY is used by amphibians, birds, reptiles, and monotremes (Kaiser 2010; Zimmerman et al. 2010). The IgA class is found in crocodilians but not in in turtles, lizards, and snakes. Humans, as a representative of mammals, apply IgM, IgD, IgG, IgE, and IgA. Lymphocytes in tetrapods can differentiate, and subpopulation may stay as long-lived memory cells. Regulation of the immune response and of immunoglobulins is performed by an extensive network of cytokines as signal molecules. Although a series of vertebrate-like cytokines have been detected in the genomes of invertebrates, it is among the fishes that the expansion becomes evident (Wang and Secombes 2013; Secombes and Zou 2017).

11.4 CONCLUSIONS

The adaptive immune system applying set-aside cells enabling the host organism to respond faster and more specifically to a second encounter with a pathogen is developed to the highest degree in mammals. Although the basic cellular outfit and responsiveness of the cells can be traced to amoebae and sponges, the most developed and sophisticated form is first seen in fishes with the origin and expansion of classes of MHC, B- and T lymphocytes, and immunoglobulins.

Although invertebrates can establish an impressive level of immune memory, it is based on innate immune response molecules and patterns and therefore not adaptive. Adaptive immunity, sensu stricto, most fully developed in mammals, is seen in its first traces in chordates (both ascidians and cephalochordates), in which lymphoid tissues are found together with genes that share similarities with *RAG1* and *RAG2*. Among vertebrates, cyclostomes have developed an alternative form of adaptive immunity, applying a series of lymphocyte subsets—but based on variable lymphocyte receptors composed of leucine-rich repeat sequences. The early gnathostomes, bony and cartilaginous fish, exhibit the basic form of set-aside cells in adaptive immunity, sensu stricto, seen in mammals. The cells in teleosts display a high level of specialization, involved in both innate and adaptive responses, leaving plenty of room for deferred cells in adaptive immunity.

REFERENCES

Abelli, L. 2016. Developmental biology of teleost lymphocytes. In *Lessons in Immunity – From Single Cell Organisms to Mammals* (Eds) Ballarin, L. and Cammarata, M., pp. 215–226. London, UK: Elsevier –Academic Press.

Abi-Rached, L., Gilles, A., Shiina, T., Pontarotti, P., and Inoko, H. 2002. Evidence of en bloc duplication in vertebrate genomes. *Nat. Genet.* 31: 100–105. doi: 10.1038/ng855.

Arizza, V., and Schillaci, D. 2016. Echinoderm antimicrobial peptides: The ancient arms of the deuterostome innate immune system. In *Lessons in Immunity – From Single Cell Organisms to Mammals* (Eds) Ballarin, L. and Cammarata, M., pp. 159–176. London, UK: Elsevier –Academic Press.

Bertheussen, K. 1983. Complement-like activity in sea urchin coelomic fluid. *Dev. Comp. Immunol.* 7: 21–31.

Beschin, A., Bilej, M., and Brys, L. 1999. Convergent evolution of cytokines. *Nature* 400: 627–628.

Betti, M., Ciacci, C., Lorusso, L. C., Canonico, B., Falcioni, T., Gallo, G., and Canesi, L. 2006. Effects of tumour necrosis factor alpha (TNFLx) on *Mytilus* haemocytes: role of stress-activated mitogen-activated protein kinases (MAPKs). *Biol. Cell* 98: 233–244. doi: 10.1242/BC20050049.

Bilej, M., De Baetselier, P., and Beschin, A. 2000. Antimicrobial defense of the earthworm. *Folia Microbiol.* 45: 283–300. doi: 10.1007/BF02817549.

Buchmann, K. 2014. Evolution of innate immunity: clues from invertebrates via fish to mammals. *Front. Immunol.* 5: 1–8. doi: 10.3389/fimmu.2014.00459.

Buckley, K. M., and Rast, J. P. 2012. Dynamic evolution of toll-like receptor multigene families in echinoderms. *Front. Immunol.* 3: 136.

Castro-Vargas, C., Linares-López, C., López-Torres, A., Wrobel, K., Torres-Guzmán, J. C., Hernández, G. A., Wrobel, K., Lanz-Mendoza, H., and Contreras-Garduño, J. 2017. Methylation on RNA: a potential mechanism related to immune priming within but not across generations. *Front. Microbiol.* 8: 473.

Cerenius, L., and Söderhäll, K. 2004. The prophenoloxidase activating system in invertebrates. *Immunol. Rev.* 198: 116–126.

Chettri, J. K., Holten-Andersen, L., Raida, M. K., Kania, P., and Buchmann, K. 2011. PAMP-induced expression of immune relevant genes in head kidney leukocytes of rainbow trout (*Oncorhynchus mykiss*). *Dev. Comp. Immunol.* 35: 476–482. doi: 10.1016/j.dci.2010.12001.

Chigasaki, J. 1925. Sur l'immunisation de *Galleria* aux differents stades de sa vie. *C R Soc. Biol.* 93: 573–574.
Cooper, E. L. 1968. Transplantation immunity in annelids. I. Rejection of xenografts exchanged between Lumbricus terrestris and Eisenia foetida. *Transplantation* 6: 322–337.
Cooper, E. L. 1969. Chronic allograft rejection in *Lumbricus terrestris*. *J. Exp. Zool.* 171: 69–73.
Cooper, E. L. 2010. Evolution of immune systems from self/not self to danger to artificial immune systems (AIS). *Phys. Life Rev.* 7: 55–78. doi: 10.1016/j.plrev.2009.12.001.
Coscia, M. R., Giacomelli, S., and Oreste, U. 2016. Teleost immunoglobulins. In *Lessons in Immunity – From Single Cell Organisms to Mammals* (Eds) Ballarin, L. and Cammarata, M., pp. 257–274. London, UK: Elsevier –Academic Press.
Cuvillier-Hot, V., Boidin-Wichlacz, C., Slomianny, C., Salzet, M., and Tasiemski, A. 2011. Characterization and immune function of two intracellular sensors, HmTLR1 and HmNLR, in the injured CNS of an invertebrate. *Dev. Comp. Immunol.* 35: 214–226.
Cytrynska, M., Wojda, I., and Jakubowicz, T. 2016. How insects combat infections. In *Lessons in Immunity – From Single Cell Organisms to Mammals* (Eds.) Ballarin, L. and Cammarata, M., pp. 117–128. London, UK: Elsevier–Academic Press.
Danchin, E. G., Abi-Rached, L., Gilles, A., and Pontarotti, P. 2003. Conservation of the MHC-like region throughout evolution. *Immunogenetics* 55: 141–148.
Desjardins, M., Houde, M., and Gagnon, E. 2005. Phagocytosis: the convoluted way from nutrition to adaptive immunity. *Immunol. Rev.* 207: 158–167. doi: 10.111/j.0105-2896.2005.00319.x.
Detournay, O., Schnitzler, C. E., Poole, A., and Weis, V. M. 2012. Regulation of cnidarian-dinoflagellate mutualisms: evidence that activation of a host TGF beta innate immune pathway promotes tolerance to the symbiont. *Dev. Comp. Immunol.* 38: 525–537. doi: 10.106/j.dci.2012.08.008.
Dzik, J. M. 2010. The ancestry and cumulative evolution of immune reactions. *Acta Biochim. Pol.* 57: 443–466.
Engelmann, P., Kiss, J., and Csöngei, V. 2004. Earthworm leukocytes kill HeLa, HEp-2, PC-12 and PA317 cells *in vitro*. *J. Biochem. Biophys. Methods* 61: 215–227.
Faulhaber, L. M., and Karp, R. D. 1992. A diphasic immune response against bacteria in the American cockroach. *Immunology* 75: 378–381.
Franchi, N., and Ballarin, L. 2016. Cytotoxic cells of compound ascidians. In *Lessons in Immunity – From Single Cell Organisms to Mammals* (Eds) Ballarin, L. and Cammarata, M., pp. 193–204. London, UK: Elsevier –Academic Press.
Franzenburg, S., Fraune, S., Kunzel, S., Baines, J. F., Domazet-Loso, T., and Bosch, T. C. G. 2012. My D88-deficient *Hydra* reveal an ancient function of TLR signaling in sensing in sensing bacterial colonizers. *Proc. Natl. Acad. Sci. U.S.A.* 109: 19374–19379. doi: 10.1073/pnas.1213110109.
Fugmann, S. D., Messier, C., Novack, L. A., Cameron, R. A., and Rast, J. P. 2006. An ancient evolutionary origin of the Rag1/2 gene locus. *Proc. Natl. Acad. Sci. U.S.A.* 103: 3728–3733.
Fuller-Espie, S. L. 2010. Vertebrate cytokines interleukin12 and gamma interferon, but not interleukin 10, enhance phagocytosis in the annelid Eisenia hortensis. *J. Invertebr. Pathol.* 104: 119–124. doi: 10.1016/j.pp.2010.02.009.
Gao, Z., Li, M., Ma, J., and Zhang, S. C. 2014. An *Amphioxus* gC1q protein binds human IgG and initiates the classical pathway: implications for a C1q-mediated complement system in the basal chordate. *Eur. J. Immunol.* 44: 3680–3695.
Garcia, A. B., Pierce, R. J., Gourbal, B., Werkmeister, E., Colinet, D., Reichart, J. M., Dissous, C., and Coustau, C. 2010. Involvement of the cytokine MIF in the snail host immune response to the parasite *Schistosoma mansoni*. *PLoS Pathog.* 6: e1001115. doi: 10.1371/journal.ppat.1001115.

Hahn, U. K., Bender, R. C., and Bayne, C. J. 2000. Production of reactive oxygen species by hemocytes of *Biomphalaria glabrata*: carbohydrate-specific stimulation. *Dev. Comp. Immunol.* 24: 531–541. doi: 10.1016/S0145-305x(00)00017-3.
Hahn, U. K., Bender, R. C., and Bayne, C. J. 2001. Involvement of nitric oxide in killing *Schistosoma mansoni* sporocysts by hemocytes from resistant *Biomphalaria glabrata. J. Parasitol.* 87: 778–785. doi: 10.1645/0022-3395(2001)087.
Hauton, C. 2012. The scope of the crustacean immune system for disease control. *J. Invertebr. Pathol.* 110: 251–260. doi: 10.1016/j.jip.2012.03.005.
Hibino, T., Loza-Coll, M., Messier, C., Majeske, A. J., Cohen, A. H., Terwilliger, D. P., Buckley, K. M., Brockton, V. et al. 2006. The immune gene repertoire encoded in the purple sea urchin genome. *Dev. Biol.* 300: 349–365.
Huang, G., Xie, X., Han, Y., Fan, L., Chen, J., Mou, C., Guo, L., Liu, H., Zhang, Q. et al. 2007. The identification of lymphocyte-like cells and lymphoid-related genes in amphioxus indicates the twilight for the emergence of adaptive immune system. *PLoS One* 2: e206. doi: 10.1371/journal.pone.0000206.
Jin, P., Hu, J., Qian, J. J., Chen, L. M., Xu, X. F., and Ma, F. 2012. Identification and characterization of a putative lipopolysaccharide-induced TNF-alpha factor (LITAF) gene from *Amphioxus* (*Branchiostoma belcheri*): an insight into the innate immunity of *Amphioxus* and the evolution of LITAF. *Fish Shellfish Immunol.* 32: 1223–1228. doi: 10.1016/j.fsi.2012.03.030.
Johansson, M. W., and Söderhäll, K. 1989. Cellular immunity in crustaceans the pro-PO system. *Parasitology* 5: 171–176.
Kaiser, P. 2010. Advances in avian immunology – prospects for disease control: a review. *Avian Pathol.* 39: 309–324. doi: 10.1080/03079457.2010.508777.
Kapitonov, V. V., and Jurka, J. 2005. RAG1 core and V(D)J recombination signal sequences were derived from transib transposons. *PLoS Biol.* 3: e181. doi: 10.1371/journal.pbio.0030181.
Kishishita, N., and Nagawa, F. 2014. Evolution of adaptive immunity: implications of a third lymphocyte lineage in lampreys. *Bioessays* 36: 244–250. doi: 10.1002/bies.201300145.
Kulski, J. K., Shiina, T., Anzai, T., Kohara, S., and Inoko, H. 2002. Comparative genomic analysis of the MHC: The evolution of Class I duplication blocks, diversity and complexity from shark to man. *Immunol Rev* 190: 95–122. doi: 10.1034/j.1600-065x.2002.
Kurtz, J., and Franz, K. 2003. Innate defense: evidence for memory in invertebrate immunity. *Nature* 425: 37–38.
Lafont, M., Petton, B., and Vergnes, A. 2017. Long-lasting antiviral innate immune priming in the Lophotrochozoan Pacific oyster, Crassostrea gigas. *Sci. Rep.* 7: 13143. doi: 10.1038/s41598-017-13564-0.
Lin, X. H., Novotny, M., Söderhäll, K., and Söderhäll, I. 2010. Ancient cytokines, the role of astakines as hematopoietic growth factors. *J. Biol. Chem.* 285. 28577–28586. doi: 10.1074/jbc.M110.138560.
Lin, X. H., and Söderhäll, I. 2011. Crustacean hematopoiesis and the astakine cytokines. *Blood* 117: 6417–6424. doi: 10.1182/blood-2010-11320614.
Litman, G. W., Rast, J. P., and Fugmann, S. D. 2010. The origins of vertebrate adaptive immunity. *Nat. Rev. Immunol.* 10: 543–553. doi: 10.1038/nri2807.
Malagoli, D., Sacchi, S., and Ottaviandi, E. 2008. Unpaired (upd)-3 expression and other immune-related functions are stimulated by interleukin-8 in *Drosophila melanogaster* SL2 cell line. *Cytokine* 44: 269–274. doi: 10.1016/j.cyto.2008.08.011.
Melillo, D., Marino, R., Italiani, P., and Boraschi, D. 2018. Innate immune memory in invertebrate metazoans: a critical appraisal. *Front. Immunol.* 9: 1–17. doi: 10.3389/fimmu.20180.01915.
Metalnikow, S. 1920. Immunité naturelle ou acquise des chenilles de *Galleria mellonella. CR Acad. Sci. Paris* 83: 817–820.

Müller, W. E., and Müller, I. M. 2003. Origin of the metazoan immune system: identification of the molecules and their functions in sponges. *Integr. Comp. Biol.* 43: 281–292. doi: 10.1093/icb/43.2.281.

Nicola, F., and Loriano, B. 2017. Morula cells as key hemocytes of the lectin pathway of complement activation in the colonial tunicate *Botryllus schlosseri. Fish Shellfish Immunol.* 63: 157–164.

Nonaka, M., and Yoshizaki, F. 2004. Primitive complement system of invertebrates. *Immunol. Rev.* 198: 203–215. doi: 10.1111/j.0105-2896.2004.00118.x.

Nydam, M. L., Netuschil, N., Sanders, E., Langenbacher, A., Lewis, D. D., Taketa, D. A., Marimuthu, A., Gracey, A. Y., and De Tomaso, A. W. 2013. The candidate histocompatibility locus of a basal chordate encodes two highly polymorphic proteins. *PLoS One* 8: e65980.

Opper, B., Bognár, A., and Heidt, D. 2013. Revising lysenin expression of earthworm coelomocytes. *Dev. Comp. Immunol.* 39: 214–218.

Peng, M. X., Niu, D. H., Wang, F., Chen, Z. Y., and Li, J. 2016. Complement C3 gene: expression characterization and innate immune response in razor clam *Sinonovacula constricta. Fish Shellfish Immunol.* 55: 223–232.

Perovic-Ottstadt, S., Adell, T., Proksch, P., Wiens, M., Korshev, M., Gamulin, V., Müller, I. M., and Müller, W. E. 2004. A (1–3) beta recognition protein from the sponge Suberites domuncula. Mediated activation of fibrinogen related protein and epidermal growth factor gene expression. *Eur. J. Biochem.* 271: 1924–1937. doi: 10.1111/j.1432-1033.2004.04102.x.

Pinaud, S., Portela, J., Duval, D., and Nowacki, F. C. 2016. A shift from cellular to humoral responses contributes innate immune memory in the vector snail *Biomphalaria glabrata. PLOS Pathog.* 12(1): e1005361. doi: 10:1371/journal.ppat.1005361.

Pujol, N., Link, E. M., Liu, L. X., Kurz, C. L., Alloing, G., Tan, M. W., Ray, K. P., Solari, R. et al. 2001. A reverse genetic analysis of components of the toll signaling pathway in *Caenorhabditis elegans. Curr. Biol.* 11: 809–882.

Rast, J. P., Smith, L. C., Loza-Coll, M., Hibino, T., and Litman, G. W. 2006. Genomic insights into the immune system of the sea urchin. *Science* 314: 952–956.

Roberts, S., Gueguen, Y., de Lorgeril, J., and Goetz, F. 2008. Rapid accumulation of an interleukin 17 homolog transcript in *Crassostrea gigas* hemocytes following bacterial exposure. *Dev. Comp. Immunol.* 32: 1099–1104. doi: 10.1016/j.dci.2008.02.006.

Rodrigues, J., Brayner, F. A., Alves, L. C., Dixit, R., and Barillasmury, C. 2010. Hemocyte differentiation mediates innate immune memory in *Anopheles gambiae* mosquitoes. *Science* 329: 1353–1355.

Roth, O., and Kurtz, J. 2009. Phagocytosis mediates specificity in the immune defence of an invertebrate, the woodlouse *Porcellio scabe* (Crustacea: Isopoda). *Dev. Comp. Immunol.* 33: 1151–1155.

Salazar-Jaramillo, L., Paspati, A., Van de Zande, L., Vermeulen, C. J., Schwander, T., and Wertheim, B. 2014. Evolution of a cellular immune response in *Drosophila*: a phenotypic and genomic comparative analysis. *Genome Biol. Evol.* 6: 273–289.

Schonhofer, C., Coatsworth, H., Caicedo, P., Ocampo, C., and Lowenberger, C. 2016. Aedes aegypti immune responses to Dengue virus. In *Lessons in Immunity – From Single Cell Organisms to Mammals* (Eds) Ballarin, L. and Cammarata, M., pp. 129–144. London, UK: Elsevier –Academic Press.

Schultz, J. H., and Adema, C. M. 2017. Comparative immunogenomics of molluscs. *Dev. Comp. Immunol.* 75: 3–15. doi: 10.1016/j.dci.2017.03.013.

Secombes, C. J., and Zou, J. 2017. Evolution of interferons and interferon receptors. *Front. Immunol.* 8: 1–10. doi: 10.3389/fimmu.2017.00209.

Skanta, F., Roubalova, R., Dvorak, J., Prochazkova, P., and Bilej, M. 2013. Molecular cloning and expression of TLR in the *Eisenia* earthworm. *Dev. Comp. Immunol.* 41: 694–702. doi: 10.1016/j.dci.2013.08.009.

Suurväli, J., Jouneau, L., Thépot, D., Grusea, S., Pontarotti, P., Pasquier, L. D., Boudinot, S. R., and Boudinot, P. 2014. The proto-MHC of placozoans, a region specialized in cellular stress and ubiquitination (proteasome pathways). *J. Immunol.* 193: 2891–2901. doi: 10.4049/jimmunol.1401177.

Tenor, J. L., and Aballay, A. 2007. A conserved toll-like receptor is required for Caenorhabditis elegans innate immunity. *EMBO Rep.* 9: 103–109. doi: 10.1038/sj.embor.7401104.

Torre, C., Abnave, P., Tsoumtsa, L. L., Mottola, G., Lepolard, C., Trouplin, V., Gimenez, G., Desrousseaux, J. et al. 2017. *Staphylococcus aureus* promotes Smed-PGRP-2/Smed-setd8-1-methyltransferase signalling in planarian neoblasts to sensitize anti-bacterial gene responses during re-infection. *EBioMed.* 20: 150–160.

Vallesi, A., Alimenti, C., and Luporini, P. 2016. Ciliate pheromones: Primordial self-/nonself-recognition signals. In *Lessons in Immunity – From Single Cell Organisms to Mammals* (Eds) Ballarin, L. and Cammarata, M., pp. 1–16. London, UK: Elsevier – Academic Press.

Venier, P., Domeneghetti, S., Sharma, N., Pallavivini, A., and Gerdol, M. 2016. Immune related signaling in mussel and bivalves. In *Lessons in Immunity – From Single Cell Organisms to Mammals* (Eds) Ballarin, L. and Cammarata, M., pp. 93–106. London, UK: Elsevier –Academic Press.

Wang, T., and Secombes, C. J. 2013. The cytokine networks of adaptive immunity in fish. *Fish Shellfish Immunol.* 35: 1703–1718. doi: 10.106/j.fsi.2013.08.030.

Weissman, I. L., Saito, Y., and Rinkevich, B. 1990. Allorecognition histocompatibility in a protochordate species: is the relationship to MHC somatic or structural? *Immunol. Rev.* 113: 227–241.

Wiens, M., Korzhev, M., Krasko, A., Thakur, N. L., Perovic-Ottstadt, S., Breter, H., Ushijima, H., Diehl-Seifert, B. et al. 2005. Innate immune defense of the sponge *Suberites domuncula* against bacteria involves a MYD88-dependent signaling pathway: induction of a perforin-like molecule. *J. Biol. Chem.* 280: 27949–27959. doi: 10.1074/jbc.M504049200.

Witteveldt, J., Cifuentes, C. C., Vlak, J. M., and van Hulten, M. C. W. 2004. Protection of *Penaeus monodon* against white spot syndrome virus by oral vaccination. *J. Virol.* 78: 2057–2061.

Yuen, B., Bayes, J. M., and Degnan, S. M. 2013. The characterization of sponge NLRs provides insight into the origin and evolution of this innate immune gene family in animals. *Mol. Biol. Evol.* 31: 106–120. doi: 10.1093/molbev/mst174.

Zeng, D. G., Lei, A. Y., and Chen, X. H. 2013. Cloning, characterization and expression of the macrophage migration inhibitory factor gene from Pacific white shrimp *Litopenaeus vannamei* (Penaeidae). *Genet. Mol. Res.* 12: 5872–5879. doi: 10.4238/2013.November.22.15.

Zhang, Y., Li, J., Yu, F., He, X. C., and Yu, Z. N. 2013. Allograft inflammatory factor-1 stimulates hemocyte immune activation by enhancing phagocytosis and expression of inflammatory cytokines in *Crassostrea gigas*. *Fish Shellfish Immunol.* 34: 1071–1077. doi: 10.1016/j.fsi.2013.01.014.

Zhang, Y.-A., Salinas, I., Li, J., Parra, D., Bjork, S., Xu, Z., LaPatra, S., Bartholomew, J., and Sunyer, J. O. 2010. IgT, a primitive immunoglobulin class specialized in mucosal immunity. *Nat. Immunol.* 11: 827–836. doi: 10.1038/ni.1913.

Zhang, Y. N., Xu, K., Deng, A. Q., Fu, X., Xu, A. L., and Liu, X. 2014. An *Amphioxus* RAG1-like DNA fragment encodes a functional central domain of vertebrate core RAG1. *Proc. Natl. Acad. Sci. U.S.A.* 11: 397–402. doi: 10.1073/pnas.1318843111.

Zhang, X., Zhuchenko, O., Kuspa, A., and Soldati, T. 2015. Social amoebae trap and kill bacteria by casting DNA nets. *Nat. Commun.* 7: 10938. doi: 10.1038/ncomms10938 (1-9).

Zimmerman, L. M., Vogel, L. A., and Bowden, R. M. 2010. Understanding the vertebrate immune system: insights from the reptilian perspective. *J. Exp. Biol.* 213: 661–671. doi: 10.1242/jeb.038315.

Zugasti, O., Bose, N., Squiban, B., Belougne, J., Kurz, C., Schroeder, F. C., Pujol, N., and Ewbank, J. J. 2014. Activation of a G protein–coupled receptor by its endogenous ligand triggers the innate immune response of Caenorhabditis elegans. *Nat. Immunol.* 15: 833–838. doi: 10.1038/ni.2957.

12 The Lack of Human Somatic Set-Aside Cells and Cancer Risks

Darryl Shibata
University of Southern California

CONTENTS

12.1 INTRODUCTION

The concept and principle of "set-aside" cells (outlined in Chapter 1) is attractive for the maintenance of human somatic tissues because humans can live for decades, and quiescent set-aside cells could potentially mitigate against the risks of cancer or serve as a reserve pool of pristine cells in case of tissue damage and the need for tissue repair. Cancer is thought to be a series of acquired genetic diseases in the sense that it is caused by somatic mutations in multiple critical oncogenes and tumor suppressor genes (collectively called *"driver" mutations*). Importantly, human cancers typically require the accumulation of multiple different driver mutations in a single cell for transformation. A relative quiescence or lack of cell division in long-lived stem cell lineages could guard against replication errors and the fixation of DNA damage.

12.2 SOMATIC STEM CELL ORGANIZATION

12.2.1 Stem Cell Hierarchy

One mechanism to prevent the accumulation of multiple driver mutations is a stem cell hierarchy (Cairns 1975), which is the organization of many human epithelial tissues (Figure 12.1). A smaller number of long-lived stem cell lineages located near the bottom or base of the epithelium divide to produce many more differentiated cells that migrate from the basal layers and are lost within days to weeks as they are shed from the surface. In this hierarchy, mutations in the more numerous nonstem cells cannot accumulate because they are quickly lost. Multiple mutations can only accumulate in the much smaller numbers of stem cell lineages, which greatly reduces the numbers of cells at risk for transformation.

12.2.2 Clonal Succession

Cell division is mutagenic because of random replication errors and the fixation of DNA damage. Hence, a further mechanism to reduce the accumulation of stem cell mutations is relative mitotic quiescence, where like "set-aside" cells, the somatic

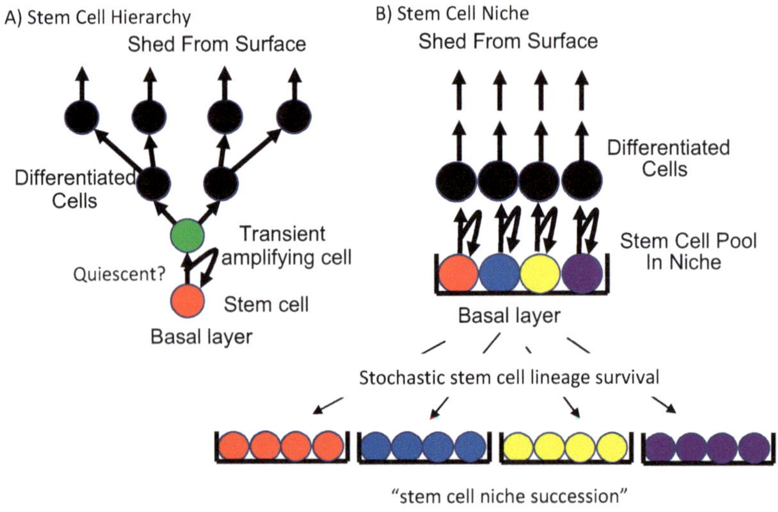

FIGURE 12.1 Epithelial tissues are maintained by a stem cell hierarchy in which a small number of long-lived stem cell lineages divide to produce many more differentiated cells (black circles) that move upward and are lost at the surface of the epithelium. (A) A schematic of quiescent stem cells (red) that seldom divide. Instead, the epithelium is maintained by mitotic transient amplifying cells (green). (B) Human mitotic epithelium is maintained by stem cell niches. The niches are located near the basal epithelial layer with multiple stem cells per niche. Niche stem cells are mitotic, and individual stem cell lineages can either be maintained as one stem cell and one differentiated daughter, expand as two stem cell daughters, or become extinct as two differentiated daughters. With random stem cell turnover, a single lineage can eventually dominate the niche.

stem cells only occasionally divide. A mechanism akin to set-aside stem cells is *clonal succession* (Kay 1965), where a large pool of quiescent stem cells periodically become active and produce differentiated cells. Clonal succession is seen with hematopoietic stem cells and hair bulge stem cells (Fuchs et al. 2004). Such individual stem cells may be mitotically active for only a small proportion of an individual's lifetime, which could protect against mutation accumulation.

12.2.3 Stem Cell Niches

Epithelia, like the skin or intestines, in which cells are shed daily, are typically highly mitotic. Potentially, there could be a further stem cell hierarchy where, similar to clonal succession, stem cells are quiescent and most differentiated cells are produced by mitotic "*transit amplifying cells*" that are imposed between the stem cell lineages and the differentiated cells (Figure 12.1A). These transit amplifying cells are not considered bona fide stem cells and are gradually replaced during aging. The transit amplifying cell concept with quiescent stem cells is exemplified by the intestinal crypts and was widely accepted (Potten and Loeffler 1990).

More recently, experimental evidence has revealed that many human somatic stem cell lineages are not quiescent but are typically highly mitotic and accumulate substantial numbers of mutations during normal aging. Single-cell fate mapping data analyzed mathematically in mice revealed both mitotic stem cells and a lack of so-called transit amplifying cells (Barker et al. 2007; Clayton et al. 2007). Moreover, mammalian stem cell lineages are typically not "immortal" (i.e., one stem always divides to produce one stem cell daughter and one nonstem cell daughter). Instead, stem cell lineages are maintained by a niche or population mechanism (Clayton et al. 2007; Lopez-Garcia et al. 2010; Snippert et al. 2010) where the numbers of stem cells are constant, but stem cell lineages can both expand and become extinct (Figure 12.1B). As noted below, often the expanding stem cell lineage is highly mutated. The net effect is that mammalian epithelia are subdivided into multiple small subclones, with each subclone defined by a small population of mitotic stem cell lineages. Multiple stem cells per niche also guard against tissue damage—the loss of any single stem cell is easily mitigated by adjacent stem cell replacement.

12.3 DNA SEQUENCING OF NORMAL HUMAN TISSUES

12.3.1 Methods

Murine fate mapping studies and their experimental manipulations are more difficult in humans, but advances in DNA sequencing technology have allowed direct measurements of somatic mutations in human tissues. DNA sequencing data provide information on the numbers of mutations, mutation allelic frequencies, and mutation spectra. Parameters important for analyzing somatic mutation data are subclone sizes and sequencing depths.

In human squamous epithelium, subclones are not well defined and vary in their sizes. Sequencing of the DNA from small regions of human epithelium ($<1\,mm^2$) reveals multiple subclones. Only the larger subclones in small epithelial patches

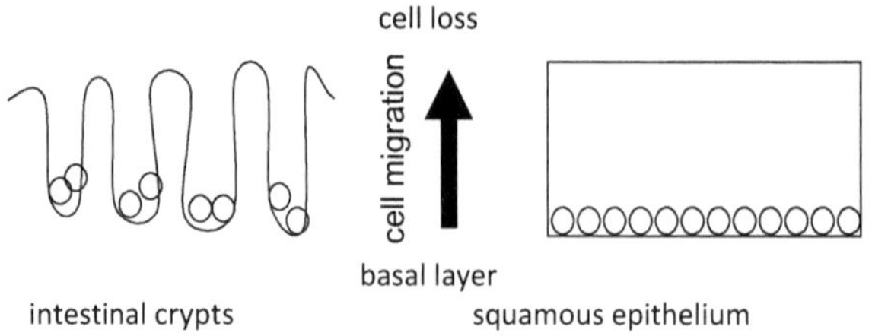

FIGURE 12.2 Physical tissue microarchitecture may modulate the spread of somatic mutations. The intestines are subdivided into millions of small invaginations called *crypts*. The stem cells (circles) at the bottom of the crypts are physically isolated from their neighbors in adjacent crypts. Consequently, the lateral spread of mutant intestinal stem cells is physically limited. By contrast, the basal layer of squamous epithelia is relatively flat, and stem cells more readily compete with their adjacent neighbors, facilitating the lateral spread of mutant stem cells with selective or driver mutations.

($<1\,mm^2$) can be detected even with deep sequencing (~1,000X). Alternatively, well-defined individual subclones (i.e., crypts) are physically present in the intestines (Figure 12.2), and individual crypts can be isolated directly (physically with individual intestinal crypts or through organoid subcloning) and then sequenced. Here, because of a stem cell hierarchy and mitotic stem cell niche turnover (Figure 12.1B), somatic mutations are typically present at clonal frequencies (0.5 allelic frequencies in a diploid cell). Current data are reviewed below for human small and large intestines, and human skin and esophagus.

12.3.2 Human Intestines

Human intestines are subdivided into millions of well-defined, subclonal cellular units called crypts (Figure 12.2). Although the human small intestine is longer than the colon (about 30 compared to 6 feet), the total numbers of crypts is about the same because the colon has a greater diameter. Small intestinal crypts are shorter and smaller than large intestinal crypts (about 400 compared with 2,000 cells per crypt). Interestingly, small intestinal adenocarcinoma is extremely rare in humans, whereas colorectal adenocarcinoma is the second most common cancer in the United States of America.

In mice, both small intestinal and colon crypt stem cells are mitotic and divide about once per day (Barker et al. 2007). The mitotic activity of human intestinal stem cells is difficult to directly measure, but several indirect studies infer that human intestinal stem cells are also mitotic (Williams et al. 1992; Yatabe et al. 2001; Greaves et al. 2006; Baker et al. 2014). A recent study by Blokzijl et al. (2016) sequenced DNA from cultured individual crypt cells and found an age-related increase in somatic mutations, which is consistent with mitotic rather than quiescent human intestinal stem cells. The absolute number of mutations was large and at older ages comparable to the numbers of mutations in human colorectal cancers.

Comparisons between the small intestine and colon are notable for the paucity of cancer in the small intestine relative to the colon, especially because mutations are generally thought to "cause" cancer. Mutations accumulate to the same extent in the small and large intestines. Moreover, mutation spectra (transitions, transversions, etc.) are similar between the small intestines and colon and largely consistent with a replication error mechanism from cell division. Hence, the sequencing data are *more consistent with mitotic rather than quiescent intestinal stem cells*, and the rates of somatic mutation accumulation are essentially identical between human small and large intestines.

12.3.3 Human Squamous Epithelium (Skin and Esophagus)

The squamous epithelium of the skin or esophagus does not have physically well-defined subclonal structures (Figure 12.2). Fate mapping studies in mice reveal stem cell lineages in squamous epithelium that also can expand laterally or become extinct, and subclonal patches derived from a single stem cell lineage can vary in sizes (Clayton et al. 2007).

A common visible manifestation of subclonal skin expansions are the variations in sizes between moles ("age spots") on sun-exposed human skin. Because of this less well-defined subclonal structure, subclones are inferred through the DNA sequencing of multiple small squamous patches ($<1\ mm^2$) from the same individual. Simplistically, subclone sizes are inferred by mutation allele frequencies, where larger subclones have more mutant reads in the sequence data. The numbers of mutations in "normal" sun-exposed skin of adults are large and comparable to the large numbers of mutations in skin cancers (Martincorena et al. 2015; Lynch et al. 2017). As might be expected, mutation spectra resemble skin cancers and are consistent with ultraviolet damage. Similarly, the numbers of mutations in normal esophagus (Martincorena et al. 2018; Yokoyama et al. 2019) are also large, and many mutations appear to be replication errors. In addition, smoking and alcohol consumption are risk factors for esophageal squamous cell carcinoma, with greater exposures associated with more mutations and mutation spectra, consistent with such environmental DNA damage in normal appearing esophagus (Yokoyama et al. 2019).

12.4 DRIVER VERSUS PASSENGER SOMATIC MUTATION ACCUMULATION

As noted above, (i) DNA sequencing data indicate that quiescent or set-aside somatic stem cells are not a feature of human epithelium, and (ii) many somatic mutations accumulate with normal human aging in normal appearing epithelium. However, not all mutations are equivalent; cancer arises when a critical number and combination of driver mutations accumulate within a single cell. Driver mutations occur in oncogenes and tumor suppressor gene and confer a selective advantage to their cells relative to surrounding cells. Interestingly, although cancer genomes have many mutations, most cancer somatic mutations (> 90%) are thought to be selectively neutral or "*passenger*" *mutations*. Among all the mutations in a cancer genome, driver mutations are relative few (Vogelstein et al. 2013) and selective genes can vary between tissue types.

Interestingly, although passenger mutations are the most common type of somatic mutation in normal tissues, driver mutations (such as in TP53 or NOTCH family genes) are relatively common in normal human squamous epithelium but are rare in human intestinal epithelium (Blokzijl et al. 2016; Martincorena et al. 2015, 2018; Lynch et al. 2017; Yokoyama et al. 2019). The relative paucity of driver mutations in normal intestines may reflect differences between the physical architecture of the intestines relative to squamous epithelium (Figure 12.2). The subdivision of the intestine into millions of small crypts results in distinct subclones, where the invaginations inherently limit the spread of mutant cells. It would be physically difficult for a mutant stem cell to climb down from the surface and colonize or take over an adjacent crypt. Hence, the physical subdivision of the intestines into millions of small subclones may be an anticancer mechanism (Cairns 1975), in lieu of the obvious antimutation mechanism of quiescent or set-aside stem cells. By contrast, the basal layer of squamous epithelia has no equivalent physical barriers to limit the lateral spread of a mutant stem cell. Therefore, competition between adjacent stem cells is less limited, and normal appearing stem cells with driver mutations can more readily spread and eliminate less fit neighboring stem cells. Given the high mitotic activity of epithelia, the lack of a suitable physical sanctuary for any potential quiescent or set-aside stem cells may preclude this mechanism as a way to limit mutation accumulation.

12.5 CONCLUSIONS

Quiescent or set-aside stem cells could potentially limit the accumulation of somatic mutations and therefore reduce cancer risks. Unfortunately, cancer is a common disease (especially with aging), and mitotic—rather than quiescent—stem cells maintain human intestinal and squamous epithelium. As a consequence, many of our epithelial cells become "riddled" with mutations as we age. In some ways, cancers may represent "bad luck" (Tomasetti and Vogelstein 2015) where cancers arise stochastically as normal mitotic stem cells accumulate age-related mutations.

Most of these somatic mutations are neutral passenger mutations, but at some frequency, a single cell could stochastically accumulate the right numbers and combinations of driver mutations and transform into a cancer. However, a degree of "skill" may be involved because only a small minority of somatic cells among billions in our bodies transform to a symptomatic tumor. There are likely systematic biological "anticancer" mechanisms that may modify progression to cancer, as manifested by the paradox of why small intestinal cancer is so rare and colon cancer is so common, even though their stem cells accumulate similar numbers of mutations (Shibata 2018). The demands of normal living and mitotic epithelium may impose a requirement for mitotic—rather than quiescent—stem cells, but perhaps many of the mechanisms of set-aside stem cells may still have interesting roles in determining the final odds of cancer.

REFERENCES

Baker, A. M., Cereser, B., Melton, S., Fletcher, A. G., Rodriguez-Justo, M., Tadrous, P. J., Humphries, A., Elia, G., et al. 2014. Quantification of crypt and stem cell evolution in the normal and neoplastic human colon. *Cell Rep.* 8: 940–947.

Barker, N., van Es, J. H., Kuipers, J., Kujala, P., van den Born, M., Cozijnsen, M., Haegebarth, A., Korving, J., et al. 2007. Identification of stem cells in small intestine and colon by marker gene Lgr5. *Nature* 449: 1003–1007.

Blokzijl, F., de Ligt, J., Jager, M., Sasselli, V., Roerink, S., Sasaki, N., Huch, M., Boymans, S., et al. 2016. Tissue-specific mutation accumulation in human adult stem cells during life. *Nature* 538: 260–264.

Cairns, J. 1975. Mutation selection and the natural history of cancer. *Nature* 255: 197–200.

Clayton, E., Doupé, D. P., Klein, A. M., Winton, D. J., Simons, B. D., Jones, P. H. 2007. A single type of progenitor cell maintains normal epidermis. *Nature* 446: 185–189.

Fuchs, E., Tumbar, T., Guasch, G. 2004. Socializing with the neighbors: stem cells and their niche. *Cell* 116: 769–778.

Greaves, L. C., Preston, S. L., Tadrous, P. J., Taylor, R. W., Barron, M. J., Oukrif, D., Leedham, S. J., Deheragoda, M., et al. 2006. Mitochondrial DNA mutations are established in human colonic stem cells, and mutated clones expand by crypt fission. *Proc. Natl. Acad. Sci. U S A*. 103: 714–719.

Kay, H. E. M. 1965. How many cell-generations? *Lancet* 2: 418.

Lopez-Garcia, C., Klein, A. M., Simons, B. D., Winton, D. J. 2010. Intestinal stem cell replacement follows a pattern of neutral drift. *Science* 330: 822–825.

Lynch, M. D., Lynch, C. N. S., Craythorne, E., Liakath-Ali, K., Mallipeddi, R., Barker, J. N., Watt, F. M. 2017. Spatial constraints govern competition of mutant clones in human epidermis. *Nat. Commun.* 8: 1119.

Martincorena, I., Fowler, J. C., Wabik, A., Lawson, A. R. J., Abascal, F., Hall, M. W. J., Cagan, A., Murai, K., et al. 2018. Somatic mutant clones colonize the human esophagus with age. *Science* 362: 911.

Martincorena, I., Roshan, A., Gerstung, M., Ellis, P., Van Loo, P., McLaren, S., Wedge, D.C., Fullam, A., et al. 2015. High burden and pervasive positive selection of somatic mutations in normal human skin. *Science* 348: 880–886.

Potten, C. S., Loeffler, M. 1990. Stem cells: attributes, cycles, spirals, pitfalls, and uncertainties. Lessons for and from the crypt. *Development* 110: 1001–1020.

Shibata, D. 2018. Evolutionary stem cell poker and cancer risks: the paradox of the large and small intestines. *Curr. Pathobiol. Rep.* 6: 193–198.

Snippert, H. J., van der Flier, L. G., Sato, T., van Es, J. H., van den Born, M., Kroon-Veenboer, C., Barker, N., Klein, A. M., et al. 2010. Intestinal crypt homeostasis results from neutral competition between symmetrically dividing Lgr5 stem cells. *Cell* 143: 134–144.

Tomasetti, C., Vogelstein, B. 2015. Cancer etiology. Variation in cancer risk among tissues can be explained by the number of stem cell divisions. *Science* 347: 7881.

Vogelstein, B., Papadopoulos, N., Velculescu, V. E., Zhou, S., Diaz, L. A. Jr., Kinzler, K. W. 2013. Cancer genome landscapes. *Science* 339: 1546–1558.

Williams, E. D., Lowes, A. P., Williams, D., Williams, G. T. 1992. A stem cell niche theory of intestinal crypt maintenance based on a study of somatic mutation in colonic mucosa. *Am. J. Pathol.* 141: 773–776.

Yatabe, Y., Tavaré, S., Shibata, D. 2001 Investigating stem cells in human colon by using methylation patterns. *Proc. Natl. Acad. Sci. U S A*. 98: 10839–10844.

Yokoyama, A., Kakiuchi, N., Yoshizato, T., Nannya, Y., Suzuki, H., Takeuchi, Y., Shiozawa, Y., Sato, Y., et al. 2019. Age-related remodelling of oesophageal epithelia by mutated cancer drivers. *Nature* 565: 312–317.

13 Microbes
New Actors in the Stem Cell Niche

Peter Nagy and Nicolas Buchon
Cornell University

CONTENTS

13.1 INTRODUCTION

The human body hosts, both internally and externally, a huge array of microorganisms that play an essential role in preserving human health. These commensal microorganisms form ecological communities, which are referred to as *microbiota*. The composition of the microbiota depends on the space they occupy and other external/internal factors.

A given host and its microbiota communicate bidirectionally via secondary metabolites, soluble factors, cytokines, and other immune system elicitors to promote tissue homeostasis. A microbiota helps tissues eliminate potentially dangerous pathogens by helping maintain cell function. Pathogen elimination is also aided by cell death and loss, which are necessary to avoid negative consequences to tissue function. Pathogen elimination injures tissues, leading to the activation of tissue-resident multipotent stem cells that replenish lost cells via proliferation and differentiation.

Tissue-resident multipotent stem cells are essential tissue constituents. They help make up the various epithelia that cover the human body, which is continuously exposed to deleterious microbes. The skin and intestinal epithelium are tightly controlled both (i) by intrinsic genetic programs (temporal expression patterns of particular subsets of genes) and (ii) by niche-derived factors originating mainly from the cell's microenvironment. The *niche* is a specific microenvironment with a specific anatomical location that houses stem cells in an undifferentiated and self-renewable state. Tissue-resident stem cells are multipotent and thereby capable of differentiating into various functional cell types within the tissue. However, they also maintain a constant pool of cells via self-renewal. These stem cells continuously communicate with components of their niche such as terminally differentiated cell types and microbes. They emit and receive instructive signals, which help them preserve their ability to participate in tissue generation, maintenance, and repair.

In this chapter, we illustrate the principles behind the interplay between stem cells and their niche as well as the contribution stem cells that make to tissue development, tissue turnover, and immune response. We discuss two ways in which microbes in the niche influence stem cell activity. They exert direct effects as well as indirect effects when different cells within the stem cell niche are regulated by microbes. After describing the anatomical structure and cellular components of tissues, we present the mechanisms of tissue homeostasis dysfunction or dysbiosis that result from cell function defects and/or a diminished niche structure. We describe how microbes influence the intestinal, respiratory, and dermal epithelia, which are continuously and directly exposed to microbes. We also discuss the influence of gut microbes on distinct stem cells in the nervous system.

13.2 INTESTINE

13.2.1 *Drosophila melanogaster*

Stem cell proliferation and differentiation occur as a result of harmful host-microbe interactions. The basic principles behind these tissue maintenance mechanisms have been described in the fruit fly *Drosophila melanogaster*. A relatively low complexity of persistent microbiota, a low redundancy of genes, powerful genetic approaches, and the discovery of tissue-resident adult stem cells in the *Drosophila* gastrointestinal (GI) tract have put this model organism to the forefront of research with the aim of understanding the mechanism of host-microbe interactions. These findings are readily transferred to mammalian systems and provide information on the complex pathomechanisms of diseases affecting the GI tract (e.g., inflammatory bowel disease, cancer), which lead to a shortened healthy life span and a serious burden to the patient, the patient's immediate family members, and society.

The GI tract is continually exposed to a variety of microbes and is consequently a primary biological defensive barrier. The intestinal epithelium undergoes constant regenerative episodes in response to stress or damage. Its functionality and a sustained variety of cell types are essential to maintaining barrier function and tissue integrity. Gut tissue maintenance is crucial for preserving both physical barrier integrity and proper immune function, both of which deteriorate in aged flies. Agedness leads to microbial *dysbiosis*, also called dysbacteriosis, when gut microbial flora are imbalanced. Agedness also leads to increased oxidative stress and intestinal stem cell (ISC) dysplasia, which indicates an abnormal development of cells within epithelia (Biteau et al. 2008; Guo et al. 2014). Recent studies revealed that alterations in intestinal microbiota are linked to the development of inflammatory bowel disease by modulating the host immune function (Manichanh et al. 2012; Sivan et al. 2015). These findings imply that the dialogue between intestinal microbes and ISCs is central to host health.

ISC proliferation must be tightly controlled to avoid the production of damaged, nonfunctional, or supernumerary cells, which are likely to contribute to pathological malformations in the GI tract. Accordingly, intrinsic ISC genetic programs are instructed via many different signaling pathways, which receive inputs from the external environment (niche) and translate these signals into an appropriate cellular response (Figure 13.1). The niche is defined as an anatomic location that produces signals that regulate stem cell behavior. Niches are considered to be dynamically changing (Scadden 2006; Lane et al. 2014).

The continuously renewing GI tract of fruit flies comprises self-renewing ISCs and their daughter cells. These cells are highly similar and functionally redundant to their mammalian counterparts in the stomach, small intestine, and colon. In fruit flies, the epithelial layer is covered by a chitinous peritrophic matrix surrounded by a basal lamina and visceral muscles (Broderick et al. 2014). In the fly gut, ISCs self-renew and, depending on division symmetry, can give rise to either (i) enteroblasts (EBs), which differentiate into mature enterocytes (ECs), or (ii) preenteroendocrine (preEE) cells, which eventually become enteroendocrine (EE) cells (Bonfini et al. 2016) without undergoing further mitotic events. Stem cell division is considered to be symmetric if the daughter cells are identical and asymmetric when the daughter cells differ.

ISCs are multipotent cells with the ability to replenish lost or damaged cells within the entire intestinal epithelium: their daughter cells may differentiate into nutrient-absorbing ECs and hormone-secreting EEs (Ohlstein and Spradling 2007) that contribute to the rebuilding of the tissue during renewal. Proper ISC renewal is controlled by various feedback signaling loops originating from the stem cell niche, which culminates in promoting their proliferation (Figure 13.1). ECs, EEs, and EBs are pivotal players of the ISC niche, which continuously responds to different luminal stimuli (e.g., microbes, chemical insults); hence, their altered activity differentially instructs ISC behavior. For instance, upon *Pseudomonas entomophila* infection or dextran sodium sulfate or bleomycin ingestion, EBs induce Wingless, JAK/STAT (Janus kinase/signal transducers and activators of transcription), cytokine, and EGF (epidermal growth factor) ligand Spitz production, which activate ISC proliferation. Hence, EBs are one of the main sources of growth factors secreted in response to bacterial challenge or chemical stress and so influence ISC proliferation greatly (Cordero et al. 2012). Upon infection with most bacteria, JAK/STAT ligand Upd3 (Unpaired-3) cytokine expression is highly upregulated in ECs via the Hippo, TGF-β (transforming growth factor ß) and SRC-MAPK (mitogen-activated protein kinase) pathways, all of which are required for ISC proliferation (Buchon et al. 2009; Jiang et al. 2009; Houtz et al. 2017).

In recent years, it has become apparent that commensal intestinal bacteria (the microbiota) influence the integrity and physiology of the gut epithelium (Broderick et al. 2014). The complexity of the fruit fly gut microbiome—both of laboratory-raised and wild-caught flies—is lower than it is in mammals and predominantly comprises five to ten bacterial species. The five most frequent bacterial

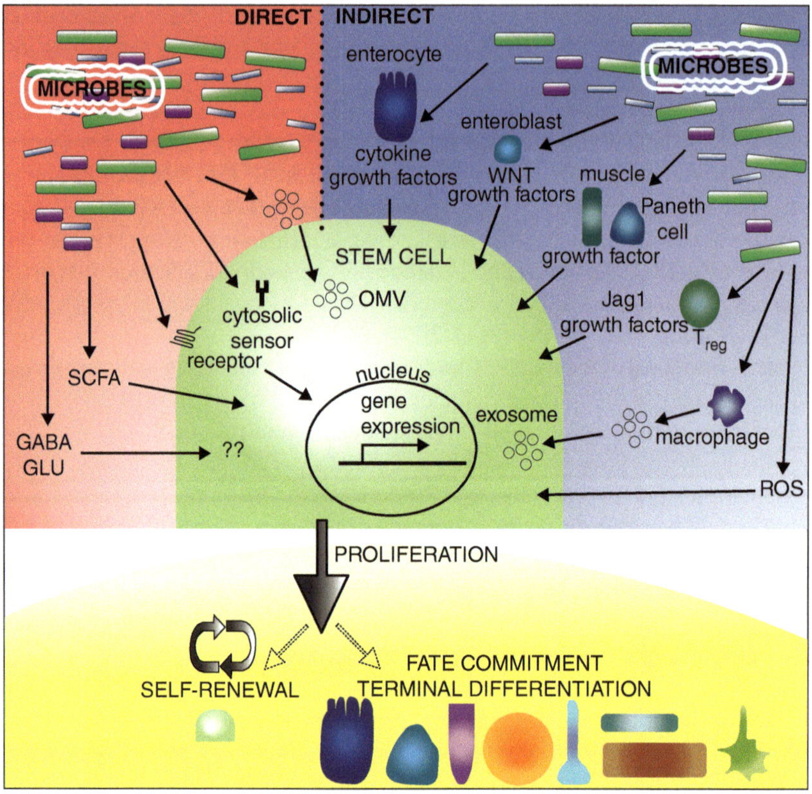

FIGURE 13.1 Microbe-derived niche factors determine stem cell activity. Microbes influence stem cell activity via direct and indirect mechanisms, each of which involves numerous signal pathways: bacterial-produced metabolites (GABA, Glu, SCFA), bacterial cell wall components (peptidoglycans), or outer membrane vesicles (OMVs). Stem cells are usually instructed by the cell types localized in their niche. These cells produce growth/differentiation factors or cytokines/exosomes as part of an inflammatory response triggered by microbes. Stem cells receive and process these niche-derived factors through receptor, cytosolic or organelle sensor-activated signaling, which lead them to change their gene expression. An altered genetic program induces their proliferative response, resulting in a heterogeneous cell progeny, including self-renewed stem cells and fate-committed intermediary cell types, which undergo terminal differentiation to produce the specific functional cell types of a given tissue. See text for further details of the pathways displayed. Abbreviations: GABA, gamma-aminobutyric acid; GLU; glutamate; OMV, outer membrane vesicle; ROS, reactive oxygen species; SCFA, short chain fatty acid; T_{reg}, regulatory T cell.

species are *Acetobacter pomorum, Acetobacter tropicalis, Lactobacillus brevis, Lactobacillus fructivorans*, and *Lactobacillus plantarum* (Broderick and Lemaitre 2012). This microbial community is acquired upon food ingestion and transmitted via defecation or regurgitation (Gilbert 1980).

It is difficult to understand the role of microbes even in conventionally reared (CR) *Drosophila* without manipulating the microbiota. This has prompted a need to develop and study germ-free (GF) and gnotobiotic flies (i.e., GF flies that are reassociated with known bacterial species by feeding them). Such studies led to the discovery that *A. pomorum* and *L. plantarum* trigger signaling pathways that regulate metabolism and cellular growth (the insulin and target of rapamycin pathways, respectively; Shin et al. 2011; Storelli et al. 2011). Interestingly, a complete lack of intestinal microbes does not impact fly survival, and these flies can be maintained in sterile conditions for generations (Bakula 1969). Aged conventionally reared flies show increased gut microbial diversity as well as higher bacterial titers and increased epithelial permeability (Broderick et al. 2014; Clark et al. 2015). Thus, emerging immunosenescence, or aging-associated deterioration of the immune system, contributes to microbial dysbiosis, resulting in an increased rate of ISC proliferation and improper differentiation, leading to tissue dysfunction (Buchon et al. 2009; Guo et al. 2014).

Transcriptomic studies of GF and CR fly guts revealed that microbiota affects intestinal immune responses and regulates epithelium physiology by activating the immune deficiency (Imd) pathway through NF-κB (nuclear factor kappa-light-chain enhancer of activated B cells) transcription factor–mediated gene expression. This leads to the induction of antimicrobial peptide expression, which limits bacterial growth (Morris et al. 2016). Bacterial cell wall–derived peptidoglycans (a polymer consisting of sugars and amino acids forming the bacterial cell wall) are recognized by peptidoglycan recognition receptors, which are dedicated receptors localized on the surface of ECs. This in turn stimulates Imd/NF-κB signaling, leading to antimicrobial peptide expression and immune tolerance to microbiota (Bosco-Drayon et al. 2012). *Lactobacilli* trigger the production of reactive oxygen species by Duox (Dual oxidase) and Nox (NADPH oxidase) oxidases, providing direct antimicrobial agents and a synergistic component for NF-κB-dependent antimicrobial peptide expression (Ryu et al. 2006; Ha et al. 2009; Lee et al. 2013).

Pathogen-activated Imd pathway in ECs also controls cell turnover in the intestinal epithelium by modulating cell shedding (Zhai et al. 2018). In addition, commensal microbes alter tissue homeostasis (i.e., ISC activity) and modulate metabolism by changing the expression of various digestive enzymes (Broderick et al. 2014; Erkosar et al. 2014) and affecting nutrition in the host (Dobson et al. 2015). Gnotobiotic larvae reassociated with *L. plantarum* or *A. pomorum* show increased protein digestion, as well as elevated amino acid levels in the body and induced growth via insulin signaling (Storelli et al. 2011). Gut microbes also provide dietary vitamin B as a supplement to promote protein nutrition and modulate nutrient acquisition (Chaston et al. 2014; Dobson et al. 2015).

13.2.2 Mice

Murine models also revealed that intestinal microbiota and their products influence ISC activity and affect host nutrition, metabolism, and intestinal epithelial barrier integrity. The structure of the GI tract and the lining epithelia is more complex than in insect guts and includes more compartmentalized regions: the esophagus, gastric region, small intestine, and colon. Most studies focus on the cellular and structural architecture of the small intestine. The intestinal epithelium folds into tubular invaginations called as Lieberkühn crypts (invaginations into the lamina propria) and into protrusions called villi (projections into the lumen), maximizing the intestinal surface dedicated to digestion and absorption. The murine intestinal epithelium has the tissue with the highest turnover rate in the body and is continuously regenerated every 2–3 days (the human intestinal epithelium is regenerated every 3–5 days) (Clevers 2013). This rapid intestinal epithelial turnover is maintained by the continuous proliferation and differentiation of crypt-based ISCs to replenish lost and damaged cells in the epithelia. This is essential to preserve tissue homeostasis, particularly in response to injury caused by pathogens or chemicals and inflammation when cytokine production is elevated. The niche helps maintain ISC function in response to several microenvironmental factors, including the microbiota, which is considered to be an essential constituent of the ecological niche of ISCs across the length of the intestine (Pedron et al. 2012; Biswas et al. 2015).

Mammalian ISCs self-renew and produce intermediate transit amplifying cells that undergo additional proliferation and differentiation, giving rise to ECs or different types of secretory cells (EEs, goblet cells, and tuft cells) as well as Paneth cells, which localize at the basal side of the Lieberkühn crypt together with ISCs (Li and Jasper 2016). Paneth cells are closely associated with stem cells and secrete antimicrobial peptides, lysozyme, and several niche factors essential for stem cells (Barker 2014). ECs absorb and transport nutrients and secrete enzymes required for the digestion of luminal content. Goblet cells are sources of protective mucus. EEs produce hormones and peptides in response to stimuli (Li and Jasper 2016).

Pioneering studies comparing GF rodents with CR animals revealed changes in the architecture of the jejunum and ileum (anatomically divided subregions of the small intestine). Villus height and crypt depth were decreased, and the mucosal surface and ISC proliferative index were reduced in GF rodents, suggesting a supportive function of microbiota on intestinal epithelia homeostasis (Gordon and Bruckner-Kardoss 1961; Khoury et al. 1969). The recolonization of GF animals with commensal microbes from healthy rodents increased proliferation in the crypt and induced morphological changes reminiscent of the guts of CR animals (Alam et al. 1994). The mice gut microbiome comprises three main bacterial phyla (namely, Firmicutes, Bacteroidetes, and Proteobacteria) according to the 16S rRNA profiling of ileocecal mucosal-associated and luminal bacteria (Rosshart et al. 2017). The importance of the microbial composition was demonstrated by the different effects of *Lactobacillus reuteri* DSM 17938 and *L. reuteri* PTA 6475 on intestinal cell proliferation (Preidis et al. 2012). Several studies carried out in gnotobiotic mice, which are a reassociation of GF mice with single or human-derived bacteria or different bacterial communities, clarified the mechanisms underlying host-microbe interactions (Clavel et al. 2016). These studies highlighted the pivotal roles of commensal intestinal microbiota in maintaining the intestinal structure, leading to a series of exciting discoveries into how microbes affect gut function and epithelial cell activity to support the health of the entire organism.

Intestinal microbes are key niche actors that directly influence gut cells. They also indirectly influence gut cell activity by stimulating cell surface receptors, secreting instructive signals, and producing secondary metabolites or neurostimulatory peptides (Peck et al. 2017). The microbial stimulation of cells in the niche of ISCs may lead to the production of stem cell regulatory signals, which in turn regulates ISC activity. Toll-like receptor (TLR) activation in Paneth cells by microbes results in the secretion of essential niche-modulating and growth-promoting signals into the crypt (lysozyme, defensins, Wnt, EGF, and Notch) (Bevins and Salzman 2011). *Lactobacillus rhamnosus*, an oral probiotic, improves crypt cell survival after radiation injury via Myd88 (adaptor protein at the intracellular domain site of TLR) and TLR-2 (Ciorba et al. 2012). In response to injury, secreted factors stimulating Wnt signaling provide stem cell–promoting signals: mesenchymal stem cells activate the Wnt pathway or macrophages secrete Wnt-containing exosomes to support ISC growth, which aids in tissue regeneration (Gong et al. 2016; Saha et al. 2016). Gram-negative bacteria produce outer membrane vesicles (OMVs) of a similar size range to exosomes, which are used for cell-cell communication in the stem cell niche (Vanaja et al. 2016). OMVs are taken up by endocytosis and may contain peptides, virulence factors, and RNA/DNA that could potentially alter the activity of the cells in the intestinal epithelia (Kunsmann et al. 2015; O'Donoghue and Krachler 2016). *Neisseria gonorrhoeae*–derived OMVs containing the bacterial outer membrane porin (PorB) targets host cell mitochondria and activates caspase-dependent apoptotic cell death in bone marrow–derived macrophage culture (Deo et al. 2018). Similar cell targeting mechanisms may exist within the dialogue between microbes and intestinal cells, leading to cell loss and thus activating stem cell proliferation and differentiation through a feedback signaling loop.

Intestinal microbiota helps in metabolization and fermentation of food in the gut lumen, and the resulting by-products can be absorbed or act as ligands for different receptors. Intestinal microbiota helps metabolize and ferment food in the gut lumen. The resulting by-products can be absorbed or act as ligands for different receptors. Short-chain fatty acids (SCFAs, 95% of which are produced by intestinal microbes) are the most studied microbe-derived metabolites affecting intestinal homeostasis (Figure 13.1). Butyrate produced in the colon by the fermentation of dietary carbohydrates suppresses colonic stem cell proliferation via Foxo3 (Forkhead box protein O3)-dependent transcription. ECs in the colon (colonocytes) take up microbe-produced butyrate and break it down by shuttling this SCFA for oxidative phosphorylation, thereby generating energy for the cells. ECs are key cellular constituents of the stem cell niche that protect ISCs in the colon from receiving restrictive signals from the niche, which contributes to balanced ISC proliferation in response to environmental cues (Kaiko et al. 2016). Interestingly, butyrate administration to young piglets increased crypt depth and villus length in the jejunum and ileum and promoted the growth of mouse intestinal enteroids in in vitro–generated cultures of ISC-derived small gut organs (Kotunia et al. 2004; Park et al. 2016). Since butyrate-producing microbes primarily reside in the colon, further research is warranted to study the SCFA effect on stem cells of the small intestine. ECs in the small intestine act, not only as a protecting metabolic barrier, like that in the colon, but may also produce proproliferative signals for stem cells in response to microbial stimuli such as Pyy (peptide YY) and Glp-1 (glucagon-like peptide 1) (Mannon 2002; Larraufie et al. 2017). Interestingly, nutrients in the intestinal lumen such as SCFAs are also sensed by EE cells via GPCR (G protein–coupled receptor)-mediated signaling. In response to activating stimuli, EEs produce small peptide hormones that are released into the intestinal microenvironment and into systemic circulation, which coordinately modulates metabolism and the coordination of local cellular responses in the epithelia (Wong et al. 2016).

Commensal microbiota also produces several neurostimulatory peptides (e.g., neuroactive amines and amino acids; Figure 13.1) that regulate the enteric nervous system, which reciprocally influences the connection between the GI tract and the central nervous system (CNS) (Cryan and Dinan 2012). *L. plantarum, Lactobacillus lactis*, and *Lactobacillus paracasei* produce glutamate (Sano 2009); *Lactobacillus bulgaricus* and *Streptococcus* produce serotonin (Mazzoli and Pessione 2016), and *Lactobacillus, Lactococcus*, and *Bifidobacterium* strains produce GABA (gamma-aminobutyric acid) (Siragusa et al. 2007), which all directly alter the secretory function of intestinal epithelial cells and stimulate the vagal nerve circuit, which in turn may indirectly affect stem cells in the GI tract or in the CNS. Moreover, it is plausible that the gut luminal microbe-derived GABA and glutamate directly act in the CNS by crossing the blood-brain barrier (BBB), although BBB permeability may be affected by stress, diet, and intestinal microbiota (Braniste et al. 2014; Kelly et al. 2015).

In addition to its role in the maturation of the innate and adaptive immune system (Garrett et al. 2010; Tremaroli and Backhed 2012), intestinal microbiota beneficially affects gut epithelial homeostasis, as directly sensed by host receptors such as Toll-like and NOD (nucleotide-binding oligomerization domain–containing protein) receptors (Rakoff-Nahoum et al. 2004; Kufer and Sansonetti 2011). The immune receptor Toll-like receptor 4 (TLR4) is expressed on ISCs where it regulates their proliferation and apoptotic death (Neal et al. 2012). The microbiota directly regulates stem cell activity via the cytosolic innate immune sensor NOD2, which is constitutively expressed at a high level in ISCs and also is involved in gut tissue regeneration after chemically induced tissue injury (Nigro et al. 2014). Intestinal dysbiosis and the deregulated expression of immune receptors are associated with colorectal cancer development (Marchesi et al. 2011; Castellarin et al. 2012).

13.3 LUNG

The mammalian lung is a multicompartmentalized tissue comprising the lower airway of the respiratory system. The lower airway can be divided into two zones: (i) the conductive zone made up of bronchioles and (ii) the respiratory zone containing the alveoli. The airways are topographically exterior to our body, and their surface is constantly exposed to inhaled particles, microbes, and toxins. The wide range of immunogens necessitates sophisticated mechanisms for the elimination of these immune elicitors. The lung epithelium is constructed of various epithelial cells (bronchiolar, lung endothelial, and alveolar), stromal cells, and multipotent stem cells. As a physical defense mechanism, ciliated cells of the epithelium propel inhaled particles and microbes upward toward the pharynx, where the surface mucus traps them. Absence of this directional flow in the airways results in accumulation of bacteria in the mucus in patients with cystic fibrosis, which can lead to serious lung infection (Hart and Winstanley 2002).

Two large discoveries have paved new routes toward understanding how mammalian lung tissue is maintained. First, microbes were identified in the lower airways, showing constitutional differences upon health status changes (Bernasconi et al. 2016). Second, the lung was found to be a slowly renewing tissue able to undergo regenerative repair in response to wounding (Snyder et al. 2009; Morrisey 2018; Yang et al. 2018). Microbes belonging to two main phyla—Bacteroidetes and Firmicutes—have thus far been identified in the healthy lung (O'Dwyer et al. 2016). Accordingly, the pulmonary epithelium expresses cell surface pattern recognition receptors (PRRs), such as TLRs, which recognize pathogen-associated molecular patterns (PAMPs) presented by viruses, bacteria, and fungi (Lambrecht and Hammad 2012).

Secretory cells within the pulmonary epithelium produce many protective agents. The first set includes defensive antimicrobial mediators (such as lysozyme, defensins, and collectins) (Iwasaki et al. 2017), which eliminate inhaled microbes. The second set includes a wide spectra of specific cytokines (interleukins), which instruct various local cell populations as niche-derived signals to adequately respond (e.g., undergo apoptosis, remodel cell-cell junctions, or proliferate) and to orchestrate the immune response (Lloyd and Marsland 2017). Studying rodent models and human lung development revealed distinct stem cell–like populations along the pulmonary epithelia: basal secretory or club cells (formerly known as Clara cells) in the bronchioles and AEC2 (alveolar epithelial type 2) cells in the alveoli. These cells can self-renew and give rise to daughter cells, thereby replenishing lost or damaged cells after exiting quiescence and subsequently reentering the cell cycle (Snyder et al. 2009; Kotton and Morrisey 2014). In response to injury triggered by infection, the basal stem cells mediate tissue repair via an EGF receptor–augmented mechanism, leading to increased RNase7, antimicrobial peptide, and EGF ligand expression (Shaykhiev 2015). Mechanisms identified thus far that regulate airway stem cell maintenance also include canonical Wnt signaling, PTEN (phosphatase and tensin homolog), GATA-6 transcription factor, and Bmi1 signaling, as well as signaling through the MAP kinase and Ras pathways (Snyder et al. 2009). Though the low mass of airway microbiota makes their characterization difficult, it has become obvious that microbes contribute to health and disease in the airways by providing tonic signals to the cells that respond to such stimuli.

13.4 SKIN

The skin epidermis provides an impermeable protective barrier that defends the body from physical, chemical, and biological insults. Our largest organ, the skin, regulates body temperature and fluid balance and is a complex ecosystem composed of microbial and physical components that occupy different topographical regions. The skin epidermis also plays an essential role in instructing the immune system. Indeed, site-specific colonization is a key feature of the human skin microbiome. According to 16S rRNA sequences, the human skin is predominantly colonized by *Propionibacterium*, *Staphylococcus* spp., *Acinetobacter* spp., *Corynebacterium*, *Proteobacteria*, and *Flavobacteriales* (Grice et al. 2009; Lange-Asschenfeldt et al. 2011).

On the surface of the epidermis, cornified cells (keratinocytes in the upper layer of the epidermis) are continuously sloughed off and replenished after epidermal differentiation of daughter cell progeny that are produced by stem cells and transient amplifying cells (Fuchs 2008). Keratinocytes guard against microbes via PRRs, TLRs, as well as mannose- and NOD-like receptors. These receptors recognize flagellin, nucleic acid, and lipopolysaccharide from Gram-negative bacteria, peptidoglycan and lipoteichoic acid from Gram-positive bacteria, as well as various fungal cell wall components. The detection of PAMPs activates AMP and cytokine expression in the keratinocytes, leading to bacterial elimination and proliferation in the basal layer (these basal cells are considered to have the ability to self-renew and to give rise to transient amplifying cells) (Grice and Segre 2011). Microbes may also contribute to the host immune response through chemokine production, which instructs epidermal cells in a niche-derived signal on how to orchestrate their response. The commensal bacterium *Staphylococcus epidermidis* eliminates *Staphylococcus aureus* and Group A *Streptococcus* by enhancing host AMP expression, hence contributing to pathogen elimination by cooperating with the host. *S. epidermidis* produces lipoteichoic acid, which induces TLR-mediated responses that inhibit skin inflammation and enable cell survival and tissue repair during infection as well as skin wound closure (Lai et al. 2009, 2010; Linehan et al. 2018).

The hair follicle, which is a specific microenvironment within the epidermis, bears only 25% of the cutaneous bacterial population (97% of these bacteria belong to Gram-positive *Micrococcaceae*; Lange-Asschenfeldt et al. 2011). The hair follicle also contains another type of skin stem cell, quiescent hair follicle stem cells (HFSCs), which reside in the bulge and do not degenerate during the hair cycle. During hair follicle regeneration, HFSCs exit their niche, proliferate, and differentiate. Interestingly, they can be recruited to wound sites to help the epidermis repair (Blanpain and Fuchs 2006). Professional immune cells called regulatory T cells (Tregs) represent an important player of the HFSC niche and contribute to HFSC activation by promoting proliferation and differentiation (Figure 13.1). Treg-derived Jag1 activates Notch signaling as well as other signals such as Wnts, fibroblast growth factors, TGFb2, and PDGFa (platelet-derived growth factor a) in HFSCs upon exit from quiescence followed by their activation (Ali et al. 2017; Horsley and Naik 2017). Importantly, Treg cell migration and hair follicle development were found to be completely dependent on microbial colonization of neonatal mice skin, emphasizing the role of microbe-derived niche signals during hair follicle morphogenesis (Scharschmidt et al. 2015, 2017; Campbell and Koch 2017).

13.5 DISTINCT STEM CELL POPULATIONS—NEURAL STEM CELLS

Here, we discuss the role of neural stem cells in tissue maintenance. This may seem strange, because neural cells, e.g., adult neural stem cells in the CNS, are localized in an immune-privileged space. However, extensive studies revealed an intricate connection between gut luminal microbes and CNS that is likely to contribute to severe neuropathologic changes.

Although the BBB is a vital and highly efficient interface that prevents the highly vulnerable nervous tissue from infections and other toxic agents, extensive studies revealed communication between the GI tract and the brain in conditions such as anxiety, depression, cognition, and autism spectrum disorder (Sharon et al. 2016). Interestingly, the bidirectional communication between the gut and the brain to regulate neurophysiological behavior is ultimately shaped by gut microbe activity via immune, endocrine, and neural pathways (Cryan and Dinan 2012). Gut microbes regulate the CNS through bacterial secondary metabolites, metabolic precursors, as well as immune and vagus nerve signaling (Sharon et al. 2016). Accordingly, GF mice display defects in working memory and reduced c-fos transcription factor levels, which regulates the cAMP response element–binding protein required for Imd- and Toll immune pathway–dependent infection tolerance and long-term memory formation via the hippocampus (Mizuno and Giese 2005; Gareau et al. 2011; Troha et al. 2018). The hippocampus is a component of the limbic system that is involved in shaping motivation and emotion and plays an essential role in memory formation and spatial navigation. Interestingly, early deleterious changes during several neurological disorders (e.g., neurodegenerative diseases) occur in the hippocampal region of the brain.

Neural progenitor (NSC) cells with hallmarks of stem cell are present in two regions of the adult mammalian brain: the subventricular zone (SVZ) lining the lateral ventricles and the subgranular zone (SGZ) within the dentate gyrus of the hippocampus (Kempermann and Gage 1999; Bonaguidi et al. 2011; Bond et al. 2015). These stem cells give rise to intermediate progenitor cells or transient amplifying progenitors before limited rounds of proliferation, thereby generating neuroblasts that differentiate into neurons and interneurons (Bond et al. 2015). As opposed to adult stem cells of epithelial origin, the main relevance of NSC progeny function is to replace cells following injury to maintain tissue function (Ma et al. 2009). The formation of new cells from adult neural stem cells in the SVZ and SGZ is finely tuned by environmental cues to ensure the integration of newly born neurons into preexisting neuronal circuits that might serve specific neuronal functions (van Praag et al. 1999; Cameron and McKay 2001). In addition, neurogenesis in the SVZ contributes to olfactory bulb maintenance, while neurogenesis in the SGZ is essential for spatial learning, memory formation, and mood regulation (Santarelli et al. 2003; Imayoshi et al. 2008; Zhang et al. 2008).

Adult neural stem cells reside in different neurogenic niches. Firstly, stem cells in the lateral ventricle are interconnected with blood vessels and ependymal cells contacting the cerebrospinal fluid. These cells play an essential role in the metabolism and detoxification of brain and spinal cord cells by circulating nutrients and removing waste (Mirzadeh et al. 2008). Secondly, SGZ neuroblasts communicate via gap junctions and direct cell-cell interactions with each other and with their niche, which consists of endothelial cells and astrocytes (Kunze et al. 2009). Both niches include astroglia, which are specialized cells consisting of astrocytes along with microglia that are sparsely localized in the brain and spinal cord. Astroglia play a central role in regulating NSC self-renewal, fate specification, migration, differentiation, and synaptic integration (Barkho et al. 2006; Jiao and Chen 2008). Microglia are resident immune cells in the brain that constantly scavenge the CNS for plaques, damage, and infectious agents. Microglia are extremely sensitive to stimuli such as lipopolysaccharide, interferon-γ, and tumor necrosis factor-α, which trigger classical activation, resulting in the production of proinflammatory cytokines and reactive oxygen species to defend nervous tissues from pathogens (Wang et al. 2015). In response to cell damage due to microglia activation, neural cells upregulate adhesion molecules and secrete trophic factors to recruit astrocytes, thereby aiding tissue repair (Gendelman 2002).

13.6 CONCLUSIONS

In this chapter, we described the emerging impact of indigenous and pathogenic microbes on stem cell niches, where they modulate stem cell activity. It is increasingly accepted that microbes regulate stem cell proliferation via both direct and indirect mechanisms, leading to the production of differentiated functional cell types in the given tissue. The influence of microbiomes on stem cells is an intriguing black box in terms of both stem cell biology and microbiota physiology. This demonstrates that the microbial environment of any given stem cell population needs to be accounted for and suggests that a balanced microbial community aids to maintain tissue function and integrity, thus contributing to sustained homeostasis of the whole organism.

REFERENCES

Alam, M., Midtvedt, T., and Uribe, A. 1994. Differential cell kinetics in the ileum and colon of germfree rats. *Scand. J. Gastroenterol.* 29: 445–451.

Ali, N., Zirak, B., Rodriguez, R. S., Pauli, M. L., Truong, H. A., Lai, K., Ahn, R., Corbin, K., et al. 2017. Regulatory T cells in skin facilitate epithelial stem cell differentiation. *Cell* 169: 1119–1129 e1111.

Bakula, M. 1969. The persistence of a microbial flora during postembryogenesis of *Drosophila melanogaster. J. Invertebr. Pathol.* 14: 365–374.

Barker, N. 2014. Adult intestinal stem cells: critical drivers of epithelial homeostasis and regeneration. *Nat. Rev. Mol. Cell Biol.* 15: 19–33.

Barkho, B. Z., Song, H., Aimone, J. B., Smrt, R. D., Kuwabara, T., Nakashima, K., Gage, F. H., and Zhao, X. 2006. Identification of astrocyte-expressed factors that modulate neural stem/progenitor cell differentiation. *Stem Cells Dev.* 15: 407–421.

Bernasconi, E., Pattaroni, C., Koutsokera, A., Pison, C., Kessler, R., Benden, C., Soccal, P. M., Magnan, A. et al. 2016. Airway microbiota determines innate cell inflammatory or tissue remodeling profiles in lung transplantation. *Am. J. Respir. Crit. Care Med.* 194: 1252–1263.

Bevins, C. L., and Salzman, N. H. 2011. Paneth cells, antimicrobial peptides and maintenance of intestinal homeostasis. *Nat. Rev. Microbiol.* 9: 356–368.

Biswas, S., Davis, H., Irshad, S., Sandberg, T., Worthley, D., and Leedham, S. 2015. Microenvironmental control of stem cell fate in intestinal homeostasis and disease. *J. Pathol.* 237: 135–145.

Biteau, B., Hochmuth, C. E., and Jasper, H. 2008. JNK activity in somatic stem cells causes loss of tissue homeostasis in the aging *Drosophila* gut. *Cell Stem Cell* 3: 442–455.

Blanpain, C., and Fuchs, E. 2006. Epidermal stem cells of the skin. *Annu. Rev. Cell Dev. Biol.* 22: 339–373.

Bonaguidi, M. A., Wheeler, M. A., Shapiro, J. S., Stadel, R. P., Sun, G. J., Ming, G. L., and Song, H. 2011. In vivo clonal analysis reveals self-renewing and multipotent adult neural stem cell characteristics. *Cell* 145: 1142–1155.

Bond, A. M., Ming, G. L., and Song, H. 2015. Adult mammalian neural stem cells and neurogenesis: five decades later. *Cell Stem Cell* 17: 385–395.

Bonfini, A., Liu, X., and Buchon, N. 2016. From pathogens to microbiota: how *Drosophila* intestinal stem cells react to gut microbes. *Dev. Comp. Immunol.* 64: 22–38.

Bosco-Drayon, V., Poidevin, M., Boneca, I. G., Narbonne-Reveau, K., Royet, J., and Charroux, B. 2012. Peptidoglycan sensing by the receptor PGRP-LE in the *Drosophila* gut induces immune responses to infectious bacteria and tolerance to microbiota. *Cell Host Microbe* 12: 153–165.

Braniste, V., Al-Asmakh, M., Kowal, C., Anuar, F., Abbaspour, A., Toth, M., Korecka, A., Bakocevic, N. et al. 2014. The gut microbiota influences blood-brain barrier permeability in mice. *Sci. Transl. Med.* 626: 263ra158.

Broderick, N. A., Buchon, N., and Lemaitre, B. 2014. Microbiota-induced changes in *Drosophila melanogaster* host gene expression and gut morphology. *MBio* 5: e01117–e01114.

Broderick, N. A., and Lemaitre, B. 2012. Gut–associated microbes of *Drosophila melanogaster. Gut Microbes* 3: 307–321.

Buchon, N., Broderick, N. A., Chakrabarti, S., and Lemaitre, B. 2009. Invasive and indigenous microbiota impact intestinal stem cell activity through multiple pathways in *Drosophila. Genes Dev.* 23: 2333–2344.

Cameron, H. A., and McKay, R. D. 2001. Adult neurogenesis produces a large pool of new granule cells in the dentate gyrus. *J. Comp. Neurol.* 435: 406–417.

Campbell, D. J., and Koch, M. A. 2017. Living in peace: host-microbiota mutualism in the skin. *Cell Host Microbe* 21: 419–420.

Castellarin, M., Warren, R. L., Freeman, J. D., Dreolini, L., Krzywinski, M., Strauss, J., Barnes, R., Watson, P. et al. 2012. *Fusobacterium nucleatum* infection is prevalent in human colorectal carcinoma. *Genome Res.* 22: 299–306.

Chaston, J. M., Newell, P. D., and Douglas, A. E. 2014. Metagenome-wide association of microbial determinants of host phenotype in *Drosophila melanogaster. MBio* 5: e01631–e01614.

Ciorba, M. A., Riehl, T. E., Rao, M. S., Moon, C., Ee, X., Nava, G. M., Walker, M. R., Marinshaw, J. M. et al. 2012. Lactobacillus probiotic protects intestinal epithelium from radiation injury in a TLR-2/cyclo-oxygenase-2-dependent manner. *Gut* 61: 829–838.

Clark, R. I., Salazar, A., Yamada, R., Fitz-Gibbon, S., Morselli, M., Alcaraz, J., Rana, A., Rera, M. et al. 2015. Distinct shifts in microbiota composition during *Drosophila* aging impair intestinal function and drive mortality. *Cell Rep.* 12: 1656–1667.

Clavel, T., Lagkouvardos, I., Blaut, M., and Stecher, B. 2016. The mouse gut microbiome revisited: from complex diversity to model ecosystems. *Int. J. Med. Microbiol.* 306: 316–327.

Clevers, H. 2013. The intestinal crypt, a prototype stem cell compartment. *Cell* 154: 274–284.

Cordero, J. B., Stefanatos, R. K., Scopelliti, A., Vidal, M., and Sansom, O. J. 2012. Inducible progenitor-derived *Wingless* regulates adult midgut regeneration in *Drosophila. EMBO J.* 31: 3901–3917.

Cryan, J. F., and Dinan, T. G. 2012. Mind-altering microorganisms: the impact of the gut microbiota on brain and behaviour. *Nat. Rev. Neurosci.* 13: 701–712.

Deo, P., Chow, S. H., Hay, I. D., Kleifeld, O., Costin, A., Elgass, K. D., Jiang, J. H., Ramm, G. et al. 2018. Outer membrane vesicles from *Neisseria gonorrhoeae* target PorB to mitochondria and induce apoptosis. *PLoS Pathol.* 14: e1006945.

Dobson, A. J., Chaston, J. M., Newell, P. D., Donahue, L., Hermann, S. L., Sannino, D. R., Westmiller, S., Wong, A. C. et al. 2015. Host genetic determinants of microbiota-dependent nutrition revealed by genome-wide analysis of *Drosophila melanogaster. Nat. Commun.* 6: 6312.

Erkosar, B., Defaye, A., Bozonnet, N., Puthier, D., Royet, J., and Leulier, F. 2014. *Drosophila* microbiota modulates host metabolic gene expression via IMD/NF-kappaB signaling. *PLoS One* 9: e94729.

Fuchs, E. 2008. Skin stem cells: rising to the surface. *J. Cell Biol.* 180: 273–284.

Gareau, M. G., Wine, E., Rodrigues, D. M., Cho, J. H., Whary, M. T., Philpott, D. J., Macqueen, G., and Sherman, P. M. 2011. Bacterial infection causes stress-induced memory dysfunction in mice. *Gut* 60: 307–317.

Garrett, W. S., Gordon, J. I., and Glimcher, L. H. 2010. Homeostasis and inflammation in the intestine. *Cell* 140: 859–870.

Gendelman, H. E. 2002. Neural immunity: friend or foe? *J. Neurovirol.* 8: 474–479.

Gilbert, D. G. 1980. Dispersal of yeasts and bacteria by *Drosophila* in a temperate forest. *Oecologia* 46: 135–137.

Gong, W., Guo, M., Han, Z., Wang, Y., Yang, P., Xu, C., Wang, Q., Du, L. et al. 2016. Mesenchymal stem cells stimulate intestinal stem cells to repair radiation-induced intestinal injury. *Cell Death Dis.* 7: e2387.

Gordon, H. A., and Bruckner-Kardoss, E. 1961. Effect of the normal microbial flora on various tissue elements of the small intestine. *Acta Anat. (Basel)* 44: 210–225.

Grice, E. A., Kong, H. H., Conlan, S., Deming, C. B., Davis, J., Young, A. C., Bouffard, G. G., Blakesley, R. W. et al. 2009. Topographical and temporal diversity of the human skin microbiome. *Science* 324: 1190–1192.

Grice, E. A., and Segre, J. A. 2011. The skin microbiome. *Nat. Rev. Microbiol.* 9: 244–253.

Guo, L., Karpac, J., Tran, S. L., and Jasper, H. 2014. PGRP-SC2 promotes gut immune homeostasis to limit commensal dysbiosis and extend lifespan. *Cell* 156: 109–122.

Ha, E. M., Lee, K. A., Park, S. H., Kim, S. H., Nam, H. J., Lee, H. Y., Kang, D., and Lee, W. J. 2009. Regulation of DUOX by the Galphaq-phospholipase Cbeta-Ca2+ pathway in Drosophila gut immunity. *Dev. Cell* 16: 386–397.

Hart, C. A., and Winstanley, C. 2002. Persistent and aggressive bacteria in the lungs of cystic fibrosis children. *Br. Med. Bull.* 61: 81–96.

Horsley, V., and Naik, S. 2017. Tregs expand the skin stem cell niche. *Dev. Cell* 41: 455–456.

Houtz, P., Bonfini, A., Liu, X., Revah, J., Guillou, A., Poidevin, M., Hens, K., Huang, H. Y. et al. 2017. Hippo, TGF-beta, and Src-MAPK pathways regulate transcription of the upd3 cytokine in *Drosophila* enterocytes upon bacterial infection. *PLoS Genet.* 13: e1007091.

Imayoshi, I., Sakamoto, M., Ohtsuka, T., Takao, K., Miyakawa, T., Yamaguchi, M., Mori, K., Ikeda, T. et al. 2008. Roles of continuous neurogenesis in the structural and functional integrity of the adult forebrain. *Nat. Neurosci.* 11: 1153–1161.

Iwasaki, A., Foxman, E. F., and Molony, R. D. 2017. Early local immune defences in the respiratory tract. *Nat. Rev. Immunol.* 17: 7–20.

Jiang, H., Patel, P. H., Kohlmaier, A., Grenley, M. O., McEwen, D. G., and Edgar, B. A. 2009. Cytokine/Jak/Stat signaling mediates regeneration and homeostasis in the *Drosophila* midgut. *Cell* 137: 1343–1355.

Jiao, J., and Chen, D. F. 2008. Induction of neurogenesis in nonconventional neurogenic regions of the adult central nervous system by niche astrocyte-produced signals. *Stem Cells* 26: 1221–1230.

Kaiko, G. E., Ryu, S. H., Koues, O. I., Collins, P. L., Solnica-Krezel, L., Pearce, E. J., Pearce, E. L., Oltz, E. M., and Stappenbeck, T. S. 2016. The colonic crypt protects stem cells from microbiota-derived metabolites. *Cell* 165: 1708–1720.

Kelly, J. R., Kennedy, P. J., Cryan, J. F., Dinan, T. G., Clarke, G., and Hyland, N. P. 2015. Breaking down the barriers: the gut microbiome, intestinal permeability and stress-related psychiatric disorders. *Front. Cell. Neurosci.* 9: 392.

Kempermann, G., and Gage, F. H. 1999. New nerve cells for the adult brain. *Sci. Am.* 280: 48–53.

Khoury, K. A., Floch, M. H., and Hersh, T. 1969. Small intestinal mucosal cell proliferation and bacterial flora in the conventionalization of the germfree mouse. *J. Exp. Med.* 130: 659–670.

Kotton, D. N., and Morrisey, E. E. 2014. Lung regeneration: mechanisms, applications and emerging stem cell populations. *Nat. Med.* 20: 822–832.

Kotunia, A., Wolinski, J., Laubitz, D., Jurkowska, M., Rome, V., Guilloteau, P., and Zabielski, R. 2004. Effect of sodium butyrate on the small intestine development in neonatal piglets fed [correction of feed] by artificial sow. *J. Physiol. Pharmacol.* 55(Suppl 2): 59–68.

Kufer, T. A., and Sansonetti, P. J. 2011. NLR functions beyond pathogen recognition. *Nat. Immunol.* 12: 121–128.

Kunsmann, L., Ruter, C., Bauwens, A., Greune, L., Gluder, M., Kemper, B., Fruth, A., Wai, S. N. et al. 2015. Virulence from vesicles: novel mechanisms of host cell injury by *Escherichia coli* O104:H4 outbreak strain. *Sci. Rep.* 5: 13252.

Kunze, A., Congreso, M. R., Hartmann, C., Wallraff-Beck, A., Huttmann, K., Bedner, P., Requardt, R., Seifert, G. et al. 2009. Connexin expression by radial glia-like cells is required for neurogenesis in the adult dentate gyrus. *Proc. Natl. Acad. Sci. U S A.* 106: 11336–11341.

Lai, Y., Cogen, A. L., Radek, K. A., Park, H. J., Macleod, D. T., Leichtle, A., Ryan, A. F., Di Nardo, A., and Gallo, R. L. 2010. Activation of TLR2 by a small molecule produced by Staphylococcus epidermidis increases antimicrobial defense against bacterial skin infections. *J. Invest. Dermatol.* 130: 2211–2221.

Lai, Y., Di Nardo, A., Nakatsuji, T., Leichtle, A., Yang, Y., Cogen, A. L., Wu, Z. R., Hooper, L. V. et al. 2009. Commensal bacteria regulate toll-like receptor 3-dependent inflammation after skin injury. *Nat. Med.* 15: 1377–1382.

Lambrecht, B. N., and Hammad, H. 2012. The airway epithelium in asthma. *Nat. Med.* 18: 684–692.

Lane, S. W., Williams, D. A., and Watt, F. M. 2014. Modulating the stem cell niche for tissue regeneration. *Nat. Biotechnol.* 32: 795–803.

Lange-Asschenfeldt, B., Marenbach, D., Lang, C., Patzelt, A., Ulrich, M., Maltusch, A., Terhorst, D., Stockfleth, E. et al. 2011. Distribution of bacteria in the epidermal layers and hair follicles of the human skin. *Skin Pharmacol. Physiol.* 24: 305–311.

Larraufie, P., Dore, J., Lapaque, N., and Blottiere, H. M. 2017. TLR ligands and butyrate increase Pyy expression through two distinct but inter-regulated pathways. *Cell. Microbiol.* 19: e12648. doi: 10.1111/cmi.12648.

Lee, K. A., Kim, S. H., Kim, E. K., Ha, E. M., You, H., Kim, B., Kim, M. J., Kwon, Y. et al. 2013. Bacterial-derived uracil as a modulator of mucosal immunity and gut-microbe homeostasis in *Drosophila*. *Cell* 153: 797–811.

Li, H., and Jasper, H. 2016. Gastrointestinal stem cells in health and disease: from flies to humans. *Dis. Model Mech.* 9: 487–499.

Linehan, J. L., Harrison, O. J., Han, S. J., Byrd, A. L., Vujkovic-Cvijin, I., Villarino, A. V., Sen, S. K., Shaik, J. et al. 2018. Non-classical immunity controls microbiota impact on skin immunity and tissue repair. *Cell* 172: 784–796 e718.

Lloyd, C. M., and Marsland, B. J. 2017. Lung homeostasis: influence of age, microbes, and the immune system. *Immunity* 46: 549–561.

Ma, D. K., Bonaguidi, M. A., Ming, G. L., and Song, H. 2009. Adult neural stem cells in the mammalian central nervous system. *Cell Res.* 19: 672–682.

Manichanh, C., Borruel, N., Casellas, F., and Guarner, F. 2012. The gut microbiota in IBD. *Nat. Rev. Gastroenterol. Hepatol.* 91: 599–608.

Mannon, P. J. 2002. Peptide YY as a growth factor for intestinal epithelium. *Peptides* 23: 383–388.

Marchesi, J. R., Dutilh, B. E., Hall, N., Peters, W. H., Roelofs, R., Boleij, A., and Tjalsma, H. 2011. Towards the human colorectal cancer microbiome. *PLoS One* 6: e20447.

Mazzoli, R., and Pessione, E. 2016. The neuro-endocrinological role of microbial glutamate and GABA signaling. *Front. Microbiol.* 7: 1934.

Mirzadeh, Z., Merkle, F. T., Soriano-Navarro, M., Garcia-Verdugo, J.-M., and Alvarez-Buylla, A. 2008. Neural stem cells confer unique pinwheel architecture to the ventricular surface in neurogenic regions of the adult brain. *Cell Stem Cell* 3: 265–278.

Mizuno, K., and Giese, K. P. 2005. Hippocampus–dependent memory formation: do memory type–specific mechanisms exist? *J. Pharmacol. Sci.* 98: 191–197.

Morris, O., Liu, X., Domingues, C., Runchel, C., Chai, A., Basith, S., Tenev, T., Chen, H. et al. 2016. Signal Integration by the IkappaB protein pickle shapes *Drosophila* innate host defense. *Cell Host Microbe* 20: 283–295.

Morrisey, E. E. 2018. Basal cells in lung development and repair. *Dev. Cell* 44: 653–654.

Neal, M. D., Sodhi, C. P., Jia, H., Dyer, M., Egan, C. E., Yazji, I., Good, M., Afrazi, A. et al. 2012. Toll-like receptor 4 is expressed on intestinal stem cells and regulates their proliferation and apoptosis via the p53 up-regulated modulator of apoptosis. *J. Biol. Chem.* 287: 37296–37308.

Nigro, G., Rossi, R., Commere, P. H., Jay, P., and Sansonetti, P. J. 2014. The cytosolic bacterial peptidoglycan sensor Nod2 affords stem cell protection and links microbes to gut epithelial regeneration. *Cell Host Microbe* 15: 792–798.

O'Donoghue, E. J., and Krachler, A. M. 2016. Mechanisms of outer membrane vesicle entry into host cells. *Cell. Microbiol.* 18: 1508–1517.

O'Dwyer, D. N., Dickson, R. P., and Moore, B. B. 2016. The lung microbiome, immunity, and the pathogenesis of chronic lung disease. *J. Immunol.* 196: 4839–4847.

Ohlstein, B., and Spradling, A. 2007. Multipotent *Drosophila* intestinal stem cells specify daughter cell fates by differential notch signaling. *Science* 315: 988–992.

Park, J. H., Kotani, T., Konno, T., Setiawan, J., Kitamura, Y., Imada, S., Usui, Y., Hatano, N. et al. 2016. Promotion of intestinal epithelial cell turnover by commensal bacteria: role of short-chain fatty acids. *PLoS One* 11: e0156334.

Peck, B. C. E., Shanahan, M. T., Singh, A. P., and Sethupathy, P. 2017. Gut microbial influences on the mammalian intestinal stem cell niche. *Stem Cells Int.* 2017: 5604727.

Pedron, T., Mulet, C., Dauga, C., Frangeul, L., Chervaux, C., Grompone, G., and Sansonetti, P. J. 2012. A crypt-specific core microbiota resides in the mouse colon. *MBio* 3: doi: 10.1128/mBio.00116-12.

Preidis, G. A., Saulnier, D. M., Blutt, S. E., Mistretta, T. A., Riehle, K. P., Major, A. M., and Venable, S. F. et al. 2012. Probiotics stimulate enterocyte migration and microbial diversity in the neonatal mouse intestine. *FASEB J.* 26: 1960–1969.

Rakoff-Nahoum, S., Paglino, J., Eslami-Varzaneh, F., Edberg, S., and Medzhitov, R. 2004. Recognition of commensal microflora by toll-like receptors is required for intestinal homeostasis. *Cell* 118: 229–241.

Rosshart, S. P., Vassallo, B. G., Angeletti, D., Hutchinson, D. S., Morgan, A. P., Takeda, K., Hickman, H. D., McCulloch, J. A. et al. 2017. Wild mouse gut microbiota promotes host fitness and improves disease resistance. *Cell* 171: 1015–1028 e1013.

Ryu, J. H., Ha, E. M., Oh, C.-T., Seol, J. H., Brey, P. T., Jin, I., Lee, D. G., Kim, J. et al. 2006. An essential complementary role of NF-kappaB pathway to microbicidal oxidants in *Drosophila* gut immunity. *EMBO J.* 25: 3693–3701.

Saha, S., Aranda, E., Hayakawa, Y., Bhanja, P., Atay, S., Brodin, N. P., Li, J., Asfaha, S. et al. 2016. Macrophage-derived extracellular vesicle-packaged WNTs rescue intestinal stem cells and enhance survival after radiation injury. *Nat. Commun.* 7: 13096.

Sano, C. 2009. History of glutamate production. *Am. J. Clin. Nutr.* 90: 728S–732S.

Santarelli, L., Saxe, M., Gross, C., Surget, A., Battaglia, F., Dulawa, S., Weisstaub, N., Lee, J. et al. 2003. Requirement of hippocampal neurogenesis for the behavioral effects of antidepressants. *Science* 301: 805–809.

Scadden, D. T. 2006. The stem-cell niche as an entity of action. *Nature* 441: 1075–1079.

Scharschmidt, T. C., Vasquez, K. S., Pauli, M. L., Leitner, E. G., Chu, K., Truong, H. A., Lowe, M. M., Rodriguez, R. S. et al. 2017. Commensal microbes and hair follicle morphogenesis coordinately drive Treg migration into neonatal skin. *Cell Host Microbe* 21: 467–477 e465.

Scharschmidt, T. C., Vasquez, K. S., Truong, H. A., Gearty, S. V., Pauli, M. L., Nosbaum, A., Gratz, I. K., Otto, M. et al. 2015. A wave of regulatory T cells into neonatal skin mediates tolerance to commensal microbes. *Immunity* 43: 1011–1021.

Sharon, G., Sampson, T. R., Geschwind, D. H., and Mazmanian, S. K. 2016. The central nervous system and the gut microbiome. *Cell* 167: 915–932.

Shaykhiev, R. 2015. Multitasking basal cells: combining stem cell and innate immune duties. *Eur. Respir. J.* 46: 894–897.

Shin, S. C., Kim, S. H., You, H., Kim, B., Kim, A. C., Lee, K. A., Yoon, J. H., Ryu, J. H., and Lee, W. J. 2011. Drosophila microbiome modulates host developmental and metabolic homeostasis via insulin signaling. *Science* 334: 670–674.

Siragusa, S., De Angelis, M., Di Cagno, R., Rizzello, C. G., Coda, R., and Gobbetti, M. 2007. Synthesis of gamma-aminobutyric acid by lactic acid bacteria isolated from a variety of Italian cheeses. *Appl. Environ. Microbiol.* 73: 7283–7290.

Sivan, A., Corrales, L., Hubert, N., Williams, J. B., Aquino-Michaels, K., Earley, Z. M., Benyamin, F. W., Lei, Y. M. et al. 2015. Commensal *Bifidobacterium* promotes antitumor immunity and facilitates anti-PD-L1 efficacy. *Science* 350: 1084–1089.

Snyder, J. C., Teisanu, R. M., and Stripp, B. R. 2009. Endogenous lung stem cells and contribution to disease. *J. Pathol.* 217: 254–264.

Storelli, G., Defaye, A., Erkosar, B., Hols, P., Royet, J., and Leulier, F. 2011. *Lactobacillus plantarum* promotes *Drosophila* systemic growth by modulating hormonal signals through TOR-dependent nutrient sensing. *Cell Metab.* 14: 403–414.

Tremaroli, V., and Backhed, F. 2012. Functional interactions between the gut microbiota and host metabolism. *Nature* 489: 242–249.

Troha, K., Im, J. H., Revah, J., Lazzaro, B. P., and Buchon, N. 2018. Comparative transcriptomics reveals CrebA as a novel regulator of infection tolerance in *D. melanogaster. PLoS Pathol.* 14: e1006847.

van Praag, H., Kempermann, G., and Gage, F. H. 1999. Running increases cell proliferation and neurogenesis in the adult mouse dentate gyrus. *Nat. Neurosci.* 2: 266–270.

Vanaja, S. K., Russo, A. J., Behl, B., Banerjee, I., Yankova, M., Deshmukh, S. D., and Rathinam, V. A. K. 2016. Bacterial outer membrane vesicles mediate cytosolic localization of LPS and Caspase-11 activation. *Cell* 165: 1106–1119.

Wang, W. Y., Tan, M. S., Yu, J. T., and Tan, L. 2015. Role of pro-inflammatory cytokines released from microglia in Alzheimer's disease. *Ann. Transl. Med.* 3: 136.

Wong, A. C., Vanhove, A. S., and Watnick, P. I. 2016. The interplay between intestinal bacteria and host metabolism in health and disease: lessons from *Drosophila melanogaster. Dis. Model Mech.* 9: 271–281.

Yang, Y., Riccio, P., Schotsaert, M., Mori, M., Lu, J., Lee, D. K., Garcia-Sastre, A., Xu, J., and Cardoso, W. V. 2018. Spatial-temporal lineage restrictions of embryonic p63(+) progenitors establish distinct stem cell pools in adult airways. *Dev. Cell* 44: 752–761 e754.

Zhai, Z., Boquete, J. P., and Lemaitre, B. 2018. Cell-specific imd-nf-kappab responses enable simultaneous antibacterial immunity and intestinal epithelial cell shedding upon bacterial infection. *Immunity* 48: 897–910 e897.

Zhang, C. L., Zou, Y., He, W., Gage, F. H., and Evans, R. M. 2008. A role for adult TLX-positive neural stem cells in learning and behaviour. *Nature* 451: 1004–1007.

14 Stem Cells in a Holobiont
Lessons from Hydra

Thomas C.G. Bosch
University of Kiel

CONTENTS

14.1 INTRODUCTION

Here, I evaluate our current understanding of the stem cells in Hydra, an apparently simple animal that shares deep evolutionary connections with all animals, including humans. Stem cells with continuous self-renewal capacity allow Hydra to have an asexual mode of reproduction by budding. I highlight growing evidence that in Hydra, stem cells interact with microbes by using the stem cell transcription factor FoxO (Forkhead box) and downstream effector genes such as antimicrobial peptides (AMPs). I also review observations in germ free animals that indicate that microbes affect development. These findings indicate that development, including stem cell behavior, can only be understood using the multiorganismic concept of the holobiont. They further suggest that stem cells in Hydra are a window to signaling pathways and strategies involved in stem cell differentiation in the bilaterian ancestor.

14.2 HYDRA'S ASEXUAL MODE OF BUDDING REQUIRES CONTINUOUSLY DIVIDING STEM CELLS AND DOES NOT ALLOW FOR SET-ASIDE CELLS

About a century ago, August Weismann observed in hydrozoans—small animals related to jellyfish and corals and belonging to the phylum Cnidaria—that germ cells are derivatives of "common embryonic tissue cells" found in a given part ("*Keimstätte*") of the tissue (Weismann 1883). Based on this observation, Weismann concluded that only certain groups of predetermined cells can differentiate as gametes and developed and published his doctrine of "*the continuity of the germ line*" (Weismann 1892). Since then, the separation of immortal germ cells from somatic cells and tissue has been documented in a wide diversity of metazoan taxa (Williamson and Lehmann 1996; McLaren 2003).

Among the best known hydrozoans are freshwater polyps of the genus *Hydra*, of which some 25 species are known (Figure 14.1A). Unlike most hydrozoans, *Hydra* do not have a jellyfish stage in their life cycle and the polyp stage is represented by single not colonial animals (Figure 14.1A). *Hydra's* asexual mode of budding is based on the presence of continuously dividing cells (Gierer et al. 1972). Diverse cell biological methods and the availability of *Hydra* mutants have revealed the lineages and corresponding cell cycle characteristics of *Hydra* cells (Figure 14.1 and see Campbell, 1967; David and Campbell 1972; Campbell and David 1974; David and Challoner 1974; David and Gierer 1974). Early statistical cloning experiments demonstrated that these continuously dividing cells have a remarkable potential for self-renewal and differentiation following strict spatiotemporal rules (David and Murphy 1977).

Single-cell cloning experiments demonstrated that interstitial stem cells in a single *Hydra* polyp are truly totipotent in the sense that they can differentiate as both somatic and—under appropriate environmental stimuli—germ line cells. Although all species of *Hydra* have been observed to go through occasional sexual phases (Ribi et al. 1985), there is no evidence for "set-aside" or "deferred-use" cells in *Hydra* (Bosch and David 1986, 1987). This also supports a hypothesis made by Buss and Green (1985; see also Buss 1987) that asexual proliferation by budding ("ramet production") requires the presence of an actively dividing multipotent cell line capable of somatic as well as germ line differentiation.

Work by Chiemi Nishimiya-Fujisawa indicates that, in addition to the continuously proliferating multipotent stem cells, *Hydra* has sperm-restricted and egg-restricted stem cells (Nishimiya-Fujisawa and Kobayashi 2012). These so-called *germ line stem cells* self-renew in a polyp and are usually transmitted to a new bud from a parental polyp during asexual reproduction. If these germ line stem cells are lost during subsequent budding or regeneration events, new ones are generated from multipotent stem cells as suggested previously (Bosch and David 1986, 1987). With the advent of "omics" and transgenesis in *Hydra* research, this admittedly surprising observation gained additional experimental support. In vivo tracing of cells of the three distinct stem cell lineages (Figure 14.1B, C) showed unquestionable that epithelial cells as well as interstitial cells located throughout the gastric region are continuously proliferating and differentiating (Wittlieb et al. 2006; Khalturin et al. 2007).

Cues from the microenvironment in which these cells exist appear to be crucial for controlling their developmental potency. Offspring of individual cells can be tracked for several years without obtaining evidence for signs of decline in proliferation rates. The biological reason behind this seemingly unique potential to escape mortality and senescence is simple: the animals have adopted a life cycle in

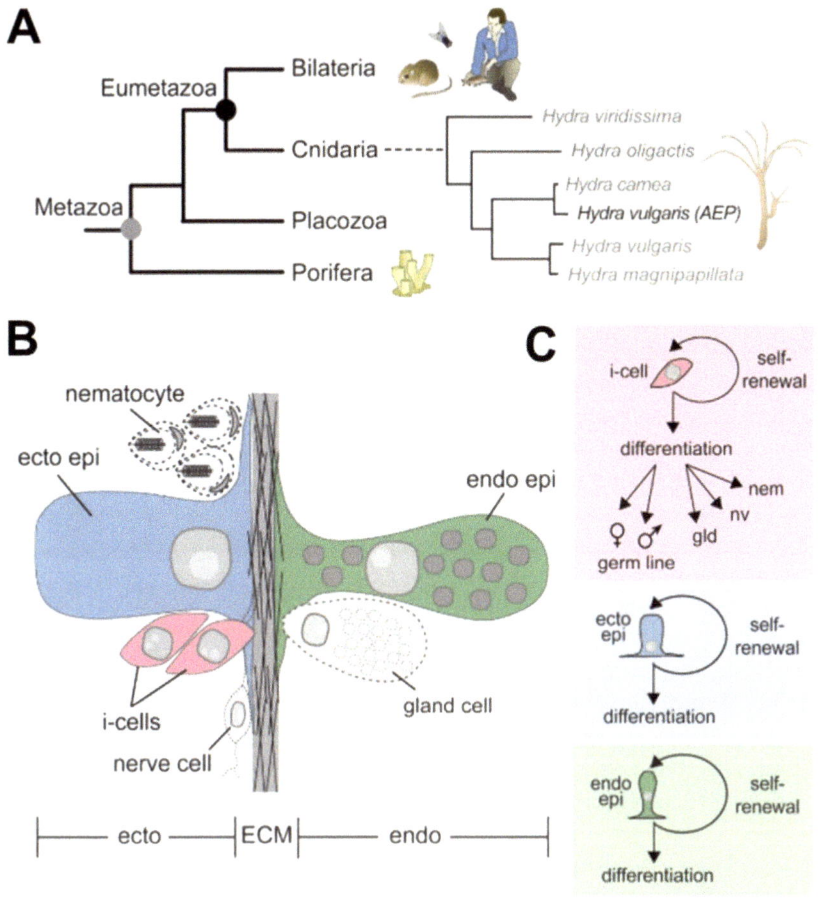

FIGURE 14.1 The stem cell lineages in *Hydra*. (A) A schematic phylogenetic tree showing the main branches in metazoan evolution. (B) The major cell types in *Hydra*. Stem cell lineages are colored; derivatives of the interstitial cell lineage are shown in gray. (C) Three stem cell systems in *Hydra*. Both epithelial cell lineages represent unipotent stem cells, whereas interstitial cells exhibit multipotent features as they are able to differentiate into various derivatives. Ecto, ectoderm; endo, endoderm; ECM, extracellular matrix; ecto epi, ectodermal epithelial cell; endo epi, endodermal epithelial cell; nv, nerve cell; gld, gland cell; nem, nematocyte.

which proliferation and population growth occur exclusively asexually by budding. This asexual mode of reproduction demands that each individual polyp maintains continuously proliferating cells. If it did not, the species would lose in number of offspring and would, sooner or later, be outcompeted by other species that did maintain continuously proliferating cells. Thus, there is strong selective constraint to equip the adult polyp's tissue with cells that are capable of continuous proliferation and differentiation.

Tracing of green fluorescent protein (GFP)–labeled transgenic ectodermal epithelial cells revealed an additional fascinating feature of *Hydra* cells: in contrast to epithelial cells in more derived animals, Hydra's epithelial cells are capable of drastic changes in structure, shape, and cell contact along the body column (Anton-Erxleben et al. 2008). It therefore seems likely that the remarkable phenotypic plasticity of epithelial cells in response to positional signals allows *Hydra* to build its body with only a limited number of different cell types.

14.2.1 Signaling Pathways and Stem Cells in Hydra

The molecular mechanisms and signal pathways governing stem cells in *Hydra* have been reviewed and summarized elsewhere (Boehm et al. 2013). In brief, each of the three stem cell populations is characterized by a specific signature set of transcriptions factors and genes playing key roles in cell type-specific function and interlineage communication (Hemmrich et al. 2012). This means that principal functions of stem cell genes, such as maintenance of stemness and control of stem cell self-renewal and differentiation, arose very early in metazoan evolution.

Patterning in *Hydra* is under the control of the Wnt pathway (Gee et al. 2010; Gufler et al. 2018; Hobmayer et al. 2000; Iachetta et al. 2018; Lengfeld et al. 2009; Nakamura et al. 2011; Petersen et al. 2015). Not too surprising, therefore, is the observation that components of the canonical Wnt signaling pathway are among the signals instructing ectodermal epithelial cells to execute different programs of terminal differentiation (Anton-Erxleben et al. 2008). In addition, signaling pathways involving Notch and glycogen synthase kinase-3 beta (GSK-3β) play a role in inducing or suppressing differentiation of stem cells in *Hydra*.

When Wnt signaling is activated experimentally by the addition of alsterpaullone—a drug that specifically inhibits GSK-3β—tracing GFP-expressing interstitial cells in vivo shows that these cells are forced to terminally differentiate into nematoblasts (Khalturin et al. 2007). In adult *Hydra*, this pathway, therefore, obviously fulfills two functions, one in patterning (Hobmayer et al. 2000; Broun et al. 2005; Guder et al. 2006) and one in interstitial cell differentiation (Khalturin et al. 2007).

An important role in controlling homeostasis of the interstitial stem cell lineage is played by the *Hydra* myc gene (*Hymyc1*) (Ambrosone et al. 2012). Inhibition of *Hymyc*1 impairs the balance between stem cell-self renewal/differentiation, as shown by the accumulation of products of stem cell intermediate and terminal differentiation cells in genetically interfered animals (Ambrosone et al. 2012). Among the regulatory proteins expressed in all Hydra stem/progenitor cells, but not in terminally differentiated cells, are two PIWI proteins, *Hydra* PIWI (Hywi) and *Hydra* PIWI-like (Hyli) (Juliano et al. 2014. *Hydra* PIWI proteins are strictly cytoplasmic and thus likely act as posttranscriptional regulators. Finding the PIWI-piRNA pathway in the somatic stem/progenitor cells of a nonbilaterian animal suggests that this pathway originated with broader stem cell functionality.

14.3 MICROBES ARE AN IMPORTANT COMPONENT OF HYDRA'S EPITHELIAL SURFACES

Hydra has one of the simplest epithelia in the animal kingdom, with only two cell layers (a mesoderm has not evolved) and few cell types derived from three distinct stem cell lineages (Figure 14.1B). *Hydra* also belongs to one of the earliest phyletic animal lineages known to form natural symbiotic relationships with both microbes and photosynthetic algae. The multipartite entity of a host and its associated microbial communities is termed "*holobiont*" (Rohwer et al. 2002) or synonymously "*metaorganism*" (Bosch and McFall-Ngai 2011).

Bacteria are an important component of the *Hydra* holobiont (Fraune and Bosch 2007; Franzenburg et al. 2013a; Bosch 2014) (Figure 14.1B). The identified bacterial phylotypes represent three different bacterial divisions with a majority of strains of Proteobacteria and Bacteroidetes (Franzenburg et al. 2013a). Even closely related species of *Hydra* are colonized by specific microbiomes that mirror the phylogenetic relatedness between their host species (*phylosymbiosis*; Brooks et al. 2016). Disturbances or shifts in any of these partners can compromise the animal's health (Franzenburg et al. 2012; Fraune et al. 2014).

The species-specific microbiomes are extraordinarily stable and still resemble the ones of *Hydra* from the wild even after years of culturing under laboratory conditions involving artificial water and feeding (Fraune and Bosch 2007; Franzenburg et al. 2013). The acquisition of bacterial symbionts occurs during early development and is established in a robust, reproducible pattern (Franzenburg et al. 2013b). Perpetuation of *Hydra*-microbe symbioses through host generations relies on symbiont transmission. Horizontally transmitted symbionts are taken up from the environment anew by each host generation, and vertically transmitted symbionts are most often transferred through the female germ line. Mixed modes also exist. In *Hydra*, we have evidence that specific bacterial taxa may not only be acquired horizontally but in some cases also via vertical transmission through the female germ line (Fraune et al. 2010). These results clearly validate the *Hydra* host as a major determinant for bacterial community composition.

14.4 MAINTENANCE OF EPITHELIAL TISSUE HOMEOSTASIS IN *HYDRA* REQUIRES DIRECT COMMUNICATION BETWEEN STEM CELLS AND MICROBES

Components of the innate immune system, such as the Toll-like receptor (TLR) signaling pathway and AMPs, as well as transcriptional regulators of stem cells, including transcription factor FoxO, are involved in maintaining homeostasis between Hydra's epithelial surfaces and its resident microbiota (Bosch 2014). A particularly interesting role is played by transcription factor *forkhead-box protein O 3* (*FoxO3*), which regulates genes involved in growth and differentiation and which has been consistently associated with human ageing and longevity (Willcox et al. 2008; Morris et al. 2015).

Since in *Hydra* the single *FoxO* gene is strongly expressed in all stem cell lineages (Boehm et al. 2012), FoxO loss-of-function mutants can provide insights into the evolutionary conserved function of this gene. FoxO's well-documented function in life span regulation in other organisms led us to speculate that in *Hydra*, FoxO might be a key driver for the continuous self-renewal capacity of stem cells. To assess this directly, we performed gain- and loss-of-function experiments. Overexpression of *FoxO* increased proliferation of stem cells (Boehm et al. 2012). Silencing of *FoxO* influenced the delicate balance between stem cells and differentiated cells by increasing numbers of cells going into terminal differentiation, accompanied by a considerable slowdown of population growth rate due to slowdown of cell division (Boehm et al. 2012).

Intriguingly, and totally unexpected at the time of the discovery, *FoxO* downregulation also caused drastic changes in the expression level of the AMPs arminin, hydramacin, and periculin2b (Boehm et al. 2012). In addition, in silico analysis revealed multiple FoxO-binding sites on the promoter sequences of the three AMPs. These unanticipated observations in epithelial FoxO loss-of-function mutants indicate that FoxO signaling leads not only to malfunctions in cell cycle progression but also to dysregulation of multiple families of genes encoding AMPs (Mortzfeld et al. 2018).

FoxO loss-of-function polyps were more susceptible to colonization by foreign bacteria and impaired in selection for bacteria, in these respects resembling the native microbiome. FoxO deficiency reduces the expression of AMPs, which results in decreased selective pressure on colonizing microbial taxa and ultimately to reduced resilience of the microbiome (Mortzfeld et al. 2018). In FoxO loss-of-function mutants, especially bacterial taxa that are rarely found in the native community can expand at the expense of bacteria resembling the naturally occurring microbiota. FoxO therefore turns out to be the direct link between tissue homeostasis, development, and the microbiota (Figure 14.2) (Mortzfeld and Bosch 2017; Mortzfeld et al. 2018).

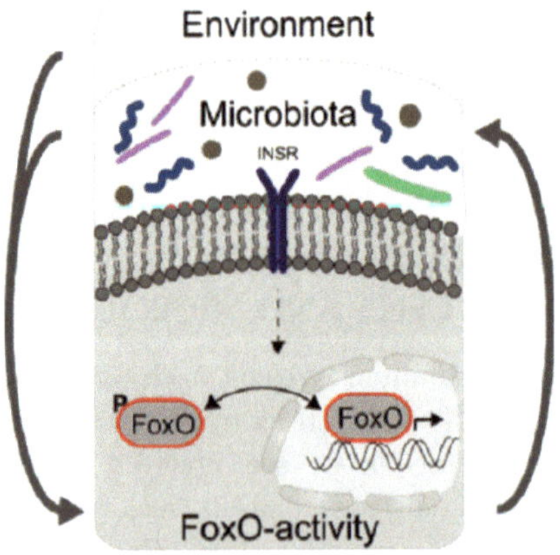

FIGURE 14.2 Hydra stem cells interact with the microbial environment (shown as colored elements outside the cell membrane) via FoxO as the key regulator of epithelial homeostasis and host-microbiome crosstalk. INSR, insulin receptor. P, phosphorylated FoxO protein. Picture credt: Benedikt Mortzfeld.

14.5 SIMILARITIES IN INTERACTIONS BETWEEN HOST CELLS AND MICROBES DURING ANIMAL DEVELOPMENT

Developmental symbiosis appears to be universal, such that the entire animal, including its persistent symbionts, may be the critical unit of anatomy, physiology, and development.

In numerous marine invertebrates, larvae will only settle in response to compounds released by environmental bacteria (McFall-Ngai et al. 2013; Alegado and King 2014). While in these species, the whole life cycle is regulated by symbiotic bacteria, in numerous other species, symbiotic bacteria are necessary for organ formation (Landmann et al. 2014). It therefore is possible that all animals form some organs through symbiosis.

The first evidence that commensal bacteria directly contribute to developmental processes came from observations revealing that bacteria provide developmental signals to intestinal epithelia. Bacteria-induced expression of mammalian genes was first demonstrated by Umesaki (1984), who noticed that a particular fucosyl transferase enzyme characteristic of mouse intestinal villi was induced by bacteria. More recent studies have shown that the intestines of germ-free mice can initiate, but not complete, their differentiation. Germ-free mice have smaller intestines, with a paucity of enteroendocrine and goblet cells (Bates et al. 2006). Microbial colonization is accompanied by profound changes in intestinal transcriptional programs (Rakoff-Nahoum et al. 2015). Moreover, microbiota also plays a key role in constructing the microvascular network during angiogenesis in the intestine (Stappenbeck et al. 2002).

In the *Drosophila* midgut, recurring damage induced by various factors is repaired by intestinal stem cells (ISCs). Microbes not only trigger cell loss and replacement, but they also modify intestinal and whole organism physiology, thus modulating ISC activity (Bonfini et al. 2016). Regulation of ISCs is integrated through a complex network of signaling pathways initiated by other gut cell populations, including enterocytes, enteroblasts, enteroendocrine, and visceral muscle cells (Bonfini et al. 2016).

Fruit flies are not the only animals whose development of gut and immune system depend on microbial symbionts. By tagging bacteria with a fluorescent molecule to observe how and when they take up residence in developing fish, University of Oregon scientists have discovered that bacteria colonize the immature gut soon after the larvae hatch and before the animal is fully mature (Bates et al. 2006). In the absence of microbiota, the zebrafish gut is arrested in specific aspects of differentiation, and the animals have trouble absorbing nutrients. All these defects can be reversed by the introduction of bacteria later in development (Bates et al. 2006).

In zebrafish, microbial symbionts use the beta-catenin signaling pathway to initiate cell division in ISCs (Rawls et al. 2004). Microbes, therefore, must be considered an integral part of this process. In comparing zebrafish gut microbe interactions to those in mice, some responses appear conserved across millions of years of evolution (Rawls et al. 2006); animals may possess a conserved program of interactions with the microbes with which they coevolved. More recently, germ-free zebrafish used in bioassays led to the discovery of novel bacterial effector proteins, such as BefA, which promotes pancreatic beta cell proliferation (Hill et al. 2016), and AimA, which reduces intestinal inflammation (Rolig et al. 2018).

Also in humans, intestinal homeostasis depends on the interaction between bacteria and the intestinal epithelium. The human gut, particularly the colon, is the host of approximately 1,000 bacterial species, the so-called *gut microbiota*. The relationship between the gut microbiota and the host is symbiotic and mutualistic, influencing many aspects of the biology of the host. This homeostatic balance can be disrupted by enteric pathogens, such as the Gram-negative bacterium *Shigella flexneri* or the anaerobic bacterium *Listeria monocytogenes*, which are able to invade the epithelial layer and consequently subvert physiological functions (Nigro et al. 2016).

The current view is that the intestinal immune system and the microbiota maintain a collaborative alliance in which the microbiota modulates the immune system and, in turn, the immune system tolerates microbiota and fights off invasive pathogenic bacteria. Dysbiosis, which is a microbial imbalance within the body, is known to play a pathogenic role. This fact is clearly manifested in a number of intestinal inflammatory diseases (Sánchez de Medina et al. 2014). In addition, dysbiosis has been related to systemic diseases such as fatty liver disease, obesity, and diabetes (Roberfroid et al. 2010; Wieland et al. 2015). The crypt culture model, known as *intestinal organoids*, has been proposed (Nigro et al. 2016) as a powerful tool to study host-microbe interactions in vitro and to examine the response of the epithelium, particularly the response of ISCs, to the presence of bacteria (Nigro et al. 2016).

As a first step toward identifying microbial factors directly influencing ISCs, Stappenbeck et al. (2002) and Kaiko et al. (2016) identified butyrate, a prominent bacterial metabolite, as a potent inhibitor of intestinal stem/progenitor proliferation at physiologic concentrations. The authors suggest that the mammalian crypt architecture protects stem/progenitor cell proliferation in part through a metabolic barrier formed by differentiated colonocytes that consume butyrate (Kaiko et al. 2016). Intriguingly, the mechanism of butyrate action depended on the transcription factor FoxO3.

This observation, in concert with many others, indicates that, as in the freshwater polyp *Hydra vulgaris*, FoxO appears to play a major role in linking tissue homeostasis to the microbial environment in numerous invertebrates and vertebrates, including humans. FoxO, for example, modulates the innate immune system in mice (Seiler et al. 2013), *Drosophila* (Becker et al. 2010), and *Caenorhabditis elegans* (Libina et al. 2003). In mice, FoxO transcription factors directly regulate TLR3-mediated innate immune responses as well as the expression of AMPs (Seiler et al. 2013). FoxO signaling reduces susceptibility to bacterial infections by reducing oxidative stress and induction of inflammatory cytokines (Joseph et al. 2016).

Functional analyses in *C. elegans* and *Drosophila* have verified that the level of FoxO expression is indeed directly linked to life span, without detectable costs for the individuals (Kenyon et al. 1993; Kenyon 2010; Hwangbo et al. 2004). Moreover, supporting the strikingly conserved components of the FoxO signaling pathway are studies of candidate genes in long-lived French population (Schächter et al. 1994) and Japanese/Okinawans in Hawaii (Willcox et al. 2008) that uncovered FoxO3 as part of the genetic ageing code in humans.

These observations, together with the functional studies in model organisms, suggest a key role for FoxO in controlling ageing and health in multicellular organisms: FoxO seems to serve as a protective gene responsible for stem cell regulation, tissue maintenance and renewal, and controlling the innate immune system. The capabilities of the FoxO transcription factor to extend life span and control effectors of the immune system demonstrate a strong and unique mechanism of cross-regulation of tissue homeostasis and innate immunity.

14.6 CONCLUSIONS

Animals are not individuals by traditional anatomical, physiological, immunological, genetic, or developmental accounts. Rather, developmental symbiosis generates holobionts, organisms composed of numerous genetic lineages whose interactions are critical for the development and maintenance of the entire organism (Gilbert et al. 2015). With the advent of novel sequencing and imaging technologies, the invisible components of animal life became visible. Today, it is very clear that symbionts are critical to stem cell development and evolution. In this newly discovered world of metaorganisms, stem cell proliferation and immunity are part of a global program, which seeks to fuse stem cell biology with ecological concepts and rules governing the interactions between an organism and its microbial environment. Documenting, comprehending, and understanding the ramifications of these phenomena are the areas of ecological evolutionary developmental biology (Eco-Evo-Devo) (Gilbert et al. 2015). The role of the master transcriptional regulator FoxO as a direct integrator of stem cell proliferation and the immune system provides a mechanistic explanation for previous diverse observations in stem cell research. Symbiotic relationships are the signature of life on earth, and stem cell biology has to include developmental symbiosis as a major component. Biology has entered a new era with the capacity to understand that an organism's genetics and fitness are inclusive of its microbiome.

ACKNOWLEDGEMENTS

I am thankful to the Wissenschaftskolleg (Institute for Advanced Studies) in Berlin for a Fellow award, which allowed me to write this review during my sabbatical year. Work in my laboratory is supported by the Deutsche Forschungsgemeinschaft (DFG) (CRC1182 "Origin and Function of Metaorganisms," DFG grant BO 848/17-1, and grants from the DFG Cluster of Excellence program "Inflammation at Interfaces"). I also gratefully appreciate support from the Canadian Institute for Advanced Research (CIFAR).

REFERENCES

Alegado, R. A., and King, N. 2014. Bacterial influences on animal origins. *Cold Spring Harbor Perspect. Biol.* 6: a016162.

Ambrosone, A., Marchesano, V., Tino, A., Hobmayer, B., and Tortiglione, C. 2012. Hymyc1 downregulation promotes stem cell proliferation in *Hydra vulgaris. PLoS One* 7: e30660.

Anton-Erxleben, F., Thomas, A., Wittlieb, J., Fraune, S., and Bosch, T. C. G. 2008. Plasticity of epithelial cell shape in response to upstream signals: a whole-organism study using transgenic *Hydra*. *Zoology* 112: 185–194.

Bates, J. M., Mittge, E., Kuhlman, J., Baden, K. N., Cheesman, S. E., and Guillemin, K. 2006. Distinct signals from the microbiota promote different aspects of zebrafish gut differentiation. *Dev. Biol.* 297: 374–386.

Becker, T., Loch, G., Beyer, M., Zinke, I., Aschenbrenner, A. C., Carrera, P., Inhester, T., Schultze, J. L., and Hoch, M. 2010. FOXO-dependent regulation of innate immune homeostasis. *Nature* 463: 369–373.

Boehm, A. M., Khalturin, K., Anton-Erxleben, F., Hemmrich, G., Klostermeier, U. C., Lopez-Quintero, J. A., Oberg H. H., Puchert, M. et al. 2012. FoxO is a critical regulator of stem cell maintenance in immortal *Hydra*. *Proc. Natl. Acad. Sci. U.S.A.* 109: 19697–19702.

Boehm, A.M., Rosenstiel, P., and Bosch, T. C. G. 2013. Stem cells and aging from a quasi-immortal point of view. *Bioessays* 35: 994–1003.

Bonfini, A., Liu, X., and Buchon, N. 2016. From pathogens to microbiota: how *Drosophila* intestinal stem cells react to gut microbes. *Dev. Comp. Immunol.* 64: 22–38.

Bosch, T. C. G. 2014. Rethinking the role of immunity: lessons from *Hydra*. *Trends Immunol.* 35: 495–502.

Bosch, T. C. G., and David, C. N. 1986. Male and female stem cells and sex reversal in *Hydra* polyps. *Proc. Natl. Acad. Sci. U.S.A.* 83: 9478–9482.

Bosch, T. C. G., and David, C. N. 1987. Stem cells of *Hydra magnipapillata* can differentiate into somatic cells and germ line cells. *Dev. Biol.* 121: 182–191.

Bosch, T. C. G., and McFall-Ngai, M. J. 2011. Metaorganisms as the new frontier. *Zoology (Jena)* 114: 185–190.

Brooks, A. W., Kohl, K. D., Brucker, R. M., van Opstal, E. J., and Bordenstein, S. R. 2016. Phylosymbiosis: relationships and functional effects of microbial communities across host evolutionary history. *PLoS Biol.* 14: e2000225.

Broun, M., Gee, L., Reinhardt, B., and Bode, H. R. 2005. Formation of the head organizer in hydra involves the canonical Wnt pathway. *Development* 132: 2907–2916.

Buss, L. W. 1987. *The Evolution of Individuality*. Princeton University Press, Princeton.

Buss, L. W., and Green, D. R. 1985. Histoincompatibility in vertebrates: the relict hypothesis. *Dev. Comp. Immunol.* 9: 191–201.

Campbell, R. D. 1967. Tissue dynamics of steady state growth in *Hydra littoralis*. I. Patterns of cell division. *Dev. Biol.* 15: 487–502.

Campbell, R. D., and David, C. N. 1974. Cell cycle kinetics and development of *Hydra attenuata*. II. Interstitial cells. *J. Cell Sci.* 16: 349–358.

David, C. N., and Campbell, R. D. 1972. Cell cycle kinetics and development of *Hydra attenuata*. I. Epithelial cells. *J. Cell Sci.* 11: 557–568.

David, C. N., and Challoner, D. 1974. Distribution of interstitial cells and differentiating nematocytes in nests in *Hydra attenuata*. *Am. Zool.* 14: 537–542.

David, C. N., and Gierer, A. 1974. Cell cycle kinetics and development of *Hydra attenuate*. III. Nerve and nematocyte differentiation. *J. Cell Sci.* 16: 359–375.

David, C. N., and Murphy, S. 1977. Characterization of interstitial stem cells in *Hydra* by cloning. *Dev. Biol.* 58: 372–383.

Fraune, S., and Bosch, T. C. G. 2007. Long-term maintenance of species-specific bacterial microbiota in the basal metazoan *Hydra*. *Proc. Natl. Acad. Sci. U.S.A.* 104: 13146–13151.

Fraune, S., Augustin, R., Anton-Erxleben, F., Wittlieb, J., Gelhaus, C., Klimovich, V. B., Samoilovich, M. P., and Bosch, T. C. G. 2010. In an early branching metazoan, bacterial colonization of the embryo is controlled by maternal antimicrobial peptides. *Proc. Natl. Acad. Sci. U.S.A.* 107: 18067–18072.

Fraune, S., Anton-Erxleben, F., Augustin, R., Franzenburg, S., Knop, M., Schröder, K., Willoweit-Ohl, D., and Bosch, T. C. G. 2014. Bacteria-bacteria interactions within the microbiota of the ancestral metazoan *Hydra* contribute to fungal resistance. *ISME J.* 9: 1543–1556.

Franzenburg, S., Fraune, S., Künzel, S., Baines, J. F., Domazet-Loso, T., and Bosch, T. C. G. 2012. MyD88-deficient *Hydra* reveal an ancient function of TLR signaling in sensing bacterial colonizers. *Proc. Natl. Acad. Sci. U.S.A.* 109: 19374–19379.

Franzenburg, S., Walter, J., Künzel, S., Wang, J., Baines, J. F., Bosch, T. C. G., and Fraune, S. 2013a. Distinct antimicrobial peptide expression determines host species-specific bacterial associations. *Proc. Natl. Acad. Sci. U.S.A.* 110: E3730–E3738.

Franzenburg, S., Fraune, S., Altrock, P. M., Künzel, S., Baines, J. F., Traulsen, A., and Bosch, T. C. G. 2013b. Bacterial colonization of Hydra hatchlings follows a robust temporal pattern. *ISME J.* 7: 781–790.

Gierer, A., Berking, S., Bode, H., David, C. N., Flick, K., Hansmann, G., Schaller, H., and Trenkner, E. 1972. Regeneration of *Hydra* from reaggregated cells. *Nat. New Biol.* 239: 98–101.

Gee, L., Hartig, J., Law, L., Wittlieb, J., Khalturin, K., Bosch, T. C. G., and Bode, H. R. 2010. Beta-catenin plays a central role in setting up the head organizer in *Hydra*. *Dev. Biol.* 340: 116–124.

Gilbert, S. F., Bosch, T. C. G., and Ledón-Rettig, C. 2015. Eco-Evo-Devo: developmental symbiosis and developmental plasticity as evolutionary agents. *Nat. Rev. Genet.* 16: 611–622.

Guder, C., Philipp, I., Lengfeld, T., Watanabe, H., Hobmayer, B., and Holstein, T. W. 2006. The Wnt code: cnidarians signal the way. *Oncogene* 25: 7450–7460.

Gufler, S., Artes, B., Bielen, H., Krainer, I., Eder, M. K., Falschlunger, J., Bollmann, A., Ostermann, T. et al. 2018. Beta-Catenin acts in a position-independent regeneration response in the simple eumetazoan *Hydra*. *Dev. Biol.* 433: 310–323.

Hemmrich, G., Khalturin, K., Boehm, A. M., Puchert, M., Anton-Erxleben, F., Wittlieb, J., Klostermeier, U. C., Rosenstiel, P. et al. 2012. Molecular signatures of the three stem cell lineages in *Hydra* and the emergence of stem cell function at the base of multicellularity. *Mol. Biol. Evol.* 29: 3267–3280.

Hill, J. H., Franzosa, E. A., Huttenhower, C., and Guillemin, K. 2016. A conserved bacterial protein induces pancreatic beta cell expansion during zebrafish development. *eLife* 5: e20145.

Hobmayer, B., Rentzsch, F., Kuhn, K., Happel, C. M., von Laue, C. C., Snyder, P., Rothbacher, U., and Holstein, T. W. 2000. WNT signalling molecules act in axis formation in the diploblastic metazoan *Hydra*. *Nature* 407: 186–189.

Hwangbo, D. S., Gershman, B., Tu, M. P., Palmer, M., and Tatar, M. 2004. *Drosophila* dFOXO controls lifespan and regulates insulin signalling in brain and fat body. *Nature* 429: 562–566.

Iachetta, R., Ambrosone, A., Klimovich, A., Wittlieb, J., Onorato, G., Candeo, A., D'andrea, C., Intartaglia, D. et al. 2018. Real time dynamics of β-catenin expression during Hydra development, regeneration and Wnt signalling activation. *Int. J. Dev. Biol.* 62: 311–318.

Joseph, J., Ametepe, E. S., Haribabu, N., Agbayani, G., Krishnan, L., Blais, A., and Sad, S. 2016. Inhibition of ROS and upregulation of inflammatory cytokines by FoxO3a promotes survival against *Salmonella typhimurium*. *Nat. Commun.* 7: 12748.

Juliano, C. E., Reich, A., Liu, N., Götzfried, J., Zhong, M., Uman, S., Reenan, R. A., Wessel, G. M. et al. 2014. PIWI proteins and PIWI-interacting RNAs function in *Hydra* somatic stem cells. *Proc. Natl. Acad. Sci. U.S.A.* 111: 337–342.

Kaiko, G. E., Ryu, S. H., Koues, O. I., Collins, P. L., Solnica-Krezel, L., Pearce, E. J., Pearce, E. L., Oltz, E. M., and Stappenbeck, T. S. 2016. The colonic crypt protects stem cells from microbiota-derived metabolites. *Cell* 165: 1708–1720.

Kenyon, C. J. 2010. The genetics of ageing. *Nature* 464: 504–512.
Kenyon, C. J., Chang, J., Gensch, E., Rudner, A., and Tabtlang, R. 1993. A *C. elegans* mutant that lives twice as long as wild type. *Nature* 366: 461–464.
Khalturin, K., Anton-Erxleben, F., Milde, S., Plötz, C., Wittlieb, J., Hemmrich, G., and Bosch, T. C. G. 2007. Transgenic stem cells in *Hydra* reveal an early evolutionary origin for key elements controlling self-renewal and differentiation. *Dev. Biol.* 309: 32–44.
Landmann, F., Foster, J. M., Michalski, M. L., Slatko, B. E., and Sullivan, W. 2014. Co-evolution between an endosymbiont and its nematode host: *Wolbachia* asymmetric posterior localization and AP polarity establishment. *PLoS Neglected Trop. Dis.* 8: e3096.
Lengfeld, T., Watanabe, H., Simakov, O., Lindgens, D., Gee, L., Law, L., Schmidt, H. A., Ozbek, S. et al. 2009. Multiple Wnts are involved in Hydra organizer formation and regeneration. *Dev. Biol.* 330: 186–199.
Libina, N., Berman, J. R., and Kenyon, C. J. 2003. Tissue specific activities of *C. elegans* DAF16 in the regulation of lifespan. *Cell* 115: 489–502.
McFall-Ngai, M., Hadfield, M. G., Bosch, T. C. G., Carey, H. V., Domazet-Lošo, T., Douglas, A. E., Dubilier, N., Eberl, G. et al. 2013. Animals in a bacterial world, a new imperative for the life sciences. *Proc. Natl. Acad. Sci. U.S.A.* 110: 3229–3236.
McLaren, A. 2003. Primordial germ cells in the mouse. *Dev. Biol.* 262: 1–15.
Morris, B. J., Willcox, D. C., Donlon, T. A., and Willcox, B. J. 2015. FOXO3: a major gene for human longevity — a mini-review. *Gerontology* 61: 515–525.
Mortzfeld, B. M., and Bosch, T. C. G. 2017. Eco-aging: stem cells and microbes are controlled by aging antagonist FoxO. *Curr. Opin. MicroBiol.* 38: 181–187.
Mortzfeld, B. M., Taubenheim, J., Fraune, S., Klimovich, A. V., and Bosch, T. C. G. 2018. Stem cell transcription factor FoxO controls microbiome resilience in *Hydra. Front. Microbiol.* 9: 629.
Nakamura, Y., Tsiairis, C. D., Özbek, S., and Holstein, T. W. 2011. Autoregulatory and repressive inputs localize Hydra Wnt3 to the head organizer. *Proc. Natl. Acad. Sci. U.S.A.* 108: 9137–9142.
Nigro, G., Hanson, M., Fevre, C., Lecuit, M., and Sansonetti, P. J. 2016. *Intestinal Organoids as a Novel Tool to Study Microbes-Epithelium Interactions.* Springer Protocols. Part of the "Methods in Molecular Biology" book series. Springer, Switzerland.
Nishimiya-Fujisawa, C., and Kobayashi, S. 2012. Germline stem cells and sex determination in *Hydra. Int. J. Dev. Biol.* 56: 499–508.
Petersen, H. O., Höger, S. K., Looso, M., Lengfeld, T., Kuhn, A., Warnken, U., Nishimiya-Fujisawa, C., Schnölzer, M. et al. 2015. A comprehensive transcriptomic and proteomic analysis of *Hydra* head regeneration. *Mol. Biol. Evol.* 32: 1928–1947.
Rakoff-Nahoum, S., Kong, Y., Kleinstein, S. H., Subramanian, S., Ahern, P. P., Gordon, J. I., and Medzhitov, R. 2015. Analysis of gene-environment interactions in postnatal development of the mammalian intestine. *Proc. Natl. Acad. Sci. U.S.A.* 112: 1929–1936.
Rawls, J. F., Mahowald, M. A., Ley, R. E., and Gordon, J. I. 2006. Reciprocal gut microbiota transplants from zebrafish and mice to germ-free recipients reveal host habitat selection. *Cell* 127: 423–433.
Rawls, J. F., Samuel, B. S., and Gordon, J. I. 2004 Gnotobiotic zebrafish reveal evolutionarily conserved responses to the gut microbiota. *Proc. Natl. Acad. Sci. U.S.A.* 101: 4596–4601.
Ribi, G., Tardent, R., Tardent, P., and Scascighini, C. 1985. Dynamics of *Hydra* populations in Lake Zürich, Switzerland, and Lake Maggiore, Italy. *Swiss J. Hydrol.* 47: 45–56.
Roberfroid, M., Gibson, G. R., Hoyles, L., McCartney, A. L., Rastall, R., Rowland, I., Wolvers, D., Watzl, B. et al. 2010. Prebiotic effects: metabolic and health benefits. *Br. J. Nut.* 104(Suppl 2): S1–S63.

Rohwer, F., Seguritan, V., Azam, F., and Knowlton, N. 2002. Diversity and distribution of coral-associated bacteria. *Mar. Ecol. Prog. Ser.* 243: 1–10.

Rolig, A. S., Sweeney, E. G., Kaye, L. E., DeSantis, M. D., Perkins, A., Banse, A. V., Hamilton, M. K., and Guillemin, K. 2018. A bacterial immunomodulatory protein with lipocalin-like domains facilitates host-bacteria mutualism in larval zebrafish. *eLife* 7: e37172.

Sánchez de Medina, F., Romero-Calvo, I., Mascaraque, C., and Martínez-Augustin, O. 2014. Intestinal inflammation and mucosal barrier function. *Inflamm. Bowel Dis.* 20: 2394–2404.

Schächter, F., Faure-Delanef, L., Guénot, F., Rouger, H., Froguel, P., Lesueur-Ginot, L., and Cohen, D. 1994. Genetic associations with human longevity at the APOE and ACE loci. *Nat. Genet.* 6: 29–32.

Seiler, F., Hellberg, J., Lepper, P. M., Kamyschnikow, A., Herr, C., Bischoff, M., Langer, F., Schäfers, H. J. et al. 2013 FOXO transcription factors regulate innate immune mechanisms in respiratory epithelial cells. *J. Immunol.* 190: 1603–1613.

Stappenbeck, T. S., Hooper, L. V., and Gordon, J. I. 2002. Developmental regulation of intestinal angiogenesis by indigenous microbes via Paneth cells. *Proc. Natl. Acad. Sci. U.S.A.* 99: 15451–15455.

Umesaki, Y. 1984. Immunohistochemical and biochemical demonstration of the change in glycolipid composition of the intestinal epithelial cell surface in mice in relation to epithelial cell differentiation and bacterial association. *J. Histochem. Cytochem.* 32: 299–304.

Weismann, A. 1883. *Die Entstehung der Sexualzellen bei Hydramedusen.* Fischer, Jena.

Weismann, A. 1892. *Das Keimplasma. Eine Theorie der Vererbung.* Fischer, Jena.

Wieland, A., Frank, D. N., Harnke, B., and Bambha, K. 2015. Systematic review: microbial dysbiosis and nonalcoholic fatty liver disease. *Aliment. Pharmacol. Ther.* 42: 1051–1063.

Willcox, B. J., Donlon, T. A., He, Q., Chen, R., Grove, J. S., Yano, K., Masaki, K. H., Willcox, D. C. et al. 2008. FOXO3A genotype is strongly associated with human longevity. *Proc. Natl. Acad. Sci. U.S.A.* 105: 13987–13992.

Williamson, A., and Lehmann, R. 1996. Germ cell development in *Drosophila. Annu. Rev. Cell Dev. Biol.* 12: 365–391.

Wittlieb, J., Khalturin, K., Lohmann, J., Anton-Erxleben, F., and Bosch, T. C. G. 2006. Transgenic *Hydra* allow *in vivo* tracking of individual stem cells during morphogenesis. *Proc. Natl. Acad. Sci. U.S.A.* 103: 6208–6211.

Subject Index

A

B

C

D

I

J

K

L

M

N

O

P

Q

R

S

Systematic Index

A

B

C

D

E